Biotechnology from A to Z

third edition

William Bains

OXFORD
UNIVERSITY PRESS

OXFORD

UNIVERSITY PRESS

Great Clarendon Street, Oxford OX2 6DP

Oxford University Press is a department of the University of Oxford.
It furthers the University's objective of excellence in research, scholarship,
and education by publishing worldwide in

Oxford New York

Auckland Bangkok Buenos Aires Cape Town Chennai
Dar es Salaam Delhi Hong Kong Istanbul Karachi Kolkata
Kuala Lumpur Madrid Melbourne Mexico City Mumbai Nairobi
São Paulo Shanghai Taipei Tokyo Toronto

Oxford is a registered trade mark of Oxford University Press
in the UK and in certain other countries

Published in the United States
by Oxford University Press Inc., New York

First published 1993
Second edition published 1998
Third edition published 2004

A Catalogue record for this title is available from the British Library

Library of Congress Cataloging in Publication Data
Data available

ISBN 0 19 852498 6

10 9 8 7 6 5 4 3 2 1

Typeset by Newgen Imaging Systems (P) Ltd., Chennai, India
Printed in Great Britain
on acid-free paper by
Biddles Ltd, www.biddles.co.uk

Biotechnology from A to Z

Foreword

Biotechnology has come of age. The industry has never been so widespread, its products have never been so successful. As a science and as an industry it promises further, extraordinary changes in our lives. And its promise, of using nature's own biochemical tools developed over 4 billion years' of evolution for our benefit, is closer to our grasp today than ever. Everyone will be affected.

Biotechnology is the application of our knowledge of biological processes. As our knowledge of the natural world grows, so it is applied to the pressing problems of our society – health, food, a safe and clean environment. Biotechnology's greatest impact remains in healthcare. 40% of the drugs now being tested come in part from biotechnology. Dozens of proteins are used every day as medicines, the product of recombinant DNA technology. Clot-busting enzymes, growth hormones, antibodies for cancer, arthritis and tissue transplant are now part of standard hospital care in the West. In the 1980s it took 5 years to find out what caused AIDS. In 2003 it took 3 weeks to find the SARS virus, thanks to the new technologies and new understanding.

Biotechnology has started to be applied in other industries. Despite widespread consumer concern in Europe, biotechnology crops now cover an area of land nearly as big as Spain. Enzymes are found in washing powders, fabric production, papermaking and food production. Biotechnology-based fuels are being tried as replacements for fossil fuels, not as scientific experiments but as real industrial projects with hard cash bottom lines. Activities as diverse as primary school education and family history research use biotech products. Our lives are being altered.

And this is only the beginning. As biotechnology merges with chemistry and physics in the world of nanotechnology, and as analysis and prediction builds up from the molecular level to understand ever more complex and fascinating forms of life, it is hard to see an area of our culture that could not be touched by biotechnology this coming century.

But with this maturity has come a new realism. The genome sequence in 2000 was found not to be the answer to all life, but just one section of a 'parts list' for life's blueprint. The collapse of share prices in biotechnology companies across the globe from late 2001 showed that fortunes could be lost as easily as won at the frontiers of technology, and that cowboys were as prone to sudden death here as in the old West. In 2000 just saying the word 'genomics' on Wall Street got you a million dollars. Today it will not get you lunch.

The 1990s also saw a naive assumption that once the public saw the benefits of biotechnology they would fall as much in love with it as its proponents. Alas, 52 million hectares of transgenic crops grew in the fields in 2002, but even some of the farmers were starting to ask 'so what?' The new maturity has realised that everyone must benefit from this new technology, and the benefit must be clear, and without unacceptable side-effects. With this, some of the more florid examples of hype have been dimmed (although there are still a few 1990s-style Bio-Barnums around). Press releases no longer claim cures for cancer in five years, or global warming solved in a decade.

Maturity brings complexity. The industry has fragmented into many sub-disciplines, each with its own jargon, each with an understanding of what is and is not possible that is often not shared by the outside world. Reproductive biologists – those involved in real cloning experiments – are amazed that every year the press go wild over claims that someone has, or will, clone a human. Surely, the scientists say, the journalists must realise that this is all nonsense? But it takes deep knowledge of the field to be able to see the details that will make cloning a human much harder than cloning a mouse, and only a few on the inside remember the years, decades of hard work that lead to a single newsworthy 'breakthrough'. Usually, only they are steeled for the decade of hard work that will be necessary before the next one.

As the industry impinges on our everyday lives, it is vital that the media, politicians, investors, bankers, lawyers, and the general public that they act for, understand what biotechnology is really about, what it can really do. Sensible decisions that will last can only be based on understanding. To make those decisions, legislators and regulators need to know what the ideas and words of biotechnology are about, what the technology can and cannot do.

That is what this book is for. Biotechnology is going to be one of the most important areas of human achievement – scientific, commercial, financial, maybe even artistic and philosophical – in the coming century. It has a vocabulary and an ideas set all its own. This opens the door into that world.

What is here, what is not

Because Biotechnology is about applying our knowledge of living things, an A to Z can cover bits of almost everything from Astronomy to Zoology, taking in environmental science, chemical industry, evolutionary theory and politics along the way. To make sure you can still lift the book, I have left a lot of things out, including basic biology or biochemistry. If you do not know what a mitochondrion is, this is probably not the place to find out – however if you can get that far then we can work together. The test is whether you are likely to hear someone say 'those biotechnologists have ...' . Whatever they have done (or tried to do, but failed), it should be in here. If it is not, please let me know.

One thing I have not left out this time is biological warfare. Publishers in the past have advised me not write about this from a factual standpoint, lest I give people ideas. Events after the autumn of 2001 confirmed that some people have ideas infinitely more horrible than those a mere scientist could dream up.

I have not tried to be inclusive about people or companies. I have mentioned a few when they are characteristic of their field of endevour, but this book is not about them, but the ideas and products they have created, so they are not indexed and do not have entries of their own, and for reasons of brevity I have mentioned only a handful. My apologies to those not included.

How to use this book

This is not a textbook. It is an extended glossary, or mini-encyclopedia, so you can dip in at any point you choose. Each entry gives a quick description of the concept, mentions related terms and ideas, and gives an indication of what the science or technology has really achieved. It is a book for dipping into quickly if you are puzzled – you do not have to read the first half to understand what the second half is about.

If what you are interested in is a major topic in biotechnology (farming, gene therapy, bioreactors), then jump straight to the entry. Or you can thumb through the book and see what catches your eye. But for most things, I would start with the index. Explaining each term in a self-contained entry would mean a lot of duplication, so different aspects of your query may be dealt with in different places. The entries are fairly extensively cross-referenced as well, so you can pick your way through the links until you feel you have enough of the background to the topic you are interested in.

ADEPT (antibody-directed enzyme prodrug therapy)

This is a way to target a drug to a specific tissue. The targeting mechanism and the drug are administered separately. The drug is administered as an inactive prodrug (*see* **Targeted drug delivery**) that does not itself have any effect. The prodrug can be converted into an active drug by an enzyme. In ADEPT, the converting enzyme can be, and indeed is preferably, one that does not occur normally in humans. Instead, it is administered with a second injection, coupled to an antibody that concentrates it in the target tissue. When the enzyme has arrived at the target tissue, the prodrug is activated there to form the active drug, while elsewhere it remains inactive.

This is being developed for tumor treatment. The prodrugs being considered are prodrugs of highly toxic anti-cancer compounds, which in their normal form have severe side-effects because they kill many cells other than the tumor cells. ADEPT can target these drugs to the tumor, sparing the rest of the body, by using an antibody that binds specifically to the tumor.

A new, closely related idea is GDEPT—gene-directed enzyme prodrug therapy. Here, the target cell is transfected with a gene which makes an enzyme, which itself converts a prodrug to a drug. A favorite is the thymidine kinase gene (TK) from herpes simplex virus, which converts gancyclovir into its active form. This is a version of using a "suicide gene" in gene therapy— a gene that kills the cell into which it is introduced. Other enzymes being considered are nitroreductase, a bacterial enzyme that converts nitrate groups to more reactive products, and CYP2B6, a natural liver enzyme, which is put into non-liver cells to enable them to convert cyclophosphamide from its inactive to its active form (a conversion usually done only in the liver).

ADMET

ADMET, sometimes called ADME/Tox, stands for absorption, disposition, metabolism, excretion, and toxicology. It summarizes all the processes that a drug undergoes when it enters the body, moves around it and leaves it again:

- Absorption is how the drug gets into the body, usually through the gut.

- Disposition is where a drug goes once it is in the body—whether it stays in the blood, gets into different tissues, into the brain, and so on.

- Metabolism is how a drug is broken down by enzymes in the body, mostly in the liver and kidneys. Drug metabolism happens in first pass metabolism (in the liver) and second pass metabolism, also sometimes called primary and secondary metabolism, respectively. The resulting molecules are called primary and secondary metabolites

- Excretion (often called "elimination" in the United States) is how the drug leaves the body, the most common routes being in the urine or via the gall bladder in the faeces.

- Toxicology is the effects the drug has on the body at each of these stages that are *not* part of its intended medical effect.

Related terms are pharmacokinetics (the practical study of how much drug there is in the body at any one time, which is the net effect of ADME), pharmacodynamics (the study of how the drug's *effect* changes with time), and DMPK (drug metabolism and pharmacokinetics), which covers most of ADME.

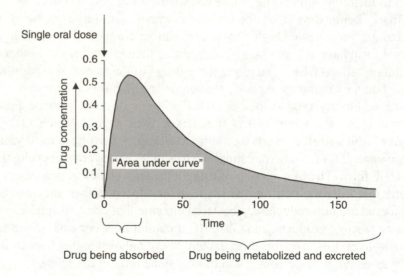

It has become increasingly topical for drug discovery groups to try to solve problems with ADMET early in the development of a new drug, as historically about 40% of drugs started in clinical trials fail because of ADMET problems. Biotechnology has therefore come up with two types of tools to try to find out what the ADMET properties of a drug will be:

1. *Tests and assays.* Many groups have attempted to make tests that mimic the complex interactions involved in ADMET *in vitro*. These include cell culture methods that seek to mimic the lining of the gut (using

CaCo-2 cells, a gut-derived cell line), the liver, and other cell types, the use of primary tissues (tissues taken from humans) and cells made from stem cell lines, and the use of artificial tissues, especially skin substitutes spun out from artificial skin technology (*see* Artificial tissues) to try to predict absorption and metabolism, and aspects of toxicity.

2. *Modelling*. ADMET is a complex property of a whole living system, not of a single molecule, and so falls within the purview of systems biology. Over the last 5 years, several companies such as Camitro and Amedis have developed mathematical models of aspects of ADMET that can be used to predict properties *in silico*. So far, this has had some success, but its main value is that computer modelling is much faster and cheaper than even *in vitro* assays, so whole chemical libraries can be screened at once.

Affinity chromatography

This is a method of separating molecules by using their ability to bind specifically to other molecules. This is particularly useful in the separation of biological molecules, because many biomolecules bind very strongly and specifically to other molecules—their substrates, inhibitors, regulators, etc. Collectively, a molecule that bind to larger molecules are called a ligand (see separate entry). In affinity chromatography, the ligand is chemically linked to a solid material (the chromatographic matrix), so that the molecule we want sticks to the matrix, and all others do not and are washed away.

There are two types of biological affinity chromatography. Either a biological molecule can be immobilized and a smaller molecule it is to bind to sticks to it, or the smaller ligand can be immobilized and the macromolecule sticks to it. (Of course, both "sticker" and "stickee" can be biological, too.) A variant is to use an antibody as the immobilized molecule and use it to "capture" its antigen.

Biological molecules used to separate smaller molecules include:

- *Enzymes*, to isolate substrates, or substrates to separate enzymes. (This only works if at least one substrate is missing from the mix, otherwise the enzyme just destroys the substrate.)

- *Antibodies* (which when immobilized on column—called an "immuno-affinity column"—can separate their antigen from a complex mixture)

- *Cyclodextrins* (for separating lipophilic materials particularly)

- *Lectins* (proteins which bind specific sugars very strongly and specifically, and which are therefore used for separating carbohydrates and anything with carbohydrates attached).

3

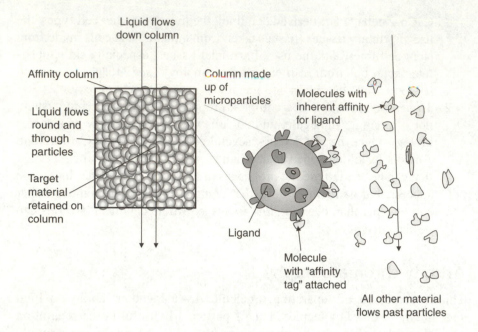

A variation is pseudoaffinity chromatography, in which a compound which is like a biological ligand is immobilized on a solid material and enzymes or other proteins bound to it (*see* **Biomimetic**). Other methods include metal affinity chromatography, where a metal ion is immobilized on a solid support: metal ions bind tightly and specifically to many biomolecules. The metal ion is bound to a chelator or chelating group, a chemical group that binds that metal extremely tightly. Such metal columns are used to purify proteins with His_6 **affinity tags** (see separate entry).

There are a wide range of support materials used in affinity chromatography. They are discussed further in the **Chromatography** entry.

Affinity chromatography is widely used in research. It is also used in production, where a particularly valuable product is to be isolated from a complex mixture of similar chemicals, of which the product is a minor component. Thus, Armour Pharmaceuticals and Baxter Healthcare, both isolate factor VIII (used to treat hemophilia A) from blood using immuno-affinity chromatography, in which an antibody to factor VIII is linked onto a "column" of solid material, and the plasma passed over it: the factor VIII sticks and all the other proteins do not, resulting in a purification of roughly 200 000-fold. Usually such techniques are too expensive to be used on anything other than high value, pharmaceutical or diagnostic products.

Affinity tag

Also sometimes called a purification tag, an affinity tail, or an affinity handle, this is a bit of the amino acid sequence of a protein engineered into the protein to make its purification easier. These can work in a number of ways.

The tag could be another protein, for example, an enzyme (making the new protein easier to detect) or a protein which binds to some other material very strongly (such as avidin, which binds to biotin very strongly), or biotin itself, (which would bind to an avidin column) which would allow the protein to be purified by affinity chromatography. Enzymes sometimes fulfil both roles, as they catalyze the reaction of substrates and bind to inhibitors very strongly. Short segments of cellulase (an enzyme which breaks down cellulose) have been used to make fusion proteins which stick to a cellulose affinity matrix. Other examples are using the last 26 amino acids of the calcium-binding protein calmodulin which binds to muscle myosin kinase, a relatively cheap protein to produce

The tag could be a short amino acid sequence, either random or selected from some other protein, which is recognized by an antibody. The antibody would then bind to the protein when it would not have done before. One such short peptide, called FLAG, has been designed so that it is particularly easy to make antibodies against it.

A short string of amino acids can also be used as a chemical tag on the protein. A string of positively charged amino acids, for example, will bind very strongly to a negatively charged filter: this could be used as the basis of a separation system. Histidine binds strongly to some metals, and short runs of six histidines (His_6 tag) can be added to the end of a protein to make it bind extremely tightly to columns of material that have nickel atoms on their surface ("nickel columns").

Aging

All vertebrates age, because it is more efficient from an evolutionary standpoint to have lots of children, and then die than to spend all your physiological energy on staying alive. As well as developing treatments for specific disease, biotechnology has sought to understand and treat aging itself.

Approaches to treatment include:

* *Organ replacements*. This suggests that, as each organ wears out, we replace it with a new one. *See* Transplant and Artificial tissues.

* *Human growth hormone*. Touted as a cure for aging, this has been found in clinical trials to reduce some of the symptoms of aging, such as the replacement of muscle by fat, but not the underlying process.

- AGE (advanced glycosylation end-products). One of the most general chemical changes that go with aging is the accumulation of proteins linked together by sugars, called advanced glycosylation end-products. These accumulate and clog up cells and the intracellular matrix. It is unclear what anyone can do to remove them, though.

- *Genes*. Some very rare genetic diseases called progerias result in what looks like greatly accelerated aging, so patients die "of old age" in their late teens. Small animals can be selected which are longer lived than their cousins, and they have identifiable genetic changes. Both lines of argument suggest that single genes can have a substantial effect on how fast we age, and imply that we could slow the rate at which we age is we understood how those genes work. Several genome companies are seeking aging-related genes. Related to the gene approach are:

- *Telomeres*. These are special DNA sequences at the ends of chromosomes, which gradually shorten with age. See separate entry.

- *Starvation*. It is a well-documented fact that small mammals who are semi-starved from birth (on a calorie restricted, or CR, diet) live up to 40% longer than their well-fed brothers and sisters. There is no convincing explanation for this, but it opens the way to modifying metabolism in a way that mimics the effect of starvation, but without the need to reduce caloric intake.

Agricultural biotechnology

Agricultural biotechnology, the application of biotechnology to food production, is far less glamorous than healthcare biotechnology and much more controversial, but is still very widespread. Biotechnology applications which the scientific community consider valuable that are discussed elsewhere include:

- Use of biological mechanisms to control pests (*see* **Biocontrol**)

- Use of transgenic plants with improved properties (*see* **Transgenic plants**)

- Creation of plants with improved nutritional, taste or other properties

- Use of plants to produce vaccines and other valuable, non-food products

- Development of rapid tests for pathogens, especially for high value crops or glasshouse environments

- Applications in environmental biotechnology.

By far, the largest application of agricultural biotechnology in terms of acres of crops planted is the use of herbicide-tolerant and pest-resistant transgenic plants.

Potential technical problems primarily are around

- the tendency towards monoculture (planting all one crop, with the resultant dependence on that one crop and vulnerability to pests and diseases)

- the use of more technologically intensive and centralized processing technologies (e.g. which have been blamed for the huge outbreak of BSE in the United Kingdom)

- the small but still significant possibility that transgenic crops will significantly affect the natural plant world.

The main political problem with agricultural biotechnology is political opposition to "GM crops."

Agrobacterium tumefaciens

Agrobacterium tumefaciens is a bacterium which causes a disease called crown gall disease in some plants. The "gall" is a hard lumpy growth on the plant, which is a benign home for the bacterium. The bacterium infects a wound, and injects a large plasmid—the Ti (tumor induction) plasmid—into the plant cells. A short region of the Ti plasmid (called T-DNA) is transferred to the plant cell, where it causes the cell to grow into a tumor-like structure.

This DNA transfer mechanism has been harnessed as a way of genetically engineering plants. The Ti plasmid is engineered so that a foreign gene is transferred into the plant cell along with the particular genes on the T-DNA (nopaline synthesis genes)—this engineering is not easy, and must be done by homologous recombination rather than conventional recombinant DNA technology. When the bacterium is cultured with isolated plant cells or with wounded plant tissue, the "new" gene is injected into the cells and ends up integrated into the chromosomes of the plant. Improved versions replace the bacterial genes entirely with the engineered ones.

A. tumefaciens usually only infects some dicotyledons, because their response to wounding is compatible with *A. tumefaciens*' DNA transfer mechanism, and so this has not been applicable to monocotyledons (grain crops) in the past. Manipulation of the plasmid and the conditions under which it transfers its DNA in culture have allowed grain crops (including rice and maize) to be transformed with Ti DNA. This is still hard to do.

A. tumefaciens has been used particularly to get DNA into trees. Trees are difficult plants to breed because of their size and long life cycles, and so genetic engineering techniques offer unusual advantages of speed and the ability to engineer millions of clones. Walnut, poplar, apple, and plum trees have all had DNA transferred to them using *A. tumefaciens*.

Airlift fermentor

Airlift fermentors or airlift reactors (ALRs) are a type of loop fermentor, and very popular in many applications. The fermentor consists of two parts, a riser and a downcomer. The liquid fermentation medium circulates between the two, driven up the riser by bubbles of air (or other gas, sometimes pure oxygen) pumped into the bottom through a sparger. There is usually a gas separator at the top of the riser which separates ("disengages") the gas from the liquid, so the bubbles of gas are not sucked back down the downcomer.

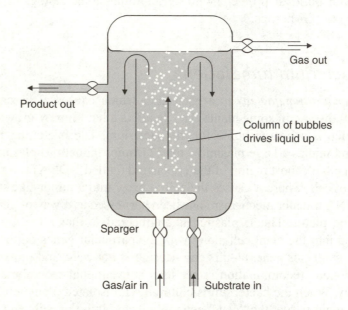

The popularity of this type of fermentor is mainly because the air drives the liquid round the fermentor quite gently: other stirring mechanisms such as paddles or plates might break open delicate mammalian cultured cells, or damage long fungal hyphae. Airlift fermentors were very popular for making monoclonal antibodies in bulk, although the trend has now swung rather towards using hollow fiber reactors instead (see separate entry) for all except very large-scale production.

A variation on the airlift fermentor is the bubble column fermentor. Here, the cells are kept suspended in solution by the bubbles rising through a tube. The difference is that in an ALR the bubble flow drives the liquid round the fermentor circuit (sometimes helped by pumps). In a bubble column fermentor, the bubbles are there to keep individual cells suspended, but not to move the bulk liquid around.

Algae: commercial uses

Algae are simple water-living plants. The fall into two types. Macroalgae, mainly seaweed, and microalgae, which are microorganisms like the ones responsible for turning ponds green and slimy.

Algae have been used for centuries in Asia to produce food ingredients and other materials, but their use elsewhere to produce biotechnology products is relatively new. Among their biotechnological uses are:

1. *Production of gums*. Many of the gums used in food, medicine (e.g. in wound dressings), and for a variety of other things such as research reagents and components of printing inks, come from seaweed. Among the more common prepared commercially from seaweed are: alginate (an acetylated polymer of mannuronic acid and glucuronic acid), carageenan, agar, and agarose.
2. *Production of chemicals*. Some chemicals are produced commercially from algae, such as the food dyes beta carotene, produced from *Dunaliella*, and astaxanthin, produced from the microscopic alga *Haematococcus*: astaxanthin is added to salmon food to give farmed fish pinker flesh (which wild fish would get from eating plankton), and to chicken feed to make egg yolks more yellow. A variety of human food ingredients are made from cultured algae, such as Omega-3 unsaturated fatty acids DHA (docosahexaenoic acid) and ARA (arachidonic acid) from dinoflagellates by Martek Biosciences.
3. Single celled algae have also been used as food—the algae *Chlorella, Scenedesmus* (in Japan), *Spirulina* (in Israel, United States, and Mexico) are used to make algal "biomass" for single celled protein (SCP) (see separate entry). Despite enthusiasm by its proponents, such microalgal food is no better for you than any other vegetable.
4. Microalgae are also used for biogas production—*see* Biogas.

Alliance

Not unique to biotechnology, this term means an alliance between two companies formalized as a legal agreement and usually aimed at developing some common area of business. Because setting up a complete biotechnology R&D department is time-consuming and expensive, companies frequently set up "strategic alliances" with each other to get access to know-how they would otherwise have to develop in-house. The essence of an alliance is that both sides benefit while remaining independent.

Alliances fall on a spectrum of "depth" from contracts to outright purchase:

• Research contracts are often called alliances, but actually are normal contracts to employ one party to do something for the other party. The only thing going from the contractor to the researcher is money.

- *Collaborations*. These are research programs where (in principle) technology, and not just money, are exchanged. Call it "strategic" if you want to emphasize how important it is.

- *Strategic alliances*. Here, two companies collaborate on a major aspect of their interests. Sometimes this can turn into a joint venture.

- *Joint venture*. The two companies actually set up a new, third company, which they both own, to carry out some business or technical development that they could not do separately.

- Mergers and acquisitions are the most intimate fusing of two company's business interests, and one partner usually loses their autonomy in the process. Probably the best-known biotechnology acquisition/alliance of all was the acquisition of 60% of Genentech by Hoffman LaRoche in 1991. Genentech was big and energetic enough to retain its own identity, so making this a strategic partnership rather than the takeover as the balance-sheet makes it appear.

Alliances are often organized around a major development program, for example, to develop a new product or technology. Such agreements are usually funded by milestone payments or royalties. The former are made when the partner doing the research achieves specific goals, for example, getting a prototype to work or getting a good phase I clinical trials result. Royalties are a percentage on the sale price of the final product, and so do not start to roll in until the joint development is completed.

Amino acids

The amino acids, key components of all living things, are produced by biotechnology in bulk using fermentation and biotransformation. Several Japanese companies dominate the world production of amino acids. They use fermentation systems in which they grow bacteria or fungi which have been selected to "over-produce" specific amino acids (i.e. make more than they need for their own metabolism), and excrete most of their production into the fermentation medium. Harvesting the medium and removing the other components yields amino acids in total amounts of hundreds or thousands of tonnes a year. Mutation and recombinant DNA techniques have increased yields of amino acids in commercial production to between 50 and 100 g/l.

Commercially produced amino acids include:

1. *Glutamic acid*. This is widely used in the food industry in the form of monosodium glutamate (MSG) as a flavor enhancer, and in the Far East as a table condiment. Several bacteria, especially *Corynebacterium glutamicum*, are used commercially to make glutamate, and can convert up to 50% of the carbon it is supplied with into glutamate. 800 000 tonnes of glutamate are produced annually using these methods.

2. *Lysine*. This is the second most abundantly produced amino acid, and is used as a feed supplement for animal feeds. (Most cereals are deficient in some amino acids, and particularly in lysine which is an essential amino acids in mammals.) Lysine biosynthesis is tightly controlled so usually it is not overproduced, but *C. glutamicum* has been mutated so that this no longer operates.

3. *Cysteine, methionine*. These are the sulfur-containing amino acids, and again are used as animal feed supplements.

4. *Phenylalanine*. As well as being used to a small amount as a feed supplement, phenylalanine is the most expensive chemical ingredient in the manufacture of aspartame.

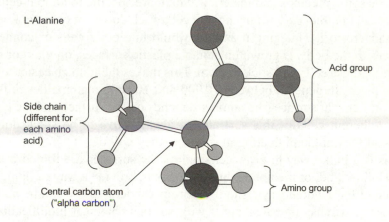

Tryptophan hit the headlines in 1990 when the tryptophan produced by a new genetically engineered *Bacillus amyloliquefaciens* made by Showa Denko KK was linked to a rare degenerative disorder called eosinophila-myalgia syndrome (EMS). Despite loud and prolonged claims that this was proof that genetic engineering was dangerous, the problem was eventually traced to a chemical generated during the (perfectly conventional) purification procedure, and had nothing to do with recombinant DNA.

Essential amino acids are ones which our bodies cannot make themselves and hence have to be eaten in our diet (*see* **Essential nutrients**). All are produced by fermentation in thousand tonne amounts. Other amino acids, not found in proteins, and especially the D-isomers, are made by biotransformations for use as chemical intermediates. The D-amino acids, for example, are used in antibiotic manufacture. (D-amino acids are ones with the opposite "handedness" to natural amino acids—*see* **Chirality**.)

Animal cell immobilization

Animal cells are used widely in biotechnology to produce natural products or genetically engineered proteins. Animal cells have the advantage that they already produce many proteins of pharmacological interest, and that genetically engineered proteins are produced by them with the post-translational modifications normal to animals. However, animal cells are much more fragile than bacterial ones, and so cannot be exposed to the high shear forces of repeated centrifugation which bacterial cells can put up with in a commercial fermentation process.

Any cell (indeed, any small particle) can be immobilized by entrapping it in some solid material, either by having it grow there or by forming the material round it after it has grown. Entrapment in some form is common and popular, from microencapsulation to growing cells inside a hollow fiber in a bioreactor (*see* **Hollow fiber reactor**). As well as these general approaches, there are some methods and materials which are specific to animal cells.

Surface adherent cells. Many animal cells stick down flat on a suitable surface (adhere to it), hugging it as they would hug other cells or connective matrices in the body. If grown on suitable plastic surfaces, on glass or many ceramics, these cells will stick to them. This makes them much easier to keep in one place. In general, between 10 000 and 100 000 mammalian cells will grow on 1 cm^2 of surface (the number depends on the surface and the type of cell). Adherent cells of this sort almost always *have* to be stuck down to grow—if they are kept floating in solution, they will die.

This is a bulky way to grow cells unless the surfaces are folded in some way. Hollow fiber or membrane bioreactors can provide a way of doing that, but one of the favored ways is to use porous carriers. These can be polysaccharides (dextrans, alginates, agar) with various chemical modifications to give them a surface charge, protein (especially collagen), plastic, or ceramic materials with microscopic holes in them a few tens or hundreds of micrometers across. Such materials are called microcarriers or microbeads. The cells grow into these holes, greatly increasing the surface area available to them while not increasing the bulk of the culture.

Animal cloning

A clone is a group of genetically identical organisms. As animals nearly all reproduce sexually, normal offspring are not genetically identical to either parent. If we want an animal that has an identical genotype to another, we must rely on naturally occurring identical twins or find an artificial approach. The latter is called "cloning."

Unlike plants, we cannot grow new animals from "cuttings." The only cells known to be capable of growing into a new mammal are fertilized egg cells or cells from very early embryos. So we can split an early embryo into eight cells and allow them all to grow into genetically identical fetuses, and hence adults. In mammals, if we split the embryo into more than eight, very few or none of the resulting clumps of cells develop into an embryo. We can also seek to inject the nucleus from another type of cell into a fertilized egg, from which we have removed the nucleus (enucleated egg), and grow up the result. This was done successfully in sheep in 1996, when Scottish researchers injected the nucleus from a cell line made from Welsh mountain sheep into eggs from Scottish Blackface Sheep mothers. Of five lambs born, only two survived more than 10 days.

Generally, the more differentiated a cell is, the harder we expect it to be to "reprogram" it to allow the development of a whole individual (*see* Stem cells for more on differentiation). Thus, it was not until a year later that two sheep (Megan and Morag) were born from nuclei from a sheep fetus that were injected into an egg cell, and a year after than when the same group announced that they had grown a sheep from a nucleus from an adult mammary gland cell (Dolly, named after Dolly Parton). Sheep have also been born from a genetically manipulated adult cell, which is the main reason for doing these experiments. Since then, pigs, cattle, goats, mice, and a cat have been cloned from adult cells.

Fish and frogs are more amenable to cloning than mammals, and frogs were cloned by nuclear injection in the early 1970s. However, even this is a very difficult endeavor.

Cloning animals is useful because it allows us to carefully make one genetically very desirable animal, by transgenic technology or decades of breeding, and then make many more with exactly the same genes. This is important when the genetic character we are seeking is determined by many alleles, which would be separated into different animals by sexual reproduction.

A rising use of cloning is in preserving pets. An anonymous US couple has donated $2.3 million to clone their mongrel pet dog Missie. In March 2002, US Savings and Clone of Texas announced that that had cloned a pet cat, called CC (for copy cat). The living cat was the only one of 86 eggs that had nuclei transplanted into them that survived to birth. If this success rate can be improved, then pet cloning is likely to grow. The group is also selling "cloning kits" to preserve your pet's DNA until cloning is less expensive, although this is very unlikely to work because the DNA will be damaged beyond repair by the extraction and storage process. (For the same reason, multiplied by about a billion, the "Jurassic Park" concept of cloning dinosaurs from preserved DNA is astonishingly improbable.)

As well as the very high rate of fetal death during gestation for fetuses cloned by nuclear transfer, there is a major issue about whether cloned fetuses that survive to birth have defects that are a result of the cloning technology. Dolly is the oldest cloned non-rodent animal, and she showed shortened telomeres (which may or may not be significant) and developed arthritis in 2002, and died of an infection-related lung disease in February 2003. She was aged $6\frac{1}{2}$ at the time, out of a maximum normal sheep lifespan 12–14 years, but how many sheep usually die when they are 6–7 years old? No one knows, because they are usually killed for meat before this.

More generally, a review at the start of 2002 found that 77% of live birth cloned animals are "apparently healthy" (i.e. no one has found anything wrong with them). This is also similar to all other highly interventive reproductive technologies, and may be related to problems with the placenta rather than the genetics: if the same techniques that are used to help premature human children were used on animals, cloned animals may appear as normal as normally conceived ones.

Alternatively, other aspects of the technology, notably the cell fusion method, or the removal of the chromosomes from the egg before nuclear injection, may be inherently damaging.

Given this, it is not surprising that cloning humans has been declared illegal in most countries, on two grounds:

- ethical/moral—many people find the idea repugnant in itself, and a violation of the idea that every person should be a unique individual.

- practical—the high failure rate in cloning mammals, both from failing to produce any viable embryo and for producing embryos that are defective and die in infancy, is not acceptable when contemplated for humans.

These arguments should not apply to **therapeutic cloning** (see separate entry), but often are as the end result is confused with the method.

See also **Embryo technology**.

Anti-idiotype antibodies

Anti-idiotype antibodies are antibodies which recognize the binding sites of other antibodies. Their binding sites are complementary to the binding site of another immunoglobulin. They are important to biotechnology in three ways.

First, they occur in normal blood. Antibodies binding to other antibodies forms a network of antibodies which can all bind to each other to various degrees, a network that helps to regulate the immune response. Thus, anti-idiotype antibodies are important to the regulation of the immune system.

If an antibody is a "key" exactly selected to fit the "lock" of a receptor molecule, then an anti-idiotype antibody is a "lock" exactly selected to fit that "key." By raising an antibody against a receptor or a hormone, and then raising an anti-idiotype antibody against the antibody, you might create an immunoglobulin with some of the characteristic shape of the original receptor, but which can be produced more easily and which is chemically quite distinct. This was planned to be a neat way of, for example, getting the body to make its own insulin replacement. In practice, antibodies only recognize a small region of the surface of a protein, so an anti-idiotype antibody can only mimic the properties or functions of part of another protein, and these functions are likely to be rather limited.

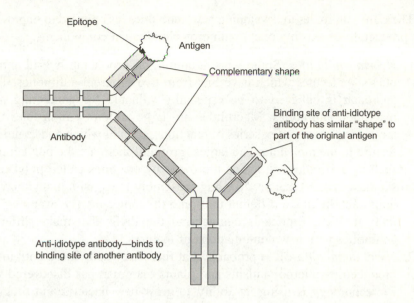

Epitope

Antigen

Complementary shape

Binding site of anti-idiotype antibody has similar "shape" to part of the original antigen

Antibody

Anti-idiotype antibody—binds to binding site of another antibody

Anti-idiotype antibodies also have potential as vaccines. We make an antibody to a bacterium, and then an anti-idiotype antibody against it, and use the anti-idiotype antibody as a vaccine instead of the bacterium. To date, the link between bacterium, antibody, anti-idiotype antibody, and immune response seems too tenuous to make this work, and the approach has been largely supplanted by using recombinant DNA techniques to clone the bacterial proteins directly.

Antibiotics

The "established" antibiotics, especially penicillins, were the first products of the industry which is now biotechnology. Produced from fungi in fermentation systems, the penicillins, streptomycins, and a host of other antibiotics were breakthrough drugs in the 1940s and 1950s, and are still major products of fermentation industries. Modified penicillins, and the closely related cephalosporins, are now produced on kilotonne scale using biotransformation (synthesis with enzymes) as well, and the antibiotic market is worth around $23 billion (2001). However, resistance to antibiotics (*see* **Resistance to anti-infectives**) means that the old antibiotics become less and less useful with time, and new ones must be discovered.

Innovation has been rather slow: nearly all new antibiotics are small (but useful) variations on well-known types ("classes") of drugs. The first new class of antibiotic for 30 years (Linezolid, an oxazolidinone) was launched spring 2000, and the first case of resistance to the drug was reported less than a year later.

There are four routes to developing new antibiotics (as opposed to improving the production of existing ones) with biotechnological components.

1. *Hybrid antibiotics*. Some work is aiming at producing hybrid antibiotics—molecules which have bits from two different antibiotics. This approach is believed to be especially valuable for generating new macrolide antibiotics. Macrolides are large antibiotic molecules that are made by a complex series of reactions, each of which adds a unique feature to the molecule. The largest group of them are the polyketides, such as erythromycin, which are made by enzymes called polyketide synthases (PKS) that are like huge assembly lines, with many enzyme units each linking a different unit onto the molecule. By swapping the units in PKS, Biotica is engineering the PKSs that make different antibiotics with new combinations of features.

2. *Novel metabolites*. It is probable that there are vastly more antibiotics produced by microorganisms and plants than man has discovered yet. Biotechnology is using its ability to grow new bacteria and fungi in large scales to screen new bacteria for compounds which have useful drug activities. This is a branch of natural products drug discovery.

3. *Animal antibacterials*. Animals, especially invertebrates (which do not have the same sophisticated, adaptable immune system as mammals) produce a wide range of materials which kill bacteria, usually proteins or peptides, which might be developed into drugs. Some, such as the defensin peptides and the bacterial permeability increasing protein (BPI), bactenecin peptides, azurocidin, and the enzyme lysozyme directly kill bacterial cells. A second group, typified by lactoferrin, inhibit bacterial growth, in this case by removing the free iron which the bacteria need from their surroundings, and binding it in a tight, inaccessible complex.

4. *Pathogenesis-targeted antibiotics*. These attempt to find out what makes a bacterium a pathogen, and then block that molecular mechanism, rather than killing the bacterium outright. It is hoped that this will get round the problems of bacterial resistance in two ways. First, the drugs will be completely different from those currently in use, so resistant bacteria will be sensitive to them. Second, they will not put as much selection pressure on the bacteria to acquire resistance—providing the bacteria do not cause disease, they will not be affected. This second point is still speculative. The bacterial genome projects—to find the DNA sequence of every gene in the genomes of several bacteria—are aimed at finding out the genes that determine pathogenesis.

Many antibiotics are semi-synthetic, as are some other drugs like steroids. Biotechnological routes are used to make the core structure, such as the beta-lactam ring structure of the penicillins, and the organic chemistry modifies it to produce the end product.

Antibodies

Antibodies are proteins made by the immune system of vertebrates to combat infection. Each antibody is made to recognize one target antigen molecule. If the antigen is a small molecule, then the antibody will recognize all of it. If the antigen is a large molecule, then the antibody will recognize only a part of the antigen, which will be called that antibody's "epitope." The binding site of the antibody latches onto this antigen very specifically and very strongly. This latching on allows the body to recognize the antigen as part of something that should not be there—a virus, a bacterium, a toxin—and so start the process of removing it.

Antibodies are made by a type of white blood cell called B-lymphocytes (B-cells). Each new B-cell makes a different antibody. If that antibody finds its target, then that fact is signalled to the B-cell which then divides, so there are more cells making that antibody. In this way the immune system develops a "memory," being able to mobilize many cells to make antibodies to enemies it has seen before.

You can get a mammal to make an antibody against almost any molecule by injecting it into the bloodstream. The immune system recognizes it as a foreign material and makes a suitable antibody. In fact, it makes a whole range of slightly different antibodies: the blood of most people usually contains a vast host of different antibody molecules targeted at the various disease agents and other foreign molecules that have got into their blood in the past. For this reason, the antibodies that are prepared from mammalian blood are called polyclonal antibodies, because they come from a large number of "clones" (i.e. identical sets) of cells. This contrasts with the synthetic Monoclonal antibodies (see separate entry).

Antibodies have been enormously useful to biotechnology because of their ability to latch tightly onto only one antigen, ignoring all others. For example, they would accurately distinguish sucrose from glucose, right-hand amino acids from left-hand (see Chirality), human blood proteins from ape proteins etc. Thus, they are the basis of many processes where great discrimination is needed. See entries on Immunotoxins, Immunodiagnostics, and Affinity chromatography.

Technically, the antibody proteins are called Immunoglobulins. There are four types usually mentioned:

- IgM—the first type made by the body when it encounters a foreign material

- IgG—the most common type, made after a prolonged encounter (as during a disease)

17

- IgE—the type responsible for allergic reactions

- IgA—a rarer type, which is present in saliva and some other non-blood fluids.

See also **Antibody structure**.

Antibody structure

Antibodies have a well-defined structure. They are made of equal numbers of "light" chains and "heavy" chains: IgG antibodies (the sort most commonly used as reagents in biology and biotechnology) contain two of each. The antigen binding region or binding site ("complementarity determining region") lies at the end of the light and heavy chains—it is therefore formed from both chains. The chains fold up into discrete blobs called domains: a "single domain antibody" (Dab) is just one domain of an antibody.

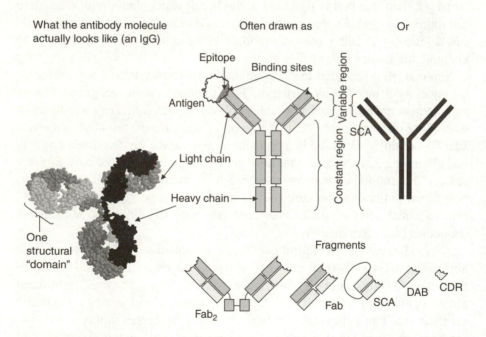

The amino terminal domains of both heavy and light chains are called the variable domains because they vary between antibodies. The other domains are constant domains, that is, they are the same between antibodies of the same class and sub-class, and not that different between any antibodies.

The antibody can be cut by proteases into several fragments known as Fab, Fab′, and Fc (for historical reasons). These also feature in some biotechnological literature.

Antisense

Antisense RNA or DNA is an approach to knocking out the function of a gene in a living organism. Chemically, it is a single-stranded nucleic acid which is complementary to the coding, or "sense," strand of a gene, and hence is also complementary to the mRNA produced from that gene. If the antisense RNA is present in the cell at the same time as the mRNA, it hybridizes to it forming a double helix. This double helical RNA cannot then be translated by ribosomes to make protein, and is broken down by RNAse H, which recognizes double-stranded RNA and cuts it.

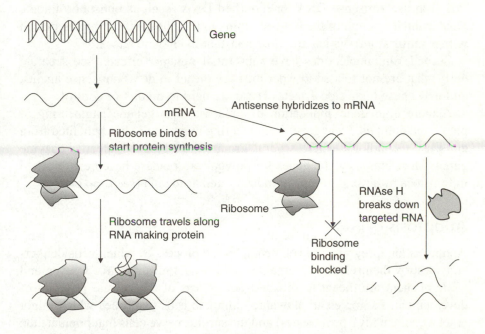

Antisense RNA is a powerful way of modifying gene activity because it is a positive genetic engineering step. Thus, rather than try to knock out all copies of a gene in, say, a plant, the genetic engineer only has to put in one gene which produces antisense RNA, and the antisense will prevent any mRNA from any copy of that gene being used by the cell.

The main use of antisense has been as a research tool—answering the question "what does this gene do?". It has two more practical applications as well.

Antisense RNA has been suggested as a potential drug, because it can block the effect of one gene without affecting any other. In particular, it could be used to block the effect of **oncogenes** (see separate entry), and so slow or prevent the development of cancer, or block the genes of viruses without affecting the patient. This was first taken all the way to a product in 1998, when, ISIS Pharmaceuticals launched Vitravene, an antisense drug to treat

CMV retinitis (infection of the retina of the eye by CMV virus). Companies like ISIS, Hybridon, and Genta have other potential antisense drugs in clinical trials.

However, developing antisense drugs has proven harder than first thought, for two reasons.

First, the pharmaceutical chemist has to be able to deliver intact antisense RNA or DNA as a chemical to all affected cells within an whole animal or patient, not just to a few in cell culture. This is doubly difficult because RNA is quite unstable, and is very easily broken down by RNAses, enzymes that are found in very many tissues and are very hard to destroy. One way round this is to use antisense DNA, or modified DNA (such as phosphorothioate DNA, which has one of the oxygen atoms in the phosphate groups replaced with a sulfur atom), which are more resistant to enzyme attack.

Second, oligonucleotides have substantial **Aptamer** effects (see separate entry) that are not related to what they are meant to do as antisense agents, and may cause toxic side-effects. These are hard to predict or control.

A more immediate application of antisense is in genetic engineering of plants and animals. Plant genetic engineering in particular has benefited from antisense technology, where several groups have blocked specific enzyme genes. Most famously, the genes for polygalacturonidase have been blocked in tomatoes by several groups in industry and academia (*see* **Crop plants**).

Apoptosis

Apoptosis, or "programmed cell death," is the process by which cells deliberately destroy themselves. Mammalian cells are not meant to live forever, and some of them are meant to self-destruct as part of the normal program of development. Examples are unwanted immune cells (e.g. ones which might react with the body's own tissues) and the surplus nerve cells that populate the brain before birth, but which must be removed to generate the functioning networks that enable us to see and remember things in postnatal life.

Apoptosis is a deliberate process—the cell dismantles itself systematically. It is unlike necrosis, where the cell is killed, often by chemicals or toxic attack. Necrotic cells burst and release their contents, causing inflammatory responses. Apoptotic cells shrink and are engulfed by cells around them. This orderly process is characterized by specific enzymes (especially a family of proteases called caspases), and specific breakdown products such as the "nucleosomal ladder," short DNA fragments generated where the DNA is cut between the nucleosomes into segments that are multiples of 200 base pairs long (*see* **Chromatin**).

Apoptosis is controlled by specific signals and genes, and mutations in those genes can be important in disease. Signals include receptors for the nerve growth factor (NGF) family of proteins, and for tissue necrosis factor.

Absence of NGF can cause cells to self-destruct—NGF is not really a growth factor, but rather a "do not die" or survival factor. (Other apparent growth factors may also act like this, rather than really promoting growth.) Other cells can also send signals that trigger apoptosis. The FAS receptor on immune system cells (which binds to the unimaginatively named "FAS ligand" on other cells) is one such signaling pathway.

Mutations in apoptotic genes are important in cell immortalization and cancer. bcl-2 is a gene which blocks apoptosis—some cancers over-produce the gene's protein and so become resistant to signals from the body which would otherwise cause it to self-destruct. bcl is an oncogene—its abnormal function allows the tumor to grow. In this case, that is because it stops apoptosis. Other genes like bax are active apoptosis triggers. An approach to cancer therapy is to give cancer cells back the ability to destroy themselves by apoptosis, by giving them normal bcl genes, or extra bax among many other possibilities.

Apoptosis is pronounced like it is spelt: "a-pop-toe-sis." Only Greek scholars and pedants insist that it should be "ay-po-toe-sis."

Aptamer

Aptamers are nucleic acid molecules, usually short lengths of DNA, which have been selected to bind to a target that is not another nucleic acid. Just as a chain of amino acids can fold to form a shape that exactly binds to some target molecule, so a nucleic acids can fold to form a specific binding shape.

It was a surprise when it was shown that nucleic acids did not only bind to each other with "classic" base pairing, but also to all sorts of other molecules in a way rather like proteins do. However, it is now well established that nucleic acids can be selected to bind to many types of molecules. Nucleic acids that bind to specific proteins (such as thrombin, an experiment done by Gilead Sciences) or small molecules such as NADPH or ATP have been developed by **Darwinian cloning** (see separate entry).

Just as there are specific structural motifs in proteins, such as alpha helices, so there are in aptamers. The most common is the G quartet—four guanosines in a row in a molecule make the molecule assemble into four-strand helices. Many other structural forms have been found.

The advantage of aptamers over proteins is that you can amplify a successful aptamer through PCR. The disadvantage is that nucleic acids are much more chemically limited than proteins.

Aptamers could be developed as drugs, like antibodies: Gilead has worked on this approach. Aptamers have the advantage that they are not immunogenic, because the body does not make antibodies against nucleic acids (except in a few rare diseases). The downside is that they are broken down quite quickly in the blood. This can be overcome by making chemical modifications of them (*see* **Antisense**). Noxxon has also developed a method for making mirror image aptamers ("speigelmers"). Because the molecules are chemically synthesized to be the "wrong" (i.e. non-natural) chiral enantiomer, the body's enzymes cannot attack them, and they can last for days in blood without breaking down at all.

Aquaculture

Aquaculture is growing water plants and animals in "farms," rather than harvesting them from wherever they happen to grow in rivers or seas. Farming of fish is called pisciculture, (*see* entry on **Fish farming**), and aquaculture using sea water is sometimes called mariculture. It is considered a part of biotechnology because, as a new commercial development, it uses the latest technologies rather than traditional ones, and often involves growing organisms in large volumes of water, which has similarities to growing large volumes of yeast or bacteria, biotechnology's "heartland."

Aquaculture is a growing industry, and produces a range of products:

- fish, especially high-value fish such as salmon and rainbow trout, see separate entry;

- crayfish, lobsters, oysters, shrimp, and other molluscs—these are farmed even more intensively (i.e. with more animal mass per cubic meter of water) than fish, being even more stupid animals;

- Macro- and micro-algae (*see* entries on **Algae** and **Biomass**).

Biotechnology comes into animal aquaculture in providing clean, well-aerated water for the animals to grow in, and food for fish farming: krill or powdered synthetic food, and as food additives such as astaxanthins (a bright red-pink pigment) to ensure the fish and prawns have the right color.

Biotechnology has been using genetic methods on aquacultured animal and plant species, especially to produce triploid and tetraploid organisms, and hybrid algae through plant cell fusions. Triploid trout, for example, are sterile, and can be used for biocontrol of weeds without the threat of being able to breed to become pests themselves. Salmon engineered with the growth hormone gene are made triploid by subjecting the eggs to pressure shocks so they are sterile, and so there is no chance that any escaping GM fish could breed with the wild population. Triploid oysters are reckoned by the US market to taste better than normal ones, and, being sterile, put more energy into muscle production and less into reproductive organs.

Artificial intelligence

Artificial intelligence is an area of computer programming techniques which attempt to capture aspects of human intelligence in programs. As no one can decide what "intelligence" is, useful AI products tend to use human intelligence as a model and then develop a specific computer version of that model. Among the applications used in biotechnology are:

1. *Expert systems*. These attempt to capture human expertise as a set of formal, explicit rules or formulae. They have been used in bioreactor control systems, where they can usefully combine quantitative and qualitative "rules" to guide a fermentor, as well as rules sets for predicting whether potential drugs may be toxic.

2. *Neural nets*. These were originally computer simulations of how nerves work. They are programs which can be "trained" to recognise patterns in large masses of data, by "training" them on previously identified examples of the patterns. Neural nets have been used in many applications, including

   • bioreactor control (where the computer is "trained" to identify the conditions in the bioreactor, derived from many sensors, under which the reactor performs optimally)

   • DNA sequence analysis (where the computer is "trained" with examples of genetic features, for example, coding regions, and then used to identify the same type of region in newly sequenced DNA.

3. *Fuzzy logic*. Normally, logic decides between statements that are true or false. Fuzzy logic, and older variants such as Baysian logic, quantify the degree of truth or falseness, or our confidence in truth, and so can handle situations where we cannot decide between absolutes.

4. *Genetic algorithms* (*including "artificial life"*). This is an AI technique inspired by the power of evolution to create the functions we see today in living systems, by evolving a solution to a problem from many possible solutions held in the computer's memory. This uses the power of natural selection to find a solution to a problem. Unfortunately, it also has the inefficiency of natural selection. A related concept is "artificial life," which is simulating very simple living things in a computer.

5. *Cellular automata*. These are computer simulations where each unit of memory or "cell" behaves according to what is going on in other cells. They have been used to simulate immune system function at a fairly primitive level. The most popular cellular automaton is a game called "Life," invented by Cambridge mathematician John Horton Conway in the 1960s.

A related topic is building better and faster computers: *see* **Bioinformatics**.

Biotechnology is also skirting the idea of artificially enhancing human intelligence. Using drugs, especially biotech-based drugs to boost intelligence in people is also a popular speculation, with a long history going back at least to the isolation of cocaine from the coca plant in 1855. Some compounds being developed for treating organic disease such as Alzheimer's disease have demonstrable effects on memory and attention in ill people, but increasing intelligence in people with no disease still seems to depend on luck, genetics, and 20 years' education.

Artificial nose

The nose works by having a large number of sensory proteins on the surface of its nerve cells, all of which bind many chemicals. The brain learns that a specific pattern of nerve cell firing is characteristic of a specific chemical, and so we associated that pattern—that smell—with that chemical. An artificial nose sensor mimics this approach. Rather than have a very specific sensor for each chemical it wants to detect, the artificial nose has a collection of relatively non-specific sensors, and uses computer software (often a neural net— *see* **Artificial intelligence**) to detect the specific pattern that is characteristic of the chemical the technologist wants to identify.

Artificial nose technology is often used when the technologist does not want to identify one single chemical alone, but rather a pattern which a human might say had a characteristic smell. Thus, artificial nose sensors have been applied in monitoring coffee quality, beer processing, and fermentations of many types. They have also been used to discriminate brands of beer, as quality control and anti-counterfeiting measure.

Common sensor elements are metal oxide-coated electrodes, MOSFETs, and piezo-electric sensors with different polymers on them that absorb different gases at different rates.

Artificial sweeteners

Many substances are used to make food taste sweeter without increasing its "calorie value." Among those of interest to biotechnology are:

Thaumatin. A protein produced by *Thaumatococcus danielli* in its fruit. Thaumatins are 3000 times as sweet as sugar, and at low concentrations enhance other flavors as well. Because they are proteins, they can and have been produced by genetic engineering in bacteria, so avoiding having to go to the tropics to harvest the fruit. Thaumatins have been produced in *E. coli, B. subtilis, Streptomyces lividans*, and *Saccharomyces cereviseae*, and the genes have been put into higher plants too.

Aspartame. Known as Nutrasweet, this is one of the most commonly used commercial artificial sweeteners. It is a dipeptide (aspartate-phenylalanine-methyl). Two aspects of its manufacture are of interest to a biotechnologist. Phenylalanine is relatively expensive, so selection, genetic engineering or other manipulation of a fermentation which produces phenylalanine more efficiently is a worthwhile goal as part of the production of aspartame. The other aspect is the synthesis of the dipeptide by enzymes, and particularly by using a protease to join the two amino acids together (rather than its more normal reaction of separating them). Both areas are undergoing continuing commercial development.

Artificial tissues

Artificial versions of some tissues can be built from the relevant cells, isolated and "expanded" (i.e. multiplied *in vitro*) and then introduced back into the body in a form that will function and not be destroyed. The simplest example is bone marrow transplant, where bone marrow cells are taken out, grown up and simply injected. However, for most other tissues, a more complicated structure is needed to support the new cells.

A related product is artificial blood, or blood substitutes, which are made chemically or biochemically. This can be hemoglobin encapsulated in artificial "red cells" (liposomes or some equivalent), hemoglobin cross-linked into large aggregates, or covalently linked to polyethylene glycol. Biotechnology produces both the hemoglobin and the encapsulation technology. Hemoglobin can also be produced as a fusion protein, with the normally separate alpha chains of hemoglobin (which are responsible for most of the toxicity) being made as a single fusion protein, which then also holds the beta chains together. This product was entered into human clinical trials in 1999.

One of the most successful artificial tissues are skin substitutes, although a true artificial skin has not been built. Keratinocytes—the cells which build up the epidermis—can be cultured in sheets from a skin biopsy: a two square centimeter biopsy will produce enough keratinocyte sheet to cover an adult in a month. This sheet can itself be used to replace skin lost in ulceration or burns. The cells are usually derived from the patient themselves, so that they are not rejected. More sophisticated grafts will be made of composites of several cell types, including the fibroblasts of the dermis, and of matrix materials to encourage the cells to grow.

This overlaps with the studies of wound healing technology, where materials are developed with biomimetic properties that can stimulate the growth of new tissue to cover a wound. Unfortunately, wound healing is stimulated by a concert of many growth regulatory factors, and so no one of them has proven to be very useful. Integrins are used widely in such product ideas, because they encourage cells to migrate and adhere to a substrate.

More complicated tissues have also been developed at an experimental level. In late 1995 experiments moulding a synthetic matrix on which to grow cartilage hit the headlines with stories of the "artificial ear" grown on a mouse. (The experiments attracted controversy as "trivial"—ironically, the scientists only used the ear shape because their previous work, on simpler shapes, was considered too uninteresting to publish in scientific journals.) There are many developments trying to get a semi-synthetic bone matrix to work as a substrate for growing new bone. A range of proteins under the general heading of bone morphogenic protein (BMP)—bone growth factors— have been tried in such matrices to enhance bone formation.

The matrix materials used in these products are important, as most of the body's cells function only when they have a complex support of proteins, growth factors, and other cells around them. The matrix in artificial tissues is meant to mimic this complex environment as closely as it can.

Work is also advanced on an "artificial pancreas" for diabetics. The transplanted cells, which would normally be destroyed by the patient's immune system, can be encapsulated in a semi-permeable membrane so that they get nutrients from the body but the immune system does not "see" them. It is quite difficult to design a system which will allow cells to get enough nutrients, especially oxygen, but will keep them immunologically isolated. Debris from dead cells can get through most semi-permeable membranes and trigger the body to mount an antibody response. Despite this, encapsulation is one of the more promising approaches to an "artificial pancreas."

An alternative approach to artificially building a tissue is to grow it in another animal, a technology known as **Xenografting** (see separate entry).

Aseptic processing

Aseptic processing is processing materials so that no microorganisms can get into them. This either means that the production process machinery must keep the material sterile, or the whole manufacturing environment must be aseptic, that is a clean-room environment. In many pharmaceutical production processes, both approaches are adopted. UHT (ultra heat treated) milk and pasteurized fruit juice is usually packaged aseptically, as opposed to being packed and then sterilized (which is what happens in traditional food canning).

Issues in aseptic processing are:

- The balance between the growth rate of bacteria and the processing. Pasteurized milk, for example, is not sterile, so it has to be processed, shipped, sold, and used fast enough so that the bacteria in it do not have time to grow. If the starting materials are sterile and can be kept sterile, then this is not a concern.

- Monitoring the microbial contamination: monitoring methods can be a slow process, slower than the rate at which the bacteria can grow in the production line.

- Cleaning and sterilizing equipment, which has to be done by **clean-in-place** technology (see separate entry).

- Validation and QC, testing to make sure that the process actually is aseptic.

Most manufacturers, even of drug products, control their process sterility using parametric release (also called parametric control), where the parameters of the production line are rigorously monitored to make sure that they fall within the range where the producer knows that contamination is kept at bay. *See* entry on **QC/QA**.

Assay

An assay is a specific chemical test for something. In biotechnology, this usually means a specific biomolecule in a mix of other ingredients. This means that an assay must be able to select the analyte molecule (the thing to be measured) from among a host of others, and measure its concentration. This in turn means that at least one of the reagents involved must be very specific for the analyte.

Lots of assays are developed by and used in biotechnology. Among the more common technologies are **immunoassays** and **enzymes**, discussed in separate entries.

Key parameters in an assay are its sensitivity and specificity. There are two meanings of these words.

Diagnostic sensitivity and specificity are slightly different. These ask how good the assay is at telling me whether or not something has happened. Any test will give true positives (i.e. tests which show up positive when the sample really is positive—P), false positives (showing up positive when the sample is negative—F), true negatives (not showing positive when there is nothing there—N), and false negatives (not showing up positive when there is something there—G). The sensitivity of a test is defined as

$$\frac{P}{P + G},$$

that is, what fraction of the test that should be positive are positive. The specificity (or selectivity) of a test is defined as

$$\frac{N}{N + F},$$

that is, what fraction of the tests that should be negative are negative. The sensitivity and selectivity of a test can be traded against each other by adjusting the threshold—for a test for explosives at airport, for example, the threshold is set very low so the slightest hint of a bomb gives a positive. F—the number of false positive alarms—is very high, but G—the number missed—is (hopefully) very low.

Sensitivity and specificity can be defined in terms of a specific parameter, or the reason you are doing the test. In medical tests, this is analytical or clinical sensitivity. The analytical sensitivity is how well the test measures the chemical analyte. The clinical (or functional, if the test is not a medical test) sensitivity measures how well it detects the disease. A thermometer has 100% sensitivity for measuring elevated temperature, but only 50% sensitivity for detecting the common cold, because temperature itself is only weakly correlated with colds.

Automation

Automation means getting machines to do something, rather than having a human do it. In biotechnology, this usually means getting machines to perform repetitive laboratory procedures. It is particularly prevalent in drug discovery, where hundreds of thousands of tests have to be run to discover one candidate drug. However, other areas of biotechnology, such as process monitoring in fermentation, environmental sample processing, even testing meat for quality are also highly automated.

Automation is an engineering discipline, not a biotechnological one, but despite this many biotechnologists think that they can build automated systems. Strong engineers can be reduced to tears by the systems that some biotechnologists call "automation." Common problems they encounter are:

- *Automating the scientist*. How a scientist does something is convenient for them, using human limbs to perform a small number of tests. It may not be the best way for a machine to do a very much larger number.

- *Not having the machines talk to each other*. This means both physically (so the output of one machine fits into the input of the next) and in handling information.

- *Not building robustly*. Having a robot that needs a PhD standing over it all day to keep it working is no better than having the PhD doing the experiments. However, building a robot that will work for 24 h without even one tiny fault is a major challenge.

- *Labs like factories*. A related idea, this says that if you are going to turn science into a factory-like process, then you must organize it like a factory, not like a research scientist's unstructured workplace. Again, this applies to both hardware layout and work-flow.

- *Reinventing the wheel*. Humble industries such as food manufacture and brewing have been automating for decades, but biotechnologists often do not realise this, and try to invent their systems from scratch.

Automation of laboratory processes is now a major industry, with companies like Tecan and Zymark making complex systems that can handle many types of reactions. Most are based around the multiwell plate (*see* **HTS**). The systems include assay readers such as fluorimiters, plate washers and stackers, pipette systems that dispense liquids into the plate, linked by moving tables and bents that move the plates around.

Autoradiograph

Radioactive labels are often detected by autoradiography (autorad for short), a technique where an X-ray film is laid against something in the dark, and the film darkens where the radiation struck it, making an image of the radio-active regions of the object. Typically, autorads produce dark patches of exposure on transparent film, and so are presented as black spots or bands on a white background. In some experiments, the film emulsion is actually laid onto the experiment directly, so that, for example, the radiation trapped in specific sections of cells can be detected.

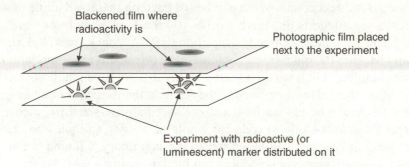

Blackened film where radioactivity is

Photographic film placed next to the experiment

Experiment with radioactive (or luminescent) marker distributed on it

Luminescent labels are now largely replacing radioactive ones for many applications, but the same approach can be used to visualize the very weak light given out by these labels—an X-ray film is laid directly on the experi-ment, and the blackened areas show where the label is. Confusingly, this technique is also often called "autoradiography" and the product an "autorad."

Patterns of radioactivity can also be captured by the electronic equivalent of an X-ray film, called an imager or phosphorimager.

Bacterial biodiversity

"Biodiversity" is usually thought of as the huge variety of animal and plant shapes we see around us. But microorganisms are also extraordinarily diverse, and this is a powerful resource for biotechnology. Unlike larger organisms, microrganisms have less diversity in shape and more in their chemistry. So, microorganisms are a vast resource of both chemicals and enzymes. Exploring this wealth is sometimes called bioprospecting.

The chemicals have been used extensively as sources of potential drugs, as well as other products (*see* **Natural products**). The enzymes can be used in a huge range of processes: indeed, enzymes are central to most of biotechnology, and are always discovered in nature before (sometimes) being "engineered" for human use (*see* **Enzymes**).

Most bacterial and fungal biodiversity remain unexplored. It is easy to find bacteria in soil by putting soil samples into culture medium and seeing what grows. But direct counts of the number of bacteria in a soil sample under the microscope suggests that there are 10–100 times as many bugs there as can be found by culture. Directly amplifying rRNA genes from soil confirms many bacteria there never seen before: these organisms are sometimes called "unculturable," but it would be better to say that we have not discovered how to culture them. However, even if the bacteria themselves cannot be grown, their DNA can be cloned from cells, either as gene-sized fragments or as larger fragments likely to code for whole pathways, and put into cells like *Escherichia coli* that can be grown in the laboratory. Cloning whole pathways is usually preferred, and this can be done using a BAC vector.

Some bacteria and fungi probably cannot be cultured on their own, because they are part of a close symbiosis, where two species' chemical lifestyles are so intertwined that they cannot live without each other. Lichens are a classic example, a symbiosis of a bacterium and a fungus that cannot survive on their own but together can survive in the harshest of conditions.

Another source of bacterial biodiversity is exotic environments, such as very cold or hot environments (*see* **Extremophile**). One of biotechnology's interests in space exploration is the faint possibility that life will be found on other planets, such as Mars or the "internal oceans" on the moons of Jupiter (particularly Europa), with chemistry utterly different from anything we have seen before. But this is too speculative for even biotech to invest in at this stage.

Bacteriophage

A bacteriophage is a virus which attacks bacteria. They have been used extensively in DNA cloning work, where they are the basis of convenient vector molecules. The bacteriophage (or "phage") used most are derived from two "wild" phages called m13 and Lambda: P1 is another, less commonly used phage.

There are two uses of phage in biotechnology, a practical one as a tool for recombinant DNA and a potential one as a therapeutic.

Both m13 and lambda are used as a basis for gene cloning vectors. Lambda is a "lytic" phage—it replicates by breaking open (lysing) its host cell. If a few phage are spread on a huge number of bacterial cells, they lyse open the cells they hit, releasing more phage, which then lyse open the neighbouring cells, releasing more phage and so on. On a bacteriological plate, this results in a small clear zone—a plaque—where each original phage landed. m13 can grow inside a bacterium as a plasmid, so that it does not destroy the cell in infects but causes it to make new phage continuously. Both these phage grow on *E. coli* as a host bacterium. Many other phages, both of *E. coli* and of other bacteria, are used in more specialist research applications.

Since they were discovered, phage have been proposed as a treatment for infectious disease, an approach called phage therapy. They look like the ideal treatment—one phage can start an infection in a bacterial population that kills millions of bacteria, and they do not affect the patient at all. However, this has never been brought to practice, for three reasons: you need a new phage for each type of bacterium (unlike antibiotics, which affect many types), the bacteria quickly develop resistance to them and then you have to start all over again, and phage are surprisingly hard to manufacture in bulk.

Baculovirus

Baculovirus is a class of insect virus which has been used to make DNA cloning vectors for gene expression in eukaryotic cells. Technically, they are *Autographica californica* (the Alfalfa looper, a small moth) Multiple Nuclear Polyhedrosis Virus (AcMNPV). The vector system was derived to enable biotechnologists to make large quantities of proteins from cloned genes in insect cells (the cells usually used are a cell line derived from the fall army-worm). Baculoviruses have a gene which expressed late on during their infection cycle at very high levels, filling the nucleus of the cell with many-sided bodies full of a protein which is not necessary to produce more viruses, but is necessary for the virus' spread in the wild. In the vector cloning system, this gene is replaced by the one we want expressed.

The baculovirus DNA is very large (100–150 kb), and so conventional recombinant DNA methods cannot be used to engineer it. Instead, plasmids containing the desired gene are recombined with the virus *in vivo* through homologous recombination. This technical difficulty meant that baculovirus was not taken up in research as quickly as some other systems, but it is now gaining popularity in several applications.

The main strength of the baculovirus system is that, as a genuine animal expression system, it produces proteins which are glycosylated like the proteins in animals. This, in combination with the relatively high levels of expression (up to 50% of the infected cells' protein content) can make this an attractive option for making proteins which are to be used as biopharmaceuticals, outweighing the problems of growing a virus in cell culture. In addition, baculovirus does not infect vertebrates, so there are no fears that any virus from the production process that gets into the final product could cause disease.

Baculovirus has recently been developed to grow in some cell lines in serum-free medium. This is to overcome the other main concerns about the safety of cell-culture-derived products, that of BSE or viruses coming from the serum that is commonly used as an additive to cell culture media.

A novel use for baculovirus systems is their use as viral insecticides. A gene is inserted into the virus which is lethal to an insect (e.g. the endotoxin gene from *B. thuringiensis*), but does not affect isolated cells. This is then used to produce infective virus, which can infect insects and kill them. There are technical problems with this, however (such as whether the virus is still infective in a real organism) as well as regulatory ones.

At a research level, baculovirus can also be used for peptide display in eukaryotic cells, analogous to **phage display** (see separate entry). The advantages of using the virus in this way are that the proteins carry the correct post-translational modification, and because other cell proteins are present as well the "new" protein can be tested for its function as well as its presence.

Binding

Biology only works because many biological molecules bind extremely tightly and specifically to other molecules—enzymes to their substrates, antibodies to their antigens, DNA strands to their complementary strands, and so on. This binding is entirely spontaneous, and depends on the chemical nature of the molecules concerned: the exact shape and chemical nature of parts of their surfaces mean that they are "complementary." A common model is "lock-and-key," much used to describe how enzymes fit around their substrate.

Binding can be characterized by a binding constant or association constant (K_a), or its inverse the dissociation constant (K_d). Mathematically, if two molecules (Mol-1 and Mol-2) link up to form a complex, then

$$K_a = \frac{[\text{complex}]}{[\text{Mol}-1] \times [\text{Mol}-2]} \qquad K_d = \frac{[\text{Mol}-1] \times [\text{Mol}-2]}{[\text{complex}]}$$

where some thing in square brackets like this—[something]—means the concentration of that something. (This is a very commonly used notation in biochemistry.) Related terms are pK_a and pK_d, which are the $\log_{10}$ of K_a and K_d respectively.

For any given concentration of Mol-1 and Mol-2, the higher the K_a (or the lower the K_d), then the more of the complex and the less free Mol-1 and Mol-2 will be around.

In general, in biotechnology when someone talks about K_a or K_d, they want a very "tight" binding, so the bigger the K_a or the smaller the K_d the better. Antibodies generally have K_a between 10^7 (bad) to 10^{10} (good). Hormones bind to the receptors with K_a from 10^4 to 10^8. Proteins such as cytokines or growth factors can bind to their receptors much more tightly, with K_a values of 10^{10}–10^{12}. The prize goes to streptavidin, the protein that binds biotin (*see* entry on **Biotin**). The K_a for the biotin–streptavidin binding is around 10^{16}, that is enough for streptavidin to be able to suck just three micrograms of biotin out of a small aircraft hanger-full of water.

K_a and K_d are thermodynamic measures. They describe what would happen if we had enough time. Often a system is kinetically limited, that is the reactions do not happen fast enough. The key rates are "on rate" (how fast the molecules stick together) and "off rate" (how fast the come apart again). In the biotin example, although in theory a few milligrams of streptavidin in a tube could suck all the biotin out of a large swimming pool, in practice it would take years for the biotin to diffuse to a single, tiny pot of streptavidin, unless we pumped the pool past the streptavidin. In recent years the kinetics of binding have been recognized as important for designing drugs as the thermodynamics. Some successful drugs, such as thiotropium, work not only because of high K_a but also because of very low off-rates, so they have a long action on the cells they are targeting. SPR-based sensors such as the BiaCore can be used to measure on- and off-rates directly (*see* **Optical sensors** and **Protein binding**).

Bioaccumulation

This is the accumulation of materials which are not critical components of an organism by that organism. Usually, it refers to the accumulation of metal. Many organisms—plants, fungi, protists, bacteria—accumulate metals when grown in a solution of them. Sometimes this is part of their defence mechanism against the poisonous effect of those metals, sometimes it is a side-effect of the chemistry of their cell walls.

In a few cases, bioaccumulation is economically important as part of the microbial mining cycle. Using this biosorption process, metals present in very low concentrations in water can be accumulated in the cell walls of living organisms and subsequently harvested. Bioaccumulation and the use of bacteria to remove toxic metals from waste water as a purification ("bioremediation") process are clearly closely related: *see* entries on **Biosorption**, **Microbial mining**.

Bioassay

A bioassay is an assay that has a living system as a key component. Usually it means a way of measuring the concentration of a chemical, although bioassays for magnetic fields (using homing pigeons or magnetic bacteria), ionising radiation (measuring mutation) or other physical effects are possible. Bioassays for chemicals are valuable when a living thing reacts extremely fast to tiny amounts of the chemicals, something that it is very hard to build physical or chemical systems to do.

Many bioassays have been in traditional use—the proverbial canary in the coal mine was a bioassay for poisonous gas, the canary being the biological element. However new bioassays are usually developed using bacteria or animal or plant cells, as these are much easier to handle than whole animals or plants, cheaper to make and keep. Thus, bacterial bioassays for BOD (*see* **Sewage treatment**) and poisons in general are in use in the water industry. Here, bacteria are mixed in with a sample of water, and how well they can metabolize (and hence use up oxygen, produce carbon dioxide or, in one case, give out light) is measured by an instrument.

Many of the cytokines and growth factors that biotechnologists are now producing using recombinant DNA methods were originally identified using bioassays in which mammalian cells were used to detect minuscule amounts of the compounds concerned through their potent effects on the cell's behavior. They are often still measured by bioassay, because this is the only reliable way to distinguish the functional, active protein from similar proteins or degraded material (i.e. they are *functional* assays). Bioassays also reflect the mode of action of the material, that is what it meant to do, rather than an arbitrary definition of chemical purity. Many very active biological substances are still measured in "international units," which are defined as the amount required to have a specific effect in a bioassay.

On the borderline between bioassays and chemical assays are **immunoassays** (see separate entry) and enzyme assays.

Bioassays can be turned into sensor systems (*see* **Immobilized cell biosensors**).

Biocatalysis

This is a broad term meaning the conversion of chemicals by a biological catalyst. This can be a whole organism (Bioconversion), or an enzyme isolated from an organism (Biotransformation).

Biocatalysis is a powerful approach, because biological systems are usually highly specific about the chemistry that has to be performed to keep them alive. Thus, biocatalysis is used to make specific changes to molecules where a "standard" chemical reaction might attack the whole molecule ("regiospecific changes"), or generate one chiral enantiomer.

Biocatalysts work at moderate temperatures and pressures, which can save energy, materials, and engineering costs. They also generally avoid the use of dangerous reagents such as heavy metals or toxic gasses, which can be very expensive to dispose of after the chemistry is complete. With growing concern about the release of chemicals into the general environment, this is becoming a more and more pressing reason to use biocatalysis.

Biocatalysis usually avoids expensive reagents as well, such as rare metal catalysts, but this advantage can be negated by the cost of the biocatalyst itself.

Biochip

A "biochip" is an integrated circuit-like device, very small and very complex, that performs some biochemical or biological process. In fact, it is not a "bio" chip at all—it is a small device that happens to do something that biologists want. However, the name has stuck. A better, and alternative, term is "lab on a chip."

The term "biochip" is also sometimes used to refer to a computer chip that uses a biological element as part of its construction. This is the more speculative field of molecular computing (see separate entry).

Biochips are related to microarrays, and microarrays are often called biochips. In particular, DNA arrays are often called "Gene Chips" (strictly, a term copyrighted by Affymetrix). (*See* Microarrays.) However, a biochip can also have some functional engineering on the chip, such as tiny tubes and reactors that materials can flow through, valves, electrodes, pumps, even light sources and detectors for analyzing the results.

Most biochips are made using MEMS, which stands for micro engineered mechanical systems. MEMS itself is a spin-out of the silicon chip industry. MEMS workers can make silicon (or sometimes glass) devices with tubes, levers, and other structures that are only microns across. The most extreme example is a tiny gas turbine, being built by MIT as a replacement for batteries. They are created by chemically etching the silicon, even though this is called "machining." Analogous devices in plastic can be made, albeit not so sophisticated, by gluing together layers of plastic sheet with dents, bumps, and holes in them.

The behavior of materials at this small size scale is quite odd. Gravity is almost irrelevant to something a micron across and weighing a millionth of a microgram, but surface tension and viscosity is extremely strong, so (if we were a micron tall) a droplet of water would look like a weightless ball of concentrated Jell-O surrounded by a Kevlar "skin." MEMS technology seeks to use these properties to the advantage of the engineer.

The advantages of doing analysis on this scale are significant. Chromatography and electrophoresis can be done at very high speed, there is no turbulence in the tiny channels. Diffusion reactions happen almost instantly in a chamber only a few microns across. And the scale of the biochip is the same as the size of many of the tiny samples that research and medical diagnostics have to work with: by keeping it in a tiny volume it remains concentrated and easier to detect.

Biochips are still fairly experimental. Agilent is the leader in applying this "lab on a chip" technology, with a system for DNA analysis. Gyros (who sold their technology to Tecan in 2002) developed a lab-on-a-CD system that uses the plastics fabrication technology of CD manufacture to make a simple analytical system.

The aim of many of the research groups in the field is the total analytical system—TAS—a system which performs a complete suite of analyses in one system, for example analyzing a set of metabolites in a blood sample or a set or proteins from a cell. Micro-TAS is a TAS system "on a chip" which only requires picoliters (10^{-12} l) of sample. So far, only prototype Micro-TAS systems for performing PCR on already clean DNA have been made.

A related idea is using MEMS technology to build a chip that performs chemical reactions. Because the amounts of chemical concerned are usually extremely small, this is not thought of as a manufacturing technology, but as a way to carry out combinatorial chemistry. The technical problems in doing this have prevented such systems being built so far.

Bioconversion

Bioconversion is the conversion of one chemical into another by living organisms. This contrasts to conversion by enzymes (which is biotransformation) or chemical processes. Synonyms are biological transformation or microbial transformation. In many cases, bioconversion is used together with more traditional organic chemistry to complete a complex synthesis.

The usefulness of bioconversion is much the same as that of biotransformation—especially its extreme specificity and its ability to work in moderate conditions of temperature and pressure. However, there are several differences.

Positive points are that bioconversions can involve several chemical steps, and the enzymes involved can be quite unstable—because the cell continuously remakes them as they break down. So, these are conversions that it would be very hard to do "in the test tube."

Negative points are concerned with efficiency. Most bacteria either convert chemicals very inefficiently, or they convert them very efficiently into more bacterium: neither is of much use, except for SCP production. Thus, to make an effective bioconversion process the bacterial strain has to be optimized so that it converts substrate to product efficiently but does not convert the product into something else.

A major commercial application is in the manufacture of steroids, such as the conversion of progesterone (which is isolated from pregnant mares' urine, a cheap and plentiful source of supply) into cortisol, using *Rhizopus* cultures. The "basic" steroid molecule is itself a very complicated molecule, and not one that it is easy to modify by normal chemical means to produce the very specific molecules needed for drug use. However, a variety of bioconversions that attack only specific bits of the molecule can achieve this.

Other applications of bioconversion are bioremediation and bacterial mining (or biohydrometallurgy), which are discussed elsewhere.

Bioconversion in organic solvents

Many chemical reactions which are to be targeted for bioconversion or biotransformation are normally carried out in organic solvents, not in water, because the reagents are insoluble in water or because water causes undesirable side-reactions. Enzymes can also be used in organic solvents, but interest is growing in the use of bacteria in solvents other than water.

Most bacteria are killed by being shaken up with organic solvents such as DMSO or chlorethylene. However, some bacterial bioconversions can be carried out in mixed phases, where the bacterium is tough enough to survive living next to droplets of the solvent. This has the advantage that a large number of enzymes, or very unstable enzymes which would not survive on their own in a bioreactor, can be used for a bioconversion. The disadvantage is that the bacterium must be kept alive, and that bacteria produce all sorts of other metabolites than the one you are after.

See also Organic phase catalysis.

Biocosmetics

A term for cosmetics with a biotechnological ingredient, or an activity or action based on biological knowledge (rather than cosmetic industry experience or belief, or marketing ploys).

Biocosmetics divide into three areas: biomaterials, biologically based ingredients, and medically rationalized products. The last class include hypoallergenic products and UV-blocking agents, whose mode of action is supported by medical research, but which are not "biotechnological" as such.

Biomaterials include the use in cosmetics of collagen and collagen hydrolysate, a wide range of lipids as "moisturisers" (including liposomes, which are claimed to carry active ingredients into the skin), fibronectin, and hyaluronic acid. These, and especially the last, are water-retention agents which are meant to prevent the skin drying and wrinkling. Lipids such as gamma linolenic acid are also claimed to have anti-inflammatory effects.

Biological ingredients include biotin, cyclodextrins, sphingosine, and a range of pigments. These are all "natural" products, that is, they are made in a living organism rather than by chemical synthesis, and so can be produced biotechnologically: however, their actual effect is debated by medical professionals.

Cosmetics cannot be too effective as any cosmetic which had a substantial effect on skin physiology would be classed as a drug, and so would have to undergo all the rigorous proof of efficacy and safety that drugs must undergo. Cosmetics which are deliberately being developed to have a drug-like effect are called cosmeceuticals. Examples include the use of retinoids for reducing skin wrinkling, and hydroxy-acid compounds called exfoliation agents, which increase the rate at which the skin cells are replaced and so give a more youthful appearance.

In reality, almost no chemicals can get across the skin to the living cells underneath. It is this "barrier function" that is what the skin is there for. So even the most effective cosmeceuticals or functional cosmetics work on the top dead layer of skin cells only.

Biodegradable materials

Biotechnology has preceded the "green" movement by some years in developing biodegradable materials. These endeavors generally fall into three areas:

1. Developing organisms which can break down normal materials, especially plastics. *See* **Bioremediation**.
2. *Developing composite materials.* Most "biodegradable" plastics are composite materials of a conventional plastic laced with a biodegradable material such as starch. They break down when soil bacteria digest the starch, leaving small granules of the plastic behind. The same approach can be used to make a "pure" biodegradable material, such as earthshell, a combination of starch, recycled fiber, some traces of minerals (such as limestone), and protective coatings such as wax.
3. *Biopolymers.* Most living things produce polymers to make cell walls or other structural materials. Some of these can be used to make things: however, most are easily wetted, and tend to fall apart if left in the rain for a long time. A few exceptions have been found, the most well-developed of which are poly-hydroxybutyrate (PHB), developed by ICI and polycaprolactone (PCL). *See* **Biopolymers**.

One strong, flexible, water-resistant, biodegradable polymeric material not usually talked about is wood. Quite a lot of plant biotechnology is aimed at trees, and biotechnology is indeed working on genetic engineering trees (*see* **Wood**).

Biodiversity

This means the diversity of life, usually either the diversity of natural organisms, or the diversity of farmed crops or animals.

The easiest measure of biodiversity is genetic diversity, a measure of how many distinct genetic species there are in an area or population. These are sometimes called genetic resources, because they can be used as the starting point for new products or processes.

In the larger world, biodiversity concerns the vast range of plants and animals alive. Many of them may produce something useful to man—a new drug, a new foodstuff, a new material. If they are allowed to be wiped out, that potential will be lost for ever. Concern over this arises from fears of habitat loss in "third world" (i.e. not rich) countries, and exploitation of those by developed countries (*see* **Bioprospecting**).

In the farm, biodiversity is considered to be a *good thing*. If a country plants only one type of crop (for example), then a disease can sweep the entire country's crop from the fields. This happened in a wave of epidemics to the US wheat farmers in the 1960s. Thus, not having a country plant just a single type (or cultivar) of a crop is protection against pandemics. This is particularly relevant to habitat destruction in Northern Europe, where there is very little wilderness left which can act as "refuges" for wild species.

The role of biotechnology in agricultural biodiversity is double-edged. If biotechnology produces a new wonder-wheat, then that will be planted in place of all other cultivars, and the wheat-growing world will end up being a monoculture—biodiversity will have been reduced. On the other hand, biotechnological methods are such that if you can transform one cereal with a gene you can transfer a lot more, so biotechnology could actually increase biodiversity by increasing the number of crops with desirable genes in them. And it has been argued that the "green revolution," including biotechnology, has been so successful that farmers are no longer under pressure to grow monocultures of the most productive crop: many farmers in Europe are paid to let fields lie fallow to reduce production, increasing overall biodiversity substantially.

On the rain forest question, biotechnology is less vociferous, but one of the key technologies in plant biotechnology is plant cloning, storage and micropropagation, which could be used to store and propagate rare or endangered species.

Bioethics

Ethics is the study of what is considered "good" (moral) behavior, and how relations between people and institutions can be arranged to advance good behavior and prevent bad behavior. This depends on what society agrees is "good," and much of the debate in biotech ethical discussion comes from a disagreement about this. Much of what is "good"; depends also on what the behaviour results in, and this is again a source of debate. For example, is DNA testing for predisposition to diseases a "good" in all cases, "good" in theory but "bad" in practice because you could not prevent the wrong people exploiting the information (so it would be ethical if *you* did it on yourself, but not if I did it on you), or basically "bad" because it panders to a genetic, deterministic view of humans that is at odds with what we really are?

At one end, ethics can be enormously useful in focussing attention on problems that need to be solved. At the other extreme it can become a name-calling argument between the "pro-biotechnology" and "anti-biotechnology" schools of thought which reduces complex arguments to clichés and abuse. The Human Genome Project has set aside around 3% of its budget solely to consider ethical issues. The medical genetics and pharmaceutical communities have employed ethical experts for years. Thus, the industry and regulators of biotechnology have a substantial concern in bioethics.

Ethical decisions can be based around one of several argument types.

"Received wisdom" uses an external reference and says that no matter what the scientists say, a particular idea is "good" or (more usually) "bad." Most religious arguments for or against specific technology are of this type, as is the "yuk factor" (see separate entry). A strong component of public opinion in bioethics is related to this received wisdom, where some things are considered as "against nature" because of the fundamental way that people understand the world.

The utilitarian argument, formalized by philosopher John Stuart Mill, says that the best ethical choice is that which creates the greatest good for the greatest number. All pragmatic arguments are of this type, where the debate becomes one of whether the benefits of doing something outweigh the costs and risks, and so this boils down to a technical argument. Most biotechnologists are Utilitarians.

Rights—it is easy to state that people "have a right to…," but who says so, and why? Usually what is meant is that people have a desire, not a right, and while that is perfectly acceptable, it is a different argument. Balanced to this is the duty argument—people ought to do something (usually "scientists," "the government," or "they"). But again, who says so, and why? And at what cost, and who should pay? The one generally accepted duty is to make the same ethical "rights" available to people without technological power as are available to those who have power, and it is this that is at the centre of the debate about whether to sell GM crops to the Third World when the same crops are not used for human food in developed countries.

The Ratchet argument—once the technology is out there it cannot be taken back, so let us not put it out until we have understood all the consequences—in fact, a variant on the Utilitarian argument, but is one that cannot be answered by facts, as in reality we never know all the effects of a new technology.

A good way to solve such conundrums is often to work out some specific examples, rather than argue generalities. Thus, much of ethics tends to shade into legal and social issues. So ethics is often grouped into the broader term "ELSI"—Ethical, Legal, and Social Issues.

Specifically, ethical issues relating to biotechnology are:

1. *Scientific conduct*. This applies everywhere, but particularly in biotech where there is the potential of substantial wealth and fame. Scientists are above all meant to report the truth as they find it (no "fudging" of results), and to respect authority (i.e. refer honestly to who originated an idea or a discovery). This has direct, practical impact on reporting bad news from your company (ethically, you have to do it), and giving credit to others for your potentially patentable invention. Usually, both are the subject of law suits rather than ethical debates in the practical world of biotechnology.

2. *Commercial* vs. *scientific motivation*. A related idea is that projects done for commercial gain are inherently more ethically suspect than those done for academic research.

3. *Genetic information*. It is generally held that someone's genetic makeup says something "fundamental" about them (that, for example, their hair color or choice of clothes does not), and so this must be treated as both valuable and private, and that an individual's consent must be given for use of that information. *See* entry on **Genetic information**.

4. *Animals*. The validity of making animal models for human diseases, for example, transgenic models of cancer. ("Animal rights" as a whole is also a major issue for some biotechnology companies, but is not really a biotechnology-specific issue.)

5. *Clinical trials*. How people should be informed for trials of experimental procedures and products, and the problem of trading off a new drug's potential side effects with a need to have patients benefit from it as quickly as possible (*see* **Treatment protocol program**).

6. *Embryo research*. The role of biotechnology in embryo and fetal research. This is not actually a biotechnology issue, but one relating to people's tremendous sensitivity to and empathy with the young.

7. The justification for patenting life forms and pieces of living material (especially genes).

8. The use of genetic resources in biotechnology exploitation (*see* **Bioprospecting**).

Biofilms

A biofilm is a layer of microorganisms growing on a surface, in a bed of polymer material which they themselves have made. Biofilms tend to form wherever a surface on which bacteria can grow is exposed to some suitable growth medium and to a supply of bacteria. Thus, biofilms form on sites as diverse as domestic plumbing, hyperbolic cooling towers in power stations, sewage plants and teeth.

The bacterial film is a community (or "consortium") of different organisms. Some corrode the surface (biocorrosion), which leaves the surface rougher and more chemically "sticky." Other bacteria synthesize polymer (mucopolysaccharides) to stick themselves, and any other bacteria nearby, to that sticky surface. The resulting films can be amazingly hard to get off. They also increase the surface's roughness, and block the diffusion of oxygen through membranes.

The process of covering surfaces with biofilm in this way is called biofouling. It is a particular problem where liquid is recirculated round a closed loop of pipework (so any bacteria washed off the film can get the chance to stick back on next time round), and in filter membranes. Unlike normal fouling of membranes by solids or large molecules, biofouling is an active process which, once under way, cannot easily be reversed by cross-filtering or reversing the flow through the membrane. The biocorrosion can also start to break down the membrane, making it leaky.

Biofouling is also a major problem in medical equipment, especially catheters (the thin tubes that are inserted into a vein to provide "intravenous" drugs, among other things). Because there are many bacterial types in a biofilm, one of them can usually break down most antibiotics—biofilms as a whole are therefore more resistant to antibiotics, and can shed bacteria into the patient.

A way of preventing biofilm build-up is to kill the bacteria with a broad-spectrum biocide, which can either be dosed into the solution (which is why chlorine generating chemicals are added to swimming pools) or impregnated into the surface.

Biofilms can also be used, in biofilters, fluidized bed reactors and some biosensor systems.

Biofilter

A biofilter is a large-scale filtering system where the surface area of the filter provides a support for organisms. Chemicals in the fluid flowing through the filter are broken down by the organisms. This is mainly used for waste disposal.

The most common form is the trickle bed sewage treatment process. Here a bed of gravel is allowed to grow a film of bacteria to assist **sewage treatment** (see separate entry).

A growing application is the use of biofilters for treatment of waste gases. Gases from food and chemical factories, especially for processes that produce a lot of organic volatiles such as acetone or hydrocarbons, can be adsorbed efficiently by a large filter mesh of microorganisms. These have to be kept moist and supplied with nitrogen and trace elements, but otherwise the filter is self-sustaining, with the organisms getting carbon from the gases and oxygen from air. This is often cheaper than conventional chemical scrubbers.

A variant on the biofilter idea are the use of reed beds as biofilter. The reeds, planted in gravel or loose soil, grow from hollow, air-containing roots (rhizomes). Oxygen passes down the plant and out through the root hairs, so the water immediately around the root is highly oxygenated—this region around the roots is called the rhizosphere (in reed beds and other plant ecosystems, too). The oxygen supports the growth of lots of bacteria, which can in turn use up the carbon pollutants in high BOD water. The bacteria can also remove nitrogen and phosphorus, pathogenic bacteria, and some metals and other toxins. If there is a high level of metals in the effluent, then the reed beds have to be cleared out and replanted from time to time, and the old, metal-saturated reeds disposed of safely.

Two types of systems are used: horizontal beds where the water flows through like a stream, and vertical flow beds where liquid is pumped over the bed like a trickle bed sewage system: this latter is used for more heavily contaminated solutions. Reedbeds are usually used for secondary treatment of farm and municipal waste—they cannot handle raw sewage. They have the advantage of being relatively cheap to build and very "green" in literal and political senses.

Biofuels

Biofuels are fuels made from bulk biological materials such as cane sugar or wood pulp. There are a range of ways of converting these rather bulky, inconvenient fuel materials into fuels which are useful for industrial or transport use, or as starting materials for the chemical industry. Converting biomass into replacements for gasoline has attracted a lot of interest in the 1980s after the "oil crisis" of the late 1970s, went out of fashion in the 1990s, but is starting to be reconsidered widely again with the rise in concern about global warming and almost continuous armed unrest in the Middle East.

Many wet biomass materials, such as starch, sugar, bagasse (the solid residue left after the juice has been squeezed from cane sugar), sewage, waste waters, etc. can be used to make ethanol or methane by fermentation. Ethanol for use as a fuel is made from cane sugar by fermentation and distillation in commercial quantities in Brazil, where the economics are unusually favorable, and "Proalcool" is a major fuel there: 14M tonnes of maize flour was fermented into 1.3 billion gallons of fuel ethanol in 1999. In the United States, fuel alcohol is made by fermentation of corn (maize) starch: ethanol made in this way is sometimes called "agricultural ethanol." Methanol has also been suggested, but is harder to make and more corrosive. Methane is widely used as a heating fuel, and some biofuel methane has been tried out for electric power generation (*see* separate entry on **Biogas**).

Another wet biomass is cellulose, which cannot be readily converted to biofuels because few organisms can break down the tough cellulose fibers into sugars. Companies such as Novozymes are working on using cellulases— enzymes that break down cellulose—in biofuel production. However, at the moment this is not economic.

A more developed version of this is biodiesel, the methyl ester of plant fatty acids (made by transesterification of plant oils). Esters are less chemically corrosive than alcohols, and so are easier on the engine. World production capacity now at about 200 000 tonnes/year, 85% of it made from canola (rapeseed). (Compare this to something like 1 billion tonnes of conventional fossil fuel used in land vehicles every year.) Biodiesel is being used experimentally in city buses in France, Italy, and Germany, with results that suggest it is economically comparable to fossil fuel diesel. Engines running on biodiesel emit less smoke (carbon particles), and almost no sulfur oxides: however, the exhaust is said to smell like French fries.

Some crops are grown specifically to provide the raw material for fuel production in this way—they are called energy crops.

The other route to making biofuels is chemical. If any dry biological matter is heated up slowly, it undergoes "pyrolysis" (literally "breakdown by fire"), generating a complex mixture of oily materials and charred polymers. These oils can be distilled in the same way that conventional mineral oil can be, to give fractions with similar properties to gasoline, diesel, lubricating oil, etc. The charred remains can themselves be burned, possibly to heat the pyrolysis reactors and stills. So far no-one has succeeded in making this sort of process competitive with mineral oil production.

Biogas

Biogas is any fuel gas generated by biological means. It usually refers to methane ("natural gas") produced by fermenting waste, and particularly sewage. Waste is incubated with suitable bacteria in a digester in the complete absence of air (an anaerobic fermentor). The organic matter in the waste is converted mainly into methane and carbon dioxide, and the methane can be burned to provide power or heat. In sewage treatment plants using anaerobic fermentation, methane is often used as a power source for the plant itself. Landfill sites also generate methane, but as it is generated in rather erratic amounts over a large area, it is usually impractical to collect it, so it is just burnt off *in situ*.

The bacteria responsible for generating methane from waste are the methanogenic bacteria, an unusual group which can turn a few carbon substrates into carbon dioxide and methane. To break the waste down into things that the methanogens can "eat" requires other bacteria. Thus, an anaerobic digester needs a specialized population of bacteria (a "consortium") to work well. In practice, waste digestion processes tend to use whatever organisms are on the waste, with consequently lower efficiency.

The other gaseous biofuel is hydrogen, made primarily by photolysis ("breakdown by light") of water. One aim of this area of biofuel research is to get organisms—probably single-celled algae—to release hydrogen gas when exposed to sunlight, rather than using the chemical energy to make sugars. In principle, algae could be used to convert water and sunlight into hydrogen fuel. In practice, this is very inefficient, and it is hard to keep such processes entirely oxygen-free: any trace of oxygen turns this process off, and destroys the enzymes involved. Routes round this are engineering the fermentation process, and making the enzyme more robust: both are being tried, particularly on the microalgae *Clamaydomonas reinhardtii*.

In theory, you can use the enzymes cellulase, glucose dehydrogenase and hydrogenase to make hydrogen from waste paper: however, there is no evidence that this would be economic.

Hydrogen made by biological systems is inevitably called biohydrogen, although it is exactly the same as any other hydrogen.

See **Solar energy**.

Biohydrometallurgy

This is the use of bacteria to perform processes involving metals. It encompasses a wide range of industrial processes, including microbial mining, oil recovery, desulfurization, and a range of physiological processes, including biosorption and the redox metabolism of bacteria. It is also the study of how bacteria corrode metal and metal-containing surfaces, a process known as **Biocorrosion** (*see* **Biofilms**).

Biohydrometallurgy includes two broad areas of bacterial activity.

Biosorption. This is the selective absorption of metal ions by bacteria and bacterial materials (such as their isolated cell walls). See separate entry.

Redox reactions. These are reactions where the bacterium uses the metal ion, or a mineral in which the metal is immobilized, for its metabolism. The area of metabolism concerned is usually "redox" metabolism, the reactions which change the oxidation state of compounds. Redox chemistry is beyond the scope of this book, but in very simple outline it is chemistry that adds or takes electrons away from atoms, thus (e.g.) turning metal salts into metals, water into hydrogen and oxygen, sulfide into sulfate, or vice versa.

A major biological use is the oxidation of sulfide to sulfate, which reaction some bacteria use as a source of energy (the reaction releases energy when carried out in air). As sulfides are frequently insoluble, and sulfates are often soluble, this can be used to release metals trapped in sulfide ores. See entry on microbial mining.

Bacteria can also oxidize or reduce metals themselves. The manganese nodules on the sea floor and the banded iron formation rock strata (laid down 1000 million years ago) are probably the result of bacterial reduction of manganese and oxidation of iron respectively.

Bioinformatics

This is the collection, storage and organization of information of biological interest. In particular, it is concerned with organizing bio-molecular data-bases, in getting useful information out of such databases, and in integrating information from disparate sources. Also sometimes called biocomputing (but *see* **Systems biology**). It is distinct from the storage of chemical information, which is called **chemoinformatics** (see separate entry).

Bioinfomatics covers three distinct types of work:

Data acquisition. This is the realm of **LIMS** systems (see separate entry)

Data collection and curation. Curation of databases is checking the information in databases, organizing it, and adding information to it (rather than just collecting it). Some databases such as dbEST (which collects expressed sequence tag sequences) have very little curation—providing the data is entered in the right format, the database managers will include it. Some such as SwissProt (protein sequence information) have a greater level of curation. This makes the data entries much more valuable to the user, but slows down the process of releasing the data and means that some data just does not appear.

The main problem with biotechnological databases is that there are dozens of different formats for many types of data, so the databases cannot be cross-referenced readily. A relevant term is interoperability, which means that data from one database can make sense in the context of another (e.g. they both define "gene" the same way).

The National Center for Biotechnology Information (NCBI, Maryland, United States) and the European Biotechnology Institute (EBI, Cambridge, United Kingdom) are major central repositories for most publicly available bioinformatics databases. They also provide suites of tools for analysis. Many large pharmaceutical companies also have big in-house databases, and some genomics companies' main business is selling access to their databases.

Analysis. Once you have got all that data, you want to analyze it. Key areas include:

- searching for similar sequence or structural data and comparing the items with each other, to give clues on function (*see* **Comparative genomics**).

- predicting structure from sequence. This is a major area of research, and is edging closer to being realistic (see separate entry).

Nearly all bioinformatics resources were developed as public domain "freeware" until the early 1990s, and much is still available free over the Internet.

In the genome era, bioinformatics is dependent on the ability to handle large databases and analyse them using time-consuming algorithms, and so this requires substantial computer power. The main providers of the computer hardware are Compaq, Sun Microsystems and Silicon Graphics Inc (SGI). SGI almost cornered the market in molecular graphics software. Other types of hardware are usually "clusters" of simpler machines harnessed together, such as the Beowulf cluster, which is a network of any type of computer (but usually PCs) which split a problem up into small parts, send a part to each computer, and then ties all the pieces of the answer together at the end. The Grid Computing initiative is a version of the same idea, but with many different types of computer at completely different sites (even different countries) connected by the Internet or a similar wide area network.

Despite its importance, bioinformatics has been a killing zone for biotech companies, because software and databases date very fast. Generally companies that started as bioinformatic companies have "evolved" into another area of business (as have Incyte Pharmaceuticals and Celera), or gone bust (as did Pangea, later named Double Twist).

Biolistics

This is a method, developed at Cornell University and commercialized by DuPont, to get DNA into cells. The DNA is mixed with small metal particles a fraction of a micron across, usually made of tungsten. These are then fired into a cell at very high speed. They puncture it and carry the DNA into the cell. In the original system a 0.22 cartridge was used to drive the particles, hence this is also called a "particle gun" system.

Biolistics has the advantage over **transfection, transduction**, etc. (see separate entry) because it can apply to any cell, or indeed to parts of a cell. Thus, biolistics has got DNA into animal, plant and fungal cells, and into mitochondria inside cells.

The force to drive the particles into the cells can also be electrical. A spark is used to vaporize a water droplet, which explodes like a small cartridge. This has the advantage that the current, and hence the energy of the explosion, can be varied at will. However, it is more complicated to set up.

As well as getting DNA into isolated cells, biolistics have been used to transfect DNA into animal tissues. Mouse skin and ears have been transfected by a suitably modified biolistic gun in whole, live mice, suggesting that this could be a route to somatic cell gene therapy in humans. The key to getting this to work is to limit the damage the gun-like propulsion causes: curiously, this is not caused by the particles themselves, but by the blast of air or gas that accompanies them. The DNA was only active for a few days, however, before the cells broke it up.

Biological containment

This is restricting how far an organism can go by arranging for their biology to allow them to grow only in some limited environments, such as inside a laboratory or a greenhouse.

Biological containment can take two forms: making the organism unable to survive in the outside environment, or making the outside environment inhospitable to the organism. The latter is rarely suitable for bacteria which, in principle, could survive almost anywhere. Thus, for bacteria and yeasts, the favored approach is to mutate the genes in the bug so that they need to have a supply of a nutrient which is usually only found in the laboratory. If they get out, they then cannot grow. Other mutants may have weakened cell walls so they fall apart outside the laboratory, or may even have "destructor genes" in them which destroy the cells if the temperature becomes lower or higher than the laboratory optimum.

Making the environment unfriendly to the organism is partly a biological control, partly a physical one. For example, some of the first genetically engineered rice strains were developed in England (which is too cold to grow rice) and tried in the field in Arizona (where it is too dry). Thus, there was no rice growing nearby to cross-pollinate with the genetically engineered rice, and if any rice "escaped," it would die. This is containment based on the biology of the plant, but without altering the plant specially.

Biological control

Also called biocontrol, this is the control of one species by another which has been deliberately introduced for that purpose. The most famous example is the introduction of myxamatosis into Australia to control rabbits, although biological control is much more ancient, dating back at least to the ancient Chinese who used pharaohs ants to combat destructive insects in their grain stores. Biocontrol is particularly useful when it is used to control a "new" species that has been introduced into a habitat by man, and can then multiply almost without control because it has no natural predators or diseases to limit its numbers. Examples of such species are rabbits in Australia, and Zebra mussels in the Great Lakes (introduced in ship's ballast water), as well as water hyacinths in much of Africa, tumbleweed in the United States, Dutch Elm Disease in many temperate climes.

Biological control particularly useful in highly unnatural environments such as intensive glasshouse cultivation, where a pest or pathogen is unlikely to have any natural enemies. For example, nematode worms are used to kill sciarid and phorid flies (*Lycoriella auripila* and *Megaselia halterata*, respectively), which attack commercial mushroom farms. Fungi, bacteria, and insects can also be used as weed-control agents: microorganisms to attack Northern Jointvetch and Milkweed vine (weeds of rice and citrus trees, respectively) are in use, and others under development. Such agents are called mycoherbicides.

The key to any biocontrol program is to isolate an effective organism— one which can spread rapidly and effectively through the target pest population but which will not spread to other species and so become a pest in its own right. The best source of a potential biocontrol agent for a foreign species is often at the original home of that species. An ideal is to find an organism that produces a short epidemic among the target species (called an epizooic) without creating a stable presence in the ecology: they can only continue to spread while there is a high population of the target around. Bacteria, such as *Bacillus popillae* (which attacks the Japanese beetle *Popilla japonicum*) have been used in this way.

Fungi, viruses, or bacteria which are known to attack a pest can be cultured in large amounts and sprayed onto a crop, there to kill that particular pest. Entomophagous fungi (fungi which infect insects) are favorites here, as they can infect insects through the cuticle (the hard external "shell"). In essence, culturing fungal pathogens is the same as culturing any other fungus, with the additional constraint that the fungi usually require very specific and unusual culture media.

Viral agents can also be found or created. Genetic engineering and advances in culturing viruses in insect cells (*see* **Baculovirus**) have enabled biotechnologists to manipulate insect viruses to be, potentially, more effective biocontrol agents, by altering the specificity of the viral proteins which bind to the cell surface, or by engineering in a toxin gene, or the "pathogenesis" genes from another virus. An example is the baculovirus engineered to contained a scorpion toxin gene that was tested on cabbages in the United Kingdom in 1993. These aims are rather hard to achieve, as viral infection is a very complex process. Model experiments where the virus is tracked with a marker gene are intermediates to discovering how to modify the infection process.

Sometimes, biocontrol agents overlap with **biopesticides** (see separate entry). For example, *B. thuringiensis* produces an anti-lepidopteric (caterpillar-killer) protein (cry proteins). *B. thuringiensis* has been used as a biocontrol agent for many years, but recently biotechnology has isolated the protein responsible to make it into a biopesticide.

A related idea is the use of sterile insects to control wild insect populations. This has been extensively tested for control of malaria and sleeping sickness. Female insects are sterilized in the laboratory, and then released into the wild in large numbers. Wild insects mate with them (simply by chance, because you have released so many), but of course no eggs result, so the next generation is much smaller. The problem here is how many insects to release—the answer is usually "as many as you can afford." A sterile pink bollworm was experimentally released in 2001 with a green fluorescent protein marker gene in to track exact ratio of pest to sterile insects, and hence be more effective in use, to address this problem.

There is a grey area between biocontrol and **biological warfare** (see separate entry), where a biocontrol agent is developed to attack an agent that some people see as a problem but others do not. An example is the US development of a *Fusarium* fungus strain that selectively attacks cannabis, as part of the American "war on drugs." The argument for this was not helped when several US environmental agencies declared the fungus too dangerous to human health to be used in the United States itself.

Biological response modifiers

A very general term, usually confined to mean proteins which affect how the immune system works. In this meaning it is almost synonymous with "cytokine." The term is widely used because of the existence of the FDA Biological Response Modifiers Advisory Committee, which oversees biopharmaceuticals which modify biological response mechanisms (i.e. all of them to date). Normally, biological response modifiers act in concert, not as isolated chemical entities. Thus, there has been much agonizing about how the components of biological response modifier drugs, cloned as pure proteins but used in combinations, can be regulated by the drug regulatory agencies, and particularly by the FDA. Cetus had well-publicised problems trying to get its IL-2 approved as an anti-cancer drug, because IL-2 was not effective on its own. Cetus wanted to use it in combination with other biopharmaceuticals, but as it was not effective enough as an isolated entity it was refused approval. (The case was not helped by Cetus' inexperience at presenting data to the FDA at the time.)

Biological warfare

A serious concern in the twenty-first century is that our increasing understanding of biology would enable the deliberate engineering of super-potent infectious agents that could be used as weapons.

Biological warfare was researched extensively in the 1940s and 1950s in the West, and into the 1980s in the former Soviet Union, but in extreme secrecy. Most countries dropped the research because biological weapons are hard to make and uncontrollable—a "biological bomb" might kill everyone in a city, might kill no-one, or might be blown back onto your own army. However, the risks were dramatically brought to light after September 11, 2001, when several deaths from cutaneous anthrax (infection of the skin by the anthrax bacterium) were shown to be because of deliberate contamination of letters with anthrax spores. Now, bioterrorism is thought the higher risk, although concerns about weapons manufacture in "rogue states" (usually political code for Iraq) remains in some countries.

The threat of biological weapons was formally recognized in the Biological Weapons Convention, first signed in 1975 and ratified by 143 states, including the United States. But the treaty contains no provisions for verification that states are complying or monitoring potential bioweapons activities, so it is powerless except as a mechanism for protest after the act. Commercial confidence is often given as the reason, but this did not prevent the FDA carrying out 22 000 very rigorous inspection visits in 2001, with no (known) security breach. Since October 2001, the United States is now much tougher on verification and enforcement, and it is possible (and to be hoped!) that the biological weapons convention will be given some real power.

A range of infectious agents have been researched as potential bioweapons:

- *Human disease*: Bacteria such as anthrax, *Yersinia pestis* (plague), and a range of viruses including the filoviruses (Marburg Virus, Ebola), the arenaviruses (Lassa Fever), and *Variola major* (Smallpox)

- *Animal and plant diseases* (to attack a country's agricultural infrastructure): Anthrax, Rinderpest, Foot and Mouth, Brucellosis for animals, Potato beetle, potato blight for plants.

Biotechnology could help to combat this threat with better drugs or improved vaccines. It can also contribute to diagnostics, where security forces and the army need rapid, specific analysis of whether a bioweapon is present "in the field." Lab-on-a-chip or biosensor technology might be able to help here. The technical problem is that you have no idea what the organism might be (when a patient comes into hospital they have symptoms, which gives a major indication): normal diagnostic methods either take time to sort through all the possibilities or test specifically for one organism, using an antibody or gene reagent.

Biotechnology could also (in principle) create new bioweapons. Known pathogens could be engineered to be resistant to the drugs used to treat them. In the 1960s Soviet scientists were trying to make antibiotic-resistant bacteria. The knowledge generated by bacterial genomics and molecular biology could make it substantially easier to do this type of engineering, although it would still take a substantial research laboratory to actually perform the work.

The other approach is to make the organism itself more lethal. Organisms that cause disease do so with a complicated array of genetic and chemical mechanisms, and it had been thought that it would be almost impossible in

51

practice to actually produce a more lethal biological weapon. However, in late 2000 this was achieved as an unexpected result from another experiment. Scientists at CSIRO (Australia) were engineering a mousepox virus to produce interleukin-4, with the aim of producing a virus that rendered the mice sterile as a pest control measure. Instead, the virus was highly lethal to mice that had previously not been susceptible, and even to mice that had been vaccinated against mousepox. This is the only publicly admitted "success" in this field, and might be a one-off, but Bob Seamark, chief scientist involved, said that the potential risk it exposed showed that the biological weapons convention should be tightened up drastically as a result.

The best way to control any bioweapons development or deployment is to control access to samples of potentially hazardous organisms. It is practically impossible to build a living organism from scratch, so the bio-arms manufacturer is reliant on getting samples from a recognized culture collection (see separate entry) and then "engineering" them. Coupled with a biological weapons convention with some verification and inspection "teeth," this would reduce the risk from biological warfare substantially.

Bioluminescence

Bioluminescence is the production of light from biological systems, either enzymes or organisms. It is of substantial interest from a technical point of view, and also very pretty.

Many organisms make light, usually ones that live in dimly lit environments (such as deep sea fish) or are active at night (such as fireflies). Algae, bacteria, and insects generate light themselves, while larger animals rely on symbiotic bacteria living in special organs to generate light for them.

Two light-generating systems are used in biotechnology applications.

Bacteria and fireflies have similar enzymes called luciferase, which acts on a molecule called luciferin together with ATP to generate light. Natural luciferase generates a yellow light, but this colour can be altered slightly by engineering the enzyme. The enzyme can be chemically linked onto a target, or produced within an organism from its gene.

The jellyfish *Aequorea victoria* contains aequorin, a protein which glows blue-white when calcium binds to it. It has to be "charged" first—a specific chemical prosthetic group (coelenterazine) is oxidized to a peroxide, and calcium binding triggers its oxidation by the peroxide group. This can be used as a label, with calcium ions as a "trigger," but can also be used as a super-sensitive calcium detector.

Biotechnologists use the light-generating systems in these organisms as molecular labels. A single molecule of a light-generating enzyme can generate dozens of photons ("packets") or light, which can be detected by a photomultiplier tube or a sufficiently sensitive CCD camera, and if this is done in a completely dark box then there is no other light around to contribute "background" signal. This is, therefore, as sensitive a detection systems as radioactive labels, but safer and in some cases easier to use.

Bioluminescence can also be used as a cell-based biosensor read-out. The amount of light that a cell produces is very dependent on its energy levels, so luminescence is used in water industry as assay for toxicity. The luciferase/luciferin combination can be used as an assay for ATP—tiny traces of ATP will cause the combination to glow. As almost all living cells contain ATP, such luminescence assays are used as hygiene tests, as they are sensitive enough to detect the tiny traces of ATP that a few bacteria on a work surface will contain.

It is also possible to link a luciferase gene to a promoter that is turned on by an environmental stimulus, such as the presence of a heavy metal, so the cells glow in the presence of that stimulus.

Bioluminescence is so spectacularly pretty that is has also generated products in its own right. Science in the Dark Inc provides the dinoflagellate pyrocyctis as an educational tool. Prolume makes bioluminescent-based squirt guns and food products, again based on marine organisms.

This is an example of the value of pure science: the study of what makes jellyfish glow spectacularly in the dark lead to a hundred million dollar product with value to industry and healthcare.

Biomass

Biomass means any bulk biological material, and by extension any large mass of biological matter. Fermentations generate biomass as well as the fermentation product, and it is important to control the amount of biomass to maintain the rate of fermentation. Active biomass—the amount of living cells in a fermentation—is most relevant. However, this can be hard to measure, as a recently dead cell looks very like a living cell.

Biomass products are ones made from the biomass of your process. Usually, in a fermentation system, the biomass is a side product, or even a waste. However, some processes grow plants, fungi or bacteria specifically for their biomass.

- **SCP (Single-cell protein).** See separate entry.

- *Algal biomass*. Single-celled plants such as *Chlorella* and *Spirulina* are grown commercially in ponds to make food materials. *Spirulina* enjoyed a vogue as a health food a few years ago, due to an unfounded belief that it was extraordinarily nutritious. Like most algae (including some seaweeds) it is quite a good food, but *Spirulina* is not outstanding. *Chorella* is grown commercially to make into fish food: it is fed to zooplankton (microscopic animals), and these in turn are harvested to feed the fish in fish farms. This is a way of concentrating sunlight into food in a more convenient and controllable way than normal farming.

- Plant biomass. Crop plants such as sugar cane have also been grown for biomass. This is usually used as the start of a chemical production process (as growing plants for food is usually called farming). Most of this is used to make fuel ethanol—*see* Biofuel.

Biomaterials

This has two meanings—materials that have a source in biology, and materials that are equivalent (especially functionally equivalent) to natural materials.

The former definition is more common. It is a general term for any biologically-derived material which is used for the sake of its material property, rather than because it is a catalyst or a pharmaceutical. Thus, DNA could be a biomaterial if you used it to make paperclips or cranes, rather than using it to store information.

The most common biomaterials are some proteins, many carbohydrates, and some specialised polymers. (*See* Biopolymers.) The proteins considered for biomaterials applications are usually proteins used as structural elements in animals or, occasionally, plants. Collagen, the protein in bone and connective tissues in a wide variety of animals, is a common candidate, and has been used (controversially) as a cosmetic biomaterial, being used as a "natural filler" for plastic surgery operations. Fibroin, the protein in silk, has been suggested as a protein with sufficient strength to rival nylon, and spider silk could challenge even Kevlar as a structural material.

Most of these structural proteins have fairly simple amino acid sequences of short blocks of amino acids repeated many times. Thus, the rigid, central sections of the collagen molecule, which give it its elastic strength, are made mostly of repeats of the three amino-acid unit glycine–X–proline (where X can be one of several amino acids). Biotechnologists have also made synthetic proteins with simple repeating patterns in the search for new biomaterials.

The spider silks illustrate some of the problems. There are dozens of different spider silk proteins (spidroins), which are adapted to the spiders that make them. All share a common general structure, consisting of rigid protein domains or modules linked by more flexible links. To make a synthetic silk that can be spun in a production plant requires that the scientist understand what those units are doing, and then assembles the right combination to work with her process, and then develop a spinning process that turns these into fibres. After 15 years of work, this has still only partly been achieved.

Such genes with the same sequence repeated many times within them tend to be "genetically unstable" in bacteria and yeast, so that on passage from generation to generation they mutate with very high frequency to lose many of the repeated sequences. Nexia have tried to overcome this by expressing spidroin genes in transgenic African goats, where genetic instability is less of a problem: the spidroin is isolated from the milk, and then spun into fibres.

Biotechnology has produced a wide range of carbohydrates with modified properties which act as lubricants for biomedical applications or as texture-modifying or bulking agents in food manufacture. These include rare but natural materials made from bacteria such as poly-dextrose, carbohydrates modified by enzymes to have improved properties, and entirely artificial polymers. *See also* entry on **Wood**.

Other polymers include "natural" plastics such as polyhydroxybutyrate (*see* **Biopolymers**), or rubbers produced by bacteria or fungi. Enzymol has developed the enzyme soybean peroxidase to make polyphenol resins, a common industrial plastic.

The properties of a polymer which are crucial in determining whether it will make a "good" biomaterial for a specific application include:

- Tensile strength (both elasticity and breaking strength)

- Hydration (How much water does it bind? How much does it need to bind to keep its properties?)

- Viscosity and visco-elastic properties. (Viscoelasticity is when a liquid seems to get thinner the more force you place on it, like ketchup.)

Biotechnology has also been active in finding alternative ways of making precursors for conventional plastics. For example, DuPont and Genencor have worked together to make propane-1,3-diol, used for making polyesters, by fermentation.

The second use of the term "biomaterial" is as a material that replaces a biological material. It is used in this sense in medical devices and implants, such as using hydroxyapatite-based materials as a bone replacement for surgery. Many of these are also biodegradable, as that they are gradually replaced by the body's own material. Other biomaterials in this context can be plastics that are implanted, such as catheters.

See also **Tissue regeneration**.

Biomimetic

Literally meaning "imitating life," this means the areas of chemistry that seek to develop reagents that perform some of the functions of biological molecules. The reason for doing this is that many biological molecules are chemically inconvenient to produce, handle or apply in large amounts and using cheap processes. By using chemical mimics of them the biotechnologist hopes to achieve a more flexible and more commercially useful way of achieving the same ends.

Areas of chemical research in the general field of biomimetics include:

Co-factor substitutes. Many enzyme co-factors are complex and labile molecules: in particular NAD and NADP are difficult to work with on a large scale. Two lines of research seek to replace them with other molecules. Triazine dyes have been used as replacements for NAD in affinity chromatography applications (called dye-ligand chromatography). Here the dye is bound to a column, and a mix of materials containing a dehydrogenase enzyme is passed over the column. The triazine dye binds to the dehydrogenase as NAD would, and so holds it onto the column—all other materials pass through. This has been used very successfully for many purifications.

Such co-factor substitutes can also be used as actual substitutes for the real molecules in enzymes, especially for NAD, NADP, and FAD (Flavine Adenine Dinucleotide) in reactions catalyzed by dehydrogenases. Here again the aim is to find a small molecule that will do the chemical work of NAD etc. with the enzyme.

Peptide biomimetics (peptidomimetics) are very valuable as drug candidates. Peptides are easy targets for manipulation through recombinant DNA techniques, but make poor drugs. A peptidomimetic is a molecule which has the same effect as a peptide (usually because it has the same critical "shape"), but which is not itself a peptide, and hence is not broken down by proteases and is cheaper to make. About 20 peptidomimetics have been developed to the point where they are potentially interesting as drugs.

DNA substitutes. Biotechnological chemists are working on altering the basic "backbone" of nucleic acids so as to make them more stable and potentially cheaper to make. In early 1992, a "DNA" substitute called poly-amide (or peptide) nucleic acid (PNA) was reported that had no sugar–phosphate backbone at all: in its place was a poly-amide chain looking more like a protein. This material bound tightly to single-stranded DNA in a base-specific way. This has applications in antisense, as such molecules could be much easier to get into cells and totally resistant to breakdown by nucleases or proteases.

Glycomimetics. These are organic molecules which have some of the properties of sugars, for potential use in affecting interactions which are usually mediated by sugars.

Synzymes. These are low molecular weight molecules that act as highly specific catalysts, like artificial enzymes. Usually, they are synthesised to deliberately copy the three-dimensional structure of the "active site" of an enzyme, but using non-peptide chemical building blocks. Unlike more usual organic chemistry catalysts, which catalyze a broad range of reactions, the aim is to make synzymes as specific as enzymes. This is still a research area at the moment.

Molecular imprinting. This is another approach to making an artificial, enzyme-like catalyst. See separate entry.

Biomineralization

This is the synthesis and breaking down of minerals by living organisms. In some applications, it is related to microbial mining (the break-up of minerals by microorganisms) and hence is part of biohydrometallurgy. However, biomineralization extends beyond this. There are two general areas of interest to the biotechnologist.

- *Microbial biomineralization.* This is how microorganisms make minerals. If the minerals are deposited inside the bacterial cell, then they are of necessity laid down as extremely small crystals or granules. Magnetic bacteria make magnetite as tiny inclusion bodies inside their cells, which can use them to point them in the right direction to swim along lines of magnetic field. (This enables them to swim towards the bottom of ponds in temperate zones.) Many larger mineral forms are also made partly by bacteria, and this has been suggested as a way that minerals could be isolated or purified using biotechnology.

- *Multicellular biomineralization.* In many plants and animals, minerals are used to give strength. Thus vertebrate bone often contains calcium phosphate, and many grasses have silica in their leaves to give them a hard cutting edge to dissuade animals from eating them.

Biomineralization is of interest to the materials scientist because organisms lay down minerals in very specific patterns, with the atoms ordered at an almost atomic scale. This contrasts to (say) concrete, where the components are mixed up in a very irregular way, although uniform on a large scale. This "nanostructuring" of the mineral can substantially alter its properties. Thus, bone and teeth are much stronger than "raw" calcium phosphate, because they are actually nanostructured materials, build on nano-scale units in hierarchical "layers." Abalone shell is made of aragonite (chemically identical to chalk) and protein in alternating layers—however, it is far stronger than either alone. As in these examples, living things achieve these feats by incorporating specific proteins into the growing mineral, to force crystal growth into the form they require or to reduce the propagation of cracks through the mineral when it is stressed.

This could be applied to synthetic materials. *Pseudomonas stutzeri* accumulates nanoparticles of silver, which have been used experimentally to make the basis of very finely structured electrodes with potential as a solar cell. The silica cases of diatoms have also been suggested as tiny units for constructing nanomachines and other exotic products, as diatoms construct enormously complex structures that are different for every species, but completely identical within each species.

The control of biomineralization is also of substantial interest in several human diseases, notably osteoporosis, a disease in which too much calcium and phosphorus is lost from the bones. The two uses of biomineralization have been combined in the use of coral as substrate for growing new bone. Coral makes calcium carbonate skeletons that combine an open structure containing many pores of the same size as bone cells with mechanical strength.

Biopesticide

A biopesticide is a pesticide, that is a compound which kills animal pests, which is based on specific biological effects, and not on broader chemical poisons. Specific types are also called bioinsecticides and biofungicides. Biopesticides are different from Biocontrol agents in that biopesticides are not living, and so cannot reproduce themselves in a target population, whereas biocontrol agents are active, living things that seek out the pest to be destroyed. Despite this, biological control agents are sometimes called biopesticides. Biopesticides are sometimes grouped with probiotics, biocontrol agents and other biologically-derived products in being called agrobiologicals.

Biotechnologists are developing a range of biopesticides from natural products, such as specific fungicides (kasugamycin), insecticides (spinosyns), antihelminthics (anti-"worm" treatments, such as avermectin). All are the product of fermentation, as these are products produced by bacteria or fungi to defend themselves against another organism or to kill their prey.

Protein anti-pest materials, such as the *Bacillus thuringiensis* (B.t., sometimes B.t.k.) toxin, are also attractive biopesticide agents. B.t.k specifically interferes with the absorption of food from the guts of some insects but is harmless to mammals. Some of these proteins (which have been used as a pesticide for some time as a bacterial suspension) have been "cloned" into more amenable bacteria. The genes for the proteins have also been inserted into a variety of crop plants, making a plant that is more resistant to pest attack, such as Monsanto's insertion of the CRY9C gene into cotton to resist the boll-worm.

The rationale behind developing biopesticides, by contrast with conventional pesticides, is threefold.

First, they are more likely to be biodegradable than chemical entities which are not normally found in nature.

Second, they are targeted at specific elements of the pest's metabolism, rather than being broad spectrum biocides. (This is often taken to mean that they are "safe" for humans, and this is usually so, although the idea that all natural products are inherently safer than artificial ones is clearly wrong, as the effects of some biological toxins show (see separate entry.)) This is an important aspect in Third world applications, where toxicity to people is a major problem. (It also means that it is not worth stealing them for use as broad-spectrum pesticides for crop protection, which is the fate of some of the more conventional pesticides used to control malaria and sleeping sickness in poor countries.)

Third, their targets are less likely to evolve resistance to them. This is speculative, and some resistance of culex mosquitoes to B.k.t. toxin has been recorded.

In 2000 there were 45 biopesticides or biocontrol agents on the market, aimed at insects, organisms which cause plant diseases, and at weeds.

Biopharmaceuticals

This has come to mean two things—a type of product and a type of company.

In its original meaning, biopharmaceuticals meant pharmaceuticals that were inherently biological, and hence pharmaceutical proteins. The more specific name for these is "biologics," derived from the FDA's classification of them. *See* entry on **Pharmaceutical proteins**. "Biopharmaceuticals" is used in this sense in this book.

Recently "biopharmaceutical" (or bio/pharmaceutical) has come to mean a company discovering and developing pharmaceuticals, but with a "biotech culture" of being small, young, having lots of neat and usually new technology, fast-moving and almost always venture capital backed and hence short of money. This is to distinguish them from "biotechnology" companies (which try to make biotech products, rather than use the tools of biotechnology like genomics to make better small molecule drugs), and from pharmaceutical companies (which everyone thinks of as being huge concerns with the resources to conduct phase III clinical trials and to manufacture and market their own drugs).

Biopolymers and bioplastics

A biopolymer can mean two things: polymers that are made from biological precursors, and polymers that are made by living organisms.

Synthetic biopolymers are made by polymerizing a simple biological monomer. An example is the hyaluronate polymers (hyaluronic acid is a sugar derivative) which are used in a range of artificial tissues as the base material on which to grow cells, and as a material in advanced wound dressings. Another is the range of dextrans, which are used for things from chromatography columns to the fillings in low calorie chocolate bars. These are synthesised chemically to have properties similar to natural polymers, refined for a specific application.

The other meaning is a polymer made by a living organism. The term is usually used to mean polymers that are storage or structural materials, such as the plastic- or rubber-like materials produced by bacteria and fungi, rather than DNA or protein. The most widely used of these are the polyhydroxyalkanoates (PHAs), such as polyhydroxybutyrate (PHB). Over 100 PHAs are now known to be made in nature. PHB on its own is too stiff and brittle to use for most purposes, so co-polymers (polymers with two different subunits linked together) are preferred. A similar type of material is polycaprolactone (PCL). The PHAs can be moulded when hot like other plastics, and are water-resistant and watertight. However, their structures can be attacked slowly by bacteria, and so after a period of months to years will be completely broken down.

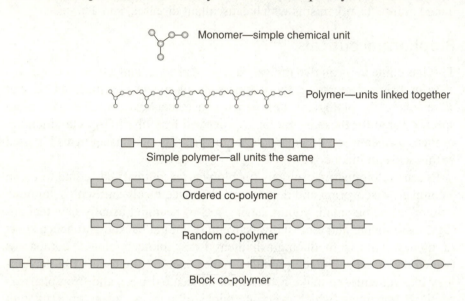

Monomer—simple chemical unit

Polymer—units linked together

Simple polymer—all units the same

Ordered co-polymer

Random co-polymer

Block co-polymer

The biochemistry of how organisms make PHAs is well understood, and the relevant genes can be cloned into more amenable bugs for fermentation, providing the organism has the ability to store such polymers in inclusion bodies (otherwise they fill up with plastic, and die). In principle this includes transgenic plants, and some initial experiments with arhabidopsis and maize show that they can produce up to 14% dry weight of PHB, and a transgenic rape has been made that stores the PHB in seeds at up to 8% of dry weight. The PHB enzymes have to be introduced into the plastids within the cells (the chloroplasts and related structures within a cell), as otherwise they compete with the other enzymes in the plant for substrates, and plant growth suffers as a result. Extraction of the polymer is difficult—the best route is to dissolve it out with acetone, which is relatively complicated and expensive.

An alternative approach is to make the PHA *in situ* in a plant. An example is expression of PHAs as 0.3% of dry weight of the fibres of cotton plants, where it significantly improved the insulating properties of the cotton.

The ICI/Zeneca Biopol project was the first to achieve a measure of commercial success. Biopol is a copolymer of PHB and polyhydroxyvalerate (PHV) is produced by fermentation on *Alcaligenes eutrophus* and *Ralstonia eutropha*, and sold by Monsanto. These materials are still expensive materials by plastics standards, so their use is limited to where these is a need for small amounts of a fully biodegradable plastic. Examples include films and fibers for biodegradable packaging, and as a latex (very fine particles) used as a water-resistant coating for paper or cardboard. Cargill-Dow has built a plant to produce polylactide plastic by fermentation of corn starch, and estimates that this will be competitive with polyethylene tetrathalate (PET) plastics used in packaging, carpeting and upholstery.

PHAs generally thought to be environmentally friendly because they are made from natural materials (not oil) and are biodegradable. However, it is not clear this is so—a study in 1999 showed that a kilogram of PHB takes 2.39 kg of fossil fuel to produce, in diesel for farm equipment, power for processing, manufacture (and then waste) of solvents, and so on.

Biopreservation

This is usually the preservation of food using biological materials. The leading example is nicin, a bacterial protein that acts as a broad spectrum antibiotic and which has been approved for food use in Japan. There are a range of other bacteriocins that could be developed, but the main barrier to using them is that they have to undergo severe regulatory tests before manufacturers are allowed to use them in food.

Often, some bacterial growth is not itself harmful, but generates undesirable side-products. The smelly amines produced by rotting fish are an example. A biopreservation approach is to use enzymes to break down these products, so the bacterial growth does not matter.

Bioprospecting

This is searching for new biological resources, usually new plant or microbial strains which could be sources of natural products and phytopharmaceuticals (*see* Natural products).

This is a disputed topic: the countries involved sometimes feel that they have got nothing in return for their genetic resources, and call such collection "biopiracy." The problem is illustrated by the 1995 row over the Neem tree. The US company W.R.Grace was granted a patent for a new process for manufacturing a pesticide extract from this tree, which has been used in India for pest control and traditional medicine for centuries. The Indian government and a range of pressure groups objected furiously, although in fact it was the manufacturing process, and not the tree, which was the subject of the patent.

Regulation of the use of "native biodiversity" (i.e. the biodiversity of your country) is the aim of the Convention on Biodiversity (CBD), signed with various delays after the Earth Summit conference in Rio de Janeiro in

1992: 175 countries have ratified or acceded to the treaty, but not the United States. It provides rules concerning access to biological material, the sustainable use of those resources, and fair distribution of the resulting benefit. It does not seek to say how that benefit should be protected, and so does not address patent or ownership issues directly.

The preferred approach is "bilateralism" (which will no doubt be called "biolateralism" in time), where the company taking the genetic resources has a long-term commitment to give back some of the value resulting to the originating country. Most companies in the field have agreements of this sort.

Microbial bioprospecting can take place closer to home, because of the huge range of environments that an industrial society provides for bacteria. The key is to look in an environment that might suit the bacteria you want to find. *See* **Strain isolation.**

Bioreactor

A bioreactor is a vessel in which a biological reaction or change takes place, usually a fermentation or biotransformation. Bioreactors, and indeed fermentation and biotransformation, are central to much of biotechnology—everything from baking bread to producing genetically engineered Interferon takes place in a fermentation, and hence uses a bioreactor.

Bioreactors are conventionally divided up into three size classes: laboratory, pilot and production systems. Laboratory bioreactors cover bench-top fermentors (up to 3 liters volume) and larger, stand-alone units (up to about 50 liters). These are used for research, and are usually used to create the

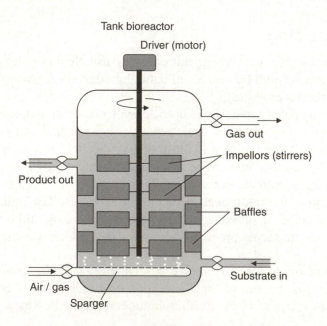

Tank bioreactor

Driver (motor)

Gas out

Impellors (stirrers)

Product out

Baffles

Air / gas

Sparger

Substrate in

fermentation process. Pilot plant fermentors are used to scale up a **fermentation process** (see separate entry), and to optimize it. They are typically between 50 and 1000 liters. Pilot plants have to be quite flexible to allow for process optimization. Production units can have any capacity, but usually hold at least 1000 liters, and can go up to the 1 000 000 liters of the ICI Pruteen plant. They generally are much more specialized than pilot plants, being designed to operate one process with maximum efficiency.

There are a number of separate entries about bioreactors. They cover different types of bioreactors:

- tank bioreactors (which is most of them)

- immobilized cell bioreactors

- fiber and membrane bioreactors

- digestors.

Other simpler types of reactors are not covered specifically. These include pond reactors and tower fermentors. Pond reactors are basically ponds: they are used mainly for growing algae. Often they are called oxidation ponds, because their large surface area allows for more rapid oxygen transfer: because of this they are useful for reducing the BOD of liquid waste. Tower reactors are relatively simple towers in which nutrient is injected at the base: the flocculent organisms settle at the base of the tower, and the product is collected at the top through a series of baffles which separate foam from bulk liquid. They are used typically for anaerobic fermentation, that is fermentation where no air is needed, as for example in brewing. Fermentation material is injected at the base of the tower.

A further general type of reactor is the plug flow reactor. Here, a substrate flows past a plug of solid support material, emerging from the end changed by the plug. Often the reactor is essentially a pipe, although it is sometimes a flat bed. This is in fact a bioreactor equivalent of column chromatography (*see* **Chromatography**). Many plug flow reactors have some stirring in the flow mechanism to even out the reaction across the "plug."

Many of the classifications of bioreactors and fermentors depend on how the materials in them are cycled and recycled. One extreme is the completely mixed bioreactor, in which the contents are mixed uniformly during their passage through the reactor, and then pass *en masse* to the next stage. Most bioreactors take some of the output, usually some of the biomass, and feed it back into the reactor again.

In almost all bioreactors, mixing the contents is important to their efficiency. Often this is achieved through the use of stirring paddles or turbines. The inflow of gas can be used to stir the fermentor contents too, as in the

airlift fermentor (see separate entry) or the deep jet fermentor, where an air jet, often starting above the liquid content of the reactor, plunges into the liquid and stirs it without the need for mixers.

Other topics covered under fermentation include:

- bioreactor control

- fermentation process

- gas transfer

- sensor systems

- substrates (what the microorganism grows on).

Bioreactor control

Very small and simple bioreactors, such as garden composts or "home brewing," can run successfully without any control on what happens. Any larger process requires control on the process to make sure that the conditions—chemical additions, pH, gas content, temperature control, and so on—remain appropriate for the reaction. Large fermentation systems can have thousands of sensors linked to hundreds of control valves. As the exact efficiency of a bioreactor can be critical to whether it is economic to run or not, such control systems are a very important topic in a range of biotechnology.

Bioreactor control is complex for two reasons.

First, most biological systems are themselves complicated. A fermentor growing a microorganism must be monitored to keep the concentration of substrate chemicals, the pH, the gas levels, the temperature, the cell mass, and the amount of product material within tight boundaries, as otherwise the yield of product will decline, and at worst the organisms will die completely. In particular, in continuous bioreactors these conditions must be monitored in real time (i.e. as they happen, also called "online"), and cannot be monitored offline (i.e. put on a laboratory bench and analysed tomorrow).

Second, many of the key parameters are hard to measure. The method used to measure something must be sterilizable, because it is going to measure it inside the bioreactor. This means, usually, that it must stand up to autoclaving, live steam, corrosive chemicals, or all three. Thermometers and pH meters can be built to do this quite easily. Chemical and biomass measurement are very much harder to do. A great deal of ingenuity has been spent on devising "non-contact" or "non-invasive" sensors for bioreactor control. Among the success stories are:

- *Capacitative and inductive sensors for biomass.* Alternating electric fields are affected in a characteristic way by biomass in the liquid between two electrodes. This can be used to find out the amount of biomass there. It does not work with only small amounts of biomass.

- *Real-time offline sensors*. These take a sample out of the bioreactor and then measure its properties. Providing the test is fast, this works OK, and the test does not have to be sterilized. However, it is expensive and complicated to engineer to ensure both speed and sterility.

- *Chemometrics* to estimate things you cannot measure directly. Chemometrics is the use of sophisticated statistical techniques to identify the concentration of one chemical species from the spectral "signature" of a complicated mixture. It is a related idea of the use of "surrogate measures," things which behave the same way as what you are trying to measure, but are not that thing. An example would be to measure cloudiness (turbidity) in a microorganism culture as a surrogate for biomass—it is not actually measuring biomass, but usually it comes out with a pretty close approximation.

Bioreactor control also needs complex algorithms to make it work. These are among the few applications of artificial intelligence techniques (see separate entry) to biotechnology. Expert systems and neural nets are among the systems that have been used with some success in controlling the more complex bioreactors.

See also Process control.

Bioremediation

Bioremediation is the use of biological systems—usually microorganisms—to clean up a contaminated site ("the environment"). The same approach (although different technologies and organisms) are also used for "bioscrubbing"—removing waste materials from flue gases. Bioremediation is an industry with a turnover around $20 billion globally, mostly treating contamination in soil, often getting rid of heavy metals (cadmium, lead, mercury, arsenic), petroleum compounds, and trichlorethylene solvent. The term "bioremediation" does not usually cover sewage treatment (see separate entry).

There are four basic approaches to cleaning up contaminated soil using biology:

Stimulation. Here we stimulate local microorganisms to metabolize the contaminants, by giving them other nutrients, oxygen, or other chemicals. This is cheap, but may not be very effective. A widely used example is the use of a combination of nitrate and phosphate as a stimulant for soil microorganisms to metabolize BTEX. (BTEX, or BTX, is a mixture of benzene, toluene, ethylbenzene and xylenes, commonly used solvents that are also a common residue from crude oil spills. They move readily into groundwater, and once there persist for years in poorly oxygenated environments.)

65

Augmentation. This is adding new organisms to the site. These can be tailored to clean up the specific contaminant involved, but this can be costly. They must also be supplied with nutrients and often with oxygen as well.

Land farming. The contaminated soil is mixed with normal surface soil (plus or minus added bugs) and aerated by tilling. This is easy to implement and effective, but you need a large area to spread the soil out. Related is *in situ* composting, where the contaminated soil is piled up with bacterial and nutrient sources.

Tank bioreactor. If a target site is very highly contaminated, or too cold or dry for bacteria to flourish in, then the soil can be placed in a tank bioreactor and the bioremediation carried out there. These bioreactors essentially large insulated tanks into which soil or waste is placed with a bacterial inoculum. Air is blown through the mass to keep it oxygenated. This is an expensive option.

A critical limiting factor in all these processes is usually supplying the microorganisms with oxygen. Oxygen can be dissolved in water which is then pumped into the site, generated locally using chemicals (like chlorate or perchlorate, coupled with organisms that can break it down to relatively harmless chloride and oxygen), or pumped in as air. Perchlorate itself is a widely found environmental contaminant, particularly from munitions manufacture: the same bacteria could break this perchlorate down, although this has not been used in environmental clean-up yet.

For many bioremediation approaches, added microorganisms that can break down a type of contaminant are needed, as such organisms, while present in the soil, would be too rare or work too slowly to clean up a contamination site or handle a waste stream on their own. Using such a microorganisms consists of selection, optimization of physiology, and inoculation.

Selection of the microorganism (*see also* **Strain isolation**)—Typically, bioremediation methods use a consortium of organisms, rather than a single organism, which can catalyze the breakdown of different components of a pollutant or can perform different parts of the breakdown of a complex molecule. Even so, some molecules are quite hard to destroy—PCBs can be dechlorinated by obligate anaerobes (bacteria killed by oxygen), and the carbon skeleton broken down by aerobes (organisms needing oxygen): however clearly these two cannot work together at the same site.

In theory, genetic engineering could produce even better bioremediation organisms, and the first patent for a genetically engineered bacterium was for an oil-eating *Pseudomonas*. However, the concern about release of GM organisms into the environment has stopped this line of research being applied.

Development of organism physiology—The scientist develops a cocktail of nutrient to boost the rate at which the soil organisms break down the target chemicals, a mixture often called an enhancer. (If the soil bacteria could do this on their own, there would not be an environmental problem.) Designs of enhancers are complex, as they must direct the bacterium's metabolism towards digesting the target chemicals, and not just feed the bug.

Inoculation of the environment—The microorganism is introduced into the site, usually with a nutrient mix and enhancer, and often nitrogen and phosphorus so that the bacterium's growth is limited only by the availability of the (carbon-based) contaminant.

The main cause of failure of practical bioremediation projects is that the organism selected cannot perform the breakdown at a useful rate at the site, despite performing well in the laboratory. Clays, for example, are particularly poorly suited to bioremediation: because they are very densely packed, water penetrates them very slowly and air hardly at all.

Typical target compounds are chlorinated aromatics (although disposing of PCBs has met with only limited success), vinyl chloride, solvent residues, gasoline fractions, and crude oil. Alpha Environmental has hit the headlines on several occasions with its oil-eating bacterial preparation, used to digest oil spills at sea into soluble molecules which other bacteria can digest. Its most public application was in the Persian Gulf in 1991. Other non-organic materials also can be metabolised if their end-product is non-toxic or volatile: selenium has been removed from soil by conversion to volatile compounds or elemental selenium, and nitrates have been removed from sewage waste by biological reduction to nitrogen gas for decades.

See also **Phytoremediation**.

Biosensors

Biosensors are devices which use a biological element to measure something. For example, an electrode could have an enzyme immobilized on its surface so that it generated a current or voltage whenever it encountered that enzyme's substrate.

There are several classes of biosensor:

- ISFET (Ion-sensitive field-effect transistor) based devices

- Physical sensors (including sensors for mass and for heat output)

- Enzyme Electrodes

- Immobilized cell biosensors

- Immunosensors

- Optical biosensors

- Ion channel sensors (*see* **Patch clamp**).

Other biosensors use DNA probes as the biological element, or even multi-cellular organisms such as daphnia (a small fresh water shrimp), or trout.

Biosensors that generate an electrical output are classified as potentiometric or ampometric. Ampometric sensors measure the current generated by a reaction, and are operated at as low a voltage as possible. Potentiometric measure the voltage generated by the reaction, and are operated with as little current flowing through them as possible.

Biosensors have the potential for being extremely sensitive and specific ways of detecting something. However, their practical application has been hampered because the biological element is hard to build into a robust engineering system:

- *Stability*—The biological element go off quite rapidly with use. Some go off in minutes when operational requirements were for days or weeks of operation.

- *Shelf life*—Even when they are not operating, biosensors decay stored in a fridge or (in extreme cases) a freezer. Papers on biosensors often claim stability for weeks of operation, but this usually means that they are used once a day and kept in the fridge in between uses, a far cry from being used in a production line 24 hour a day.

- *Manufacturability*—Most biosensors are difficult to make, and constructing an assembly line to make in commercial quantities requires a well defined way of making them. Even commercially successful sensors are hard to make sufficiently reliable for such a method to be defined.

The most prominent exception is the glucose biosensor, an enzyme electrode based on glucose oxidase and commercialised by several companies, notably Exactech, as a test for blood glucose levels. These work where others fail because the amounts of glucose being measured are large (so the electrode does not have to be very sensitive) and the enzyme glucose oxidase is unusually stable.

A related idea is the biochip, where the sensor element carries out chemical reactions as well as sensing: see separate entry.

Biosorption

Biosorption is the sequestering (i.e. capture from solution) of chemicals (usually metals) by materials of biological origin. Biosorption is widely talked about and little used as a method for removing materials from waste or purifying rare metals. Potential uses include:

- removing heavy metals from industrial waste water, especially nuclear waste streams, where the metals are present at low concentrations but are the most hazardous element in the water

- purification of precious metals such as silver and gold from very low grade ores, by washing the metal out of the ore and then concentrating it from the leachate using biosorption. *See* **Gold and Uranium extraction**

Many organisms have components which bind metal ions: bone matrix material, for example, binds strontium rather well. In some cases this is an active process—the organism uses energy to take the metal ions inside and trap them in an insoluble form. Among the ways in which organisms actively accumulate metal ions is by "pumping" them into special sections of the cell and then precipitating them as phosphates or sulfides (which are usually insoluble). In others, the process is passive—the metals stick of their own accord to a material that the organism makes. "Passive" systems include proteins which bind the metal, specifically metallothioneins (sulfur-containing proteins found in many organisms), lignin (from wood), chitin, chitosan, and some cellulose derivatives.

Biosorpting organisms can be stuck on filters or in pellets which are then put in a reactor and a solution washed over them, or the biosorptive material can be extracted from the organism and used on its own. This second option is useful for non-microbial biosorption systems: chitin, produced from waste prawn shells, absorbs a number of metal ions, and has been extensively investigated as a biosorption medium. Organisms can be selected which can accumulate more of the "target" metal, or which accumulate one metal specifically. For industrial use, bacteria or yeast are almost always the organisms used, although many other organisms such as protozoa, simple plants, even trees can accumulate substantial amounts of metals.

To be useful, biosorption must be specific and efficient. For metal removal from waste streams, removal must be at least 90% efficient to be of any industrial use, and the organisms or polymers must be able to remove at least 15% of their own weight in metal. Any less efficient system costs more to use than conventional removal systems (such as "ion-exchange" materials). The efficiencies for precious metal extraction can be lower, depending on how valuable the metal is, but must be very specific: there is no point purifying gold if you purify a lot of iron along with it. In general, these properties rely on strain selection. Strain improvement can do little to alter the basic properties of the organisms, and we do not know enough about the metal binding materials involved to engineer them "rationally" using genetic engineering.

Biotechnology

Biotechnology is the application of knowledge of living systems to use those systems or their components for industrial purposes. The work "biotechnology" was first used by the Hungarian agricultural economist Karoly (Karl) Ereky in 1919, to mean "all lines of work by which products are produced from raw materials with the aid of living organisms." The definition has been broadened slightly to include producing things with the aid of materials from living organisms (such as enzymes) and some raw materials that are produced from living organisms themselves (such as alginates or biomass), and narrowed to focus on new technologies, rather than traditional production processes. The UK government in *Developments in Biotechnology* (1992) focussed more on the product than the production process, with "... the production of innovative products, devices and organisms by exploitation of biological processes." Either way, farming is not considered part of biotechnology.

So biotechnology is the pragmatic combination of science and technology to make use of our knowledge of living systems for practical application. This includes a wide variety of applied biological science, but also includes aspects of chemistry, chemical technology, engineering, and specialist disciplines in specific industries such as pharmaceutical, environmental treatment or agricultural industries. There are about 1300 biotechnology companies in the United States, with 260 being public companies (whose shares are traded on the stock markets), and a total revenue of about $25 billion. There are around the same number in Europe, but only around 100 public, those mostly in the United Kingdom, reflecting the breadth of ground covered by the industry.

Some argue that biotechnology is just an extension of brewing and baking. Usually, this is a prelude to an argument about how safe biotechnology is. The record shows that biotechnology really is amazingly safe—Orsen Welles' famous radio broadcast of "The War of the Worlds" caused more deaths than biotechnology has done, suggesting that radio plays are more dangerous than genetic engineering. However, saying that it is an extension of traditional brewing is wrong. The difference is that biotechnology seeks to use rational approaches to developing its technologies and products, rather than traditional craft or trial-and-error. (Modern brewing, of course, is a highly technical and scientific branch of biotechnology.)

Note also that biotechnology is not the same as "applied molecular biology." Genes, DNA and all the science associated with them get a lot of press nowadays, but equally important for the end users of biotechnology (i.e. you) are the food, beer, vitamins, and other products through the wide range of technologies outlined in this book.

Biotin

Biotin, a vitamin that is required as a co-enzyme for some enzymes, turns up in some unexpected places in biotechnology as a "label" system. Biotin can be linked onto many different macromolecules by chemical reaction, a process called biotinylation. The protein avidin (usually made from egg white) or its bacterial counterpart streptavidin binds to biotin extremely tightly—far tighter than an antibody binds to its antigen. The avidin can be labelled with an enzyme, a fluorescent group, a colored bead etc. This will then seek out and recognize the biotinylated molecules, and not stick to any others. This can be preferable to trying to link the enzyme, fluorescent tag or other label onto the target macromolecule directly because (i) you can get more biotins onto a macromolecule that enzyme molecules, and (ii) the biotin is very stable, and so can be treated with extreme pH, boiled, or irradiated, whereas an enzyme would be destroyed by these conditions.

Biotransformation

Biotransformation is the conversion of one chemical into another using a biological catalyst: a near synonym is biocatalysis, and hence the catalyst used is called a biocatalyst. Usually the catalyst is an enzyme, or whole, dead microorganisms that contain an enzyme or several enzymes. The advent of catalytic antibodies and ribozymes will broaden the definition somewhat. Conversion of one material into another using whole living organisms is sometimes called bioconversion.

Biotransformation is one of the largest areas of industrial biotechnology: around 5% by volume of the enzymes used industrially are used for biotransformation (*see* **Enzymes**). A wide range of materials are made by biotransformation, from commodity items such as high-fructose corn syrup to speciality chemicals for the pharmaceutical industry. Some biotransformation processes, such as that producing vitamin C, produce thousands of tonnes of product per annum. The advantage of biotransformation over conventional chemistry is the specificity of enzymes. Reactions can be

- Stereospecific—they produce only one optical isomer of a chiral compound

- Regiospecific—they change only one part of a large and rather homogeneous molecule (analogous to only digging up one particular stretch of a motorway.)

A key use for biotransformation is in "resolution." This is a biotransformation which takes a racemic mix of a chiral compound and converts one optical isomer into another compound. Conventional chemistry or separation techniques can now take this and separate the two, different chemicals. (*See* **Chirality**).

The most commonly used biotransformations involve:

- Acylases (to resolve chemically synthesised amino acids)

- Esterases and lipases (to make a range of esters, lipids, and to resolve fatty acids and alcohols).

- Beta lactamases and penicillin acylase (to make penicillins and cephalosporins)

- Peptidases and proteases (to make peptides)

71

- Steroid transforming enzymes (to make steroid derivatives). These are always used in whole organisms, as many enzymes are involved in each biotransformation.

Biotransformations involving the use of proteases, amylases, lipases are discussed in separate entries.

A version of biotransformation is directed synthesis. This is feeding an organism a chemical that would not normally be present in its environment so that its normal metabolic pathways include that chemical into a new product. An example is the production of benzylpenicillin. Penicillin is produced by fermentation, during which an acetate group is linked onto the growing penicillin molecule by an enzyme in the *Penicillium* organism. If the fungus is fed phenylacetate at high levels, then it links this on instead and produces benzylpenicillin.

Blood products

Originally these were biopharmaceutical products made from human blood, such as the blood clotting factor VIII used to treat hemophiliacs. Such extracted products are usually made by a series of filtrations and solvent extractions. The major "blood products" in this category are HSA—human serum albumin, used to produce blood substitutes and extenders for transfusion, and human gamma globulins, immunoglobulins used medically to give people an extra high level of antibodies when they might be exposed to specific, unusual diseases.

The term blood products is also used to refer to biopharmaceuticals which act on blood or the cells which make blood. They are also made by those cells, but in such tiny amounts that extracting them from blood itself is impractical. So they are made by genetic engineering instead.

Among the "blood products" category of biopharmaceuticals are:

- Thrombolytics—drugs such as tissue plasminogen activator (tPA) produced by Genentech, streptokinase, eminase (made by GlaxoSmithKline). These dissolve blood clots in the arteries and hence are used as treatments for heart attacks. An enormous clinical trial spreading over several years in the 1990s (GUSTO) showed that tPA gave a slightly better clinical outcome in heart attack patients than did streptokinase: however, it costs ten times as much. The protein hirudin, originally extracted from leeches, is also being developed as an anti-thrombin compound, which prevents further clotting rather than breaking up clots that already exist.

- Clotting agents—Factors VIII and IX to treat hemophilia, a disease where these proteins are missing. Baxter Healthcare and Mile Inc are developing recombinant Factor VIII.

- Erythropoietin (EPO)—stimulates the bone marrow to make more red blood cells (*see* **Growth factors**).

- G-CSF (marketed by Amgen as Neupogen), GMCSF, etc.—These are **cytokines**, materials made by the immune cells to regulate the immune system's function. *See* **cytokines**.

A related product is animal serum (i.e. the fluid from blood after the cells and clotting factors have been removed), especially from fetal and newborn calves. Serums ("Sera" to purists) are used very widely as a supplement to the growth medium used for mammalian cell culture. Finding reliable sources of serum is a substantial issue, as they have to be very consistent: how cells grow in serum can vary from batch to batch, for no clear reason. In addition, serum has to be from a source that can be proved to be free of BSE (*see* **Transmissible Encephalopathies**).

Blots

A range of molecular biological techniques are called "blots." They are all based on transferring biological material from a gel onto a porous membrane in a way that preserves how they were separated in the gel. There are a variety of ways of doing the transfer—passive diffusion, wicking, suction (vacuum blotting), or electric field (electroblotting). The figure shows the "classic" blot arrangement, which uses wicking to draw buffer through a gel. Once the

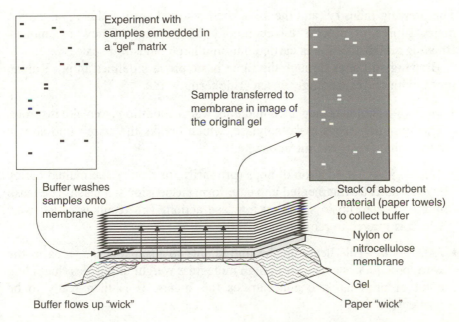

Experiment with samples embedded in a "gel" matrix

Sample transferred to membrane in image of the original gel

Buffer washes samples onto membrane

Stack of absorbent material (paper towels) to collect buffer

Nylon or nitrocellulose membrane

Gel

Buffer flows up "wick"

Paper "wick"

sample molecules are on the membrane, some new analysis can be done on them that could not be done in the gel.

The variations on this theme depend on the molecules:

- Southern Blot—Named after Professor Edwin Southern, the molecules are DNA, separated by electrophoresis.

- Northern Blot—Almost the same as the southern blot, but the molecules are RNA.

- Western Blot—Here the molecules are protein, also separated by gel electrophoresis.

- Dot Blot—Here DNA, RNA or protein is dotted directly onto the membrane support, so that they form discrete spots. Also called slot blots (if the spots are not round).

- Colony Blot—Here the molecules (usually DNA) are from colonies of bacteria or yeast growing on a bacteriological plate. A variation (called the plaque lift) can also be used for viruses.

With the advent of PCR, which is a more sensitive way of detecting DNA and RNA than hybridisation methods, the most commonly seen blot now is the western blot, where proteins are separated according to size and then identified by reacting them with an antibody (*see* **Gel electrophoresis**).

Brewing

The brewing industry (and the associated distilled spirit industry) is both a major source of process biotechnology and a user of new techniques. Brewing and distillation is carried out on a large industrial scale.

Brewing still goes through the same basic process defined in pre-Roman times, with correspondingly unscientific terms.

- *Malting*—Barley is incubated in damp, warm conditions to make it sprout. The sprouting seeds make amylase, which breaks the starch into simple sugars that yeasts can metabolize.

- *Mashing*—After addition of hops (primarily for flavor), the malted barley is mechanically fragmented into a uniform suspension—a mash. The mash is then heated up to speed up amylase action. The result is called wort which is filtered to remove most of the solids.

- *Kettle boil*—This heats up the ingredients to inactivate enzymes in the wort, precipitates a number of high molecular weight impurities (including a lot of proteins) and concentrates the sugars. It is then ready to be fermented.

- *Fermentation*—Yeast performs fermentation. The two most common types of yeast used in beer brewing—*Saccharomyces carlsbergensis* and *Saccharomyces cereviseae*—are also mainstays of both molecular genetics research and other biotechnological production processes. The yeasts grow rapidly, and the ethanol slows down any competing yeasts or bacteria that survive the boiling. As fermentation nears completion yeast "flocculates" (sticks together in clumps) and either sinks to the bottom or floats with the foam on the top, where it can be removed.

- *Clarification*—This is the removal of remaining proteins and other high molecular weight material, mostly polyphenols, that can cause "haze" in the beer. They are removed by precipitation (started by adding a polymer which binds to them, making them stick together), or cooling (which can alter their chemistry, again making them sticky). The polymer mixtures are called "finings," and can be polyvinylpyrolidone, various forms of silica, gelatin, or a number of traditional preparations such as "Isinglass" (a mixture made from the swim bladders of sturgeons).

Many other carbohydrate sources can be used for alcoholic fermentation, such as rice, potatoes, swede. The result is often concentrated by distillation. The entire broth is distilled to separate an alcohol-containing liquid from the yeast and remaining solids from the grain. The liquid distilled off usually is almost pure ethanol and water. In whiskey distillation, this is called grain spirit or grain neutral spirit. Some drinks such as vodka and tequila use the primary distillate as a drink. Others make it more potable by "aging" it to diffuse woody flavors into it (whiskey), or using it to extract the flavor from other ingredients, such as juniper berries (gin).

The solid material left after distillation of the alcohol from grain-based fermentations is called "distillers grains," and is often used as an animal feed component. "Dark grains" is the same thing, but with some of the spend liquor still absorbed into it.

New biotechnologies are applied to all aspects of this process. Particularly high profile is the use of recombinant yeasts with enzymes that degrade more of the sugars (producing "lite" beers with little sugar left), or which can be separated more easily.

Cancer

One of the most frequent causes of death, cancer is a disease where molecular biology has made enormous advances in understanding, and many—maybe the majority—of biotechnology companies working in drug discovery are trying to apply that knowledge. Most pre-biotech drug treatment was based on drugs which are poisonous to cancer cells but slightly less poisonous to normal cells. With a few exceptions, treatment was therefore a race between whether the drug killed the cancer, the cancer killed the patient, or the drug killed the patient.

Cancer is a disease where the normal controls on cell division fail (*see* Cell cycle). This is because some of the genes involved in control of the cell cycle have mutated in those cells. As the cancer progresses, more and more genes mutate, resulting in "wilder" cells. So, in general, the earlier cancer can be diagnosed and treated the better. Biotechnology has sought to use this understanding of the disease to develop more sophisticated strategies for therapy.

Some of the major applications are:

Drug discovery. A huge range of targets for drug discovery have been uncovered in cancer research. Among the more successful are the tyrosine kinases (*see* Kinase), which is the target for Gleevec (or Glivec), launched in 1998 by Novartis. Other kinase inhibitors have been less successful in clinical trials, illustrating the difficulty of this approach even when you know a lot about the disease.

Other drug discovery targets in cancer in which biotechnology plays a large part are neovascularization (also called angiogenesis), and metastasis. The former is how blood vessels grow into the tumor, to provide the tumor with oxygen and nutrients. Metastasis is how a "primary" tumor (the first one that develops) sheds cells around the body where they can lodge and grow new tumors of their own ("secondary" tumors). Stopping this process is therefore very valuable. It is also very hard—British Biotechnology Ltd almost went bankrupt trying to pursue an anti-metastasis drug.

Drug targeting. Cancer drugs could be targeted to the molecular peculiarities of cancer cells. A major success for this approach is Herceptin (from Genentech), launched in 1998. This is an antibody drug that targets a specific protein present on some breast cancers, called HER/neu. If tests show that HER/neu is present, then Herceptin is quite effective at pushing back the tumor and prolonging life. It has minimal side-effects, unlike "normal" anti-cancer agents, because it only targets HER/neu. (*See also* Immunotoxin.)

Cancer vaccines. A related approach to the Herceptin one is to treat cancer as an invasion by dangerous cells, and to vaccinate the patient against them. As always, the trick is to find a suitable antigen (*see* Vaccine).

Current targets include

- HER/neu (the same target as for Herceptin, but as a vaccine)

- MUC-1, a protein that is common on pancreatic cancers as well as some others

- A range of antigens present on melanoma, which are being developed by Corixa and Onyvax.

Success in the cancer vaccine field has been most notable with melanomas. Cancer vaccines can be single proteins, killed or engineered cancer cells, or immune cells such as dendritic cells which are involved in "presenting" the antigen to the immune system (basically, saying "hey, look at this foreign protein!" to the immune system).

Immunotherapeutics. A related idea is to boost the immune system to recognise cancers on its own. Drugs such as interleukin-2 (IL-2) and interferons have been tried with this, with indifferent success.

Gene therapy. As cancer is a disease of the genes, it would be attractive to develop a gene therapy to cure it. There are three general approaches:

- *Gene replacement.* Putting correct versions of the mutant genes into cancer cells. It has turned out to be impractical to get the gene into enough cancer cells to have a significant effect.

- *Immunotherapy.* Rather than inject IL-2 to stimulate the immune system, you inject the gene for IL-2 so that it stimulates the immune system continuously in the area of the tumor. This has had rather poor results to date.

- *Suicide gene therapy.* This is another name for GDEPT—*see* **ADEPT**. It has the advantage that you only need to get the gene into a few tumor cells to kill off all the surrounding ones.

- Oncolytic viruses. These are viruses that selectively grow in cancer cells. A range of such viruses based on adenovirus (Onyx Inc), herpes viruses and a polio-rhinovirus hybrid have been tried.

There has been a huge amount of work in these fields since the 1970s, but the death rate from cancer has actually increased in that time. This is partly because death rates from other causes have declined, and partly because it takes 10 years minimum to get a new drug researched and tested, so the breakthroughs of the 1980s are only today being seen as new medicines (such as Glivec). Even so, this is not as impressive a record as the company brochures of the 1980s promised.

Diagnostics. An area that has been successful is diagnosis. Use of antibodies, DNA probes, and PCR have revolutionized cancer diagnosis, to find out what the molecular problems are behind each disease. The problem is that, at the moment, the drugs are not here to do much about it. The exception is Herceptin, where a specific diagnostic test can be used to tell whether a breast cancer patient has the HER/neu gene or not. If they do, then Herceptin can be used.

Capillary electrophoresis

Also capillary zone electrophoresis or capillary gel electrophoresis, this is a widely used technique in many biochemical and biotechnological fields.

Capillary electrophoresis (CE) is electrophoresis—the separation of molecules by moving them in an electric field—in a very fine capillary tube (a tube with an internal diameter of less than 1 mm). This can be "free zone" electrophoresis, that is, when a solution fills the tube, or gel electrophoresis when the tube is filled with a polymer. *See* Electrophoresis.

The advantages of capillary electrophoresis are:

- *Speed*. The electrophoresis can be "run" much faster than conventional electrophoresis. The speed of molecules through a gel is dependent on the voltage. The mass of liquid in a capillary tube is so small that even very high voltages produce tiny currents, and the heat produced can be radiated away from the tube rapidly. So the electrophoresis can either be run very fast or it can be run on a very long capillary tube, so increasing resolution.

- *Higher resolution*. A major cause of poor resolution (how well the method separates molecules) is unevenness across the electrophoretic medium due to uneven current flow, heating or convection. This is minimized when the tube diameter is very small (less than 1 mm).

As well as electric forces, CE systems can also use electroendosmosis: the movement of a solution of ions in an electric field, when they are carried in a charged tube. In such small tubes, this causes the solution to move, allowing separation to be a balance between fluid flow and electrophoresis.

Capillary gel electrophoresis is widely used for large-scale DNA sequencing, especially genome programs where applied biosystems capillary electrophoresis machines are almost an industrial standard. They can also be used for very high throughput SNP detection (see separate entry).

Capital (money)

Most biotechnology companies are based on research or development, and do not make a profit as soon as they start up. Some do not expect to make a profit for years, and so they need a lot of money invested in them. Investors and biotechnologists have quite different ideas about the nature of the universe, so this can be a difficult process.

78

There are generally three types of investors. They put money into the company in return for shares (stock, equity) in its future.

- *Rich people.* There are different names for these—private investors, business angels, high net worth individuals—but they are people with sufficient money of their own to make a substantial contribution to the running cost of a company in its early years.

- *Venture capital companies.* These are professional groups who invest in risky companies in the hopes that their value will rocket. Traditionally regarded as very fickle mercenary groups (the standing joke is that biotechnology venture capitalists have no capital, show no liking for adventure and know nothing about biotechnology), they have been the main funders behind the rise in the numbers of biotech companies. They are looking for a very high return on investment (ROI) and an exit route, that is some way that they can sell their share in your company and get hard cash back in 3–5 years.

- *Public investors.* When a company is stable enough, shares in it can be traded on a public stock exchange. Usually, the most important investors here are big institutions (called institutional investors), like pension funds, who will have specialists who invest just in healthcare companies, others in environmental companies, etc.

The path to these riches is rocky. It usually involves

- *Seed funding.* Less than $1 million, to get you started.

- *Private financing.* Any investment that gets substantial funds in, but which is not on a public stock market. There may be several "rounds" of financing, that is, the company may run out of money and need to ask more investors for more. The key here is usually to find a "lead investor"—someone who will say "yes, I am going to put my money in." Others will then follow, sheep-like, if you are lucky. The shares given out ("issued") at each round are usually given letters—A-series, B-series, etc., and the rights of the different classes of shareholder can differ substantially.

- *IPO.* Initial Public Offering, the Rubicon of funding, when the company is "taken public" and its shares are traded on a stock exchange where anyone can buy or sell them. In the United States there is a special stock market called NASDAQ for "high risk" (almost always technologically-based) investments. Companies will sometimes say that they are "quoted on NASDAQ" if their shares are being bought and sold in that market.

There can then be a quite bewildering range of methods of issuing more shares on a public stock exchange, which it is beyond the scope of this book to discuss.

If a company is not making money, then a key issue is its "burn rate"—how fast the company is using up its supply of money. Burn rates of $2–10 million a year are normal for a start-up, Burn rates of $10–40 million a year are typical for a medium size company.

There are several vexed issues here.

The value of technology. To know how much a company should be paid in return for a share, you have to know how much the whole company is worth, and that means valuing its most valuable asset, its science and technology. There is no "right answer" to this.

Once you are a public company, your value is defined by your share price, which is influenced by public news about you: analysts scrutinize your financial figures every 3 months ("quarterly returns") and mark your shares down by 70% if your clinical trial fails. But it is also influenced by casual remarks made by brokers over a beer, the performance of unrelated companies that are seen to be similar to you, the economy as a whole, even the phase of the moon. (This latter really is a well documented effect.) Many scientists feel that this is a poor way to manage the long-term development of new technology, and in 1996 a survey of biopharmaceutical CEOs found that 50% of them thought that they could improve their clinical trials results "by ignoring Wall Street."

Catalytic antibodies

Catalytic antibodies, also called abzymes, are antibodies whose binding sites catalyze a reaction in their target "antigen," rather than just passively binding to it. Antibodies do not normally possess catalytic activity.

In the 1940s, Linus Pauling suggested than an enzyme was simply a protein which bound to and stabilised the transition state of a reaction. (A transition state is what the reaction would look like half way through, with some chemical bonds "half broken" and others "half formed." *See* **Enzyme mechanism.**) In the 1960s several workers suggested that an antibody which bound to the transition state of a reaction would catalyze that reaction.

Because the transition state is not a stable chemical, you have to make an antibody against something similar to it, called a transition state analog: a chemical that has the same shape, size, charge, etc. of the transition state, but, because it is made of different atoms, is actually a stable molecule.

Catalytic antibodies can also work through reducing the entropy of reaction, that is, bringing together two molecules in the right orientation to allow them to react. This can apply to two substrates for a reaction, or a substrate and a co-factor. Catalytic antibodies have been made which catalyse reactions through both these mechanisms.

A lot of research work has generated catalytic antibodies which have some small catalytic effect. However, in 1996 it was found that albumin, a completely non-antibody-like protein from blood, can also catalyze reactions at similar rates to catalytic antibodies, just by providing hydrophobic "pockets" on the protein's surface. So, the jury is still out on whether the abzyme approach will produce genuine, powerful catalysts, or whether it is just detecting the very poor catalysis that any protein could do.

cDNA

cDNA is copy-DNA (or complementary-DNA). It is a DNA copy of an RNA, and is made from the RNA using reverse transcriptase. This is a gene cloning technology.

There are a variety of reasons for doing this: the DNA gene itself may be unknown, but we know the cells that are making RNA, or the scientist might not want the "original" gene, but the spliced version of it with no introns in (*see* **Genetic code**). However, the main driver behind them is that RNA is much less convenient to handle than DNA. DNA can be readily replicated, can be "cloned" in many organisms, and is chemically robust. RNA is very easy to break down, and while it can be replicated directly using RNA-directed RNA polymerase, this is a less established enzymatic technique.

cDNA is also made when you want to use PCR to amplify an RNA. PCR does not work on RNA, so you copy it to DNA using reverse transcriptase and then PCR amplify the cDNA. This is called RT-PCR. This can be used as a method for measuring how much RNA was there, even when the amounts are tiny (a few molecules per cell).

Cell adhesion molecules (CAM)

Also called Intracellular adhesion molecules (ICAMs), these molecules are present on a wide range of human cells, and are part of the mechanism used by cells to recognize each other. They are glycoproteins, and the sugar residues can be crucial to their function: the differences between some blood groups, for example, are the result of variation in the sugar residues on some ICAM molecules.

There are several families of mammalian cell adhesion molecules, with many variants of each family being seen in different cell and disease states. The major families are:

Integrins. These are cell-surface receptors for a range of other cell adhesion molecules, and for molecules found in the intracellular matrix (which is the material that mechanically supports cells, and is secreted by them). Integrins bind to proteins with the three amino acid sequence arginine-glycine-aspartic acid ("RGD" in the one-letter amino acid code) in them.

Immunoglobulin superfamily. These are molecules with a structural similarity to the antibody (immunoglobulin) molecules. They include ICAM-1, VCAM-1, LFA-1, and Mac-1 (found on leukocytes only).

Selectins. They bind to carbohydrate cell surface molecules, and are important in the initiation of blood clots and inflammatory responses among other processes.

Fibronectin. A cell surface glycoprotein which links the cell to other proteins in the intercellular matrix, notably to collagen. It is also sometimes called large external transformation-sensitive protein (LETS), because it is often absent from cancer cells.

Cell adhesion molecules are important to drug companies, and hence to biotechnology companies, because they are the molecules through which cells control their interactions with each other. This is especially relevant for drugs for inflammatory diseases and blocking some viral infections.

Cell culture

This is the cultivation of cells from a multicellular organism, usually mammalian cells although insect cells are also commonly used. The term "tissue culture" is often used as a synonym, which strictly it is not (see separate entry): however, the intuitive nature of cell culture science does not lend itself to such pedantry.

Many biotechnological production processes and research programs depend on the ability to grow animal cells outside their parent animal. Monoclonal antibodies are manufactured in bulk using cell culture, and a range of biopharmaceutical products are produced in genetically engineered mammalian cells, as these synthesize the correct glycoforms of the proteins.

Because animal cells are not free-living organisms like yeast or bacteria, they do not function well when separated from their animal. They need to be treated with special care to enable them to proliferate, with careful control of pH, temperature, and the chemical ingredients of the liquid in which they grow ("culture medium"). This medium is also often supplemented with animal serum, which provides growth factors and some other minor nutrients. Usually fetal calf serum (FCS, also called fetal bovine serum, FBS) or newborn calf serum (NCS, NBS) is reckoned to be rich in growth factors. Pregnant mare serum (PMS) and horse serum is also sometimes used. There is an on-going debate about the quality and reliability of serum for cell culture, especially if it is to be used to grow cells that are making a biotherapeutic product such as a therapeutic antibody. This is a substantial economic issue for biotechnology, as about 500 000 litres of FBS are traded worldwide each year at $200/liter for commercial cell culture.

Some animal cells are "anchorage dependent," which means that the cells have to stick down onto a surface to grow. Usually the surface has to be specially treated with polymers such as collagen or poly-lysine. Glass is a good surface for many cells. Anchorage independent cells can survive and grow drifting free in solution. Sometimes anchorage independent cells will stick onto things anyway, but they do not need to in order to survive. White blood cells are an example of an anchorage-independent cell, nerve and muscle cells are anchorage dependent.

Most mammalian cells also do not grow well in isolation from all other cells, so they needed to be grown in fairly dense or "crowded" conditions. The needs of some cells are so specialized that they will only grow on other, living cells. Such a layer of support cells is called a "feeder layer."

A key condition for cell culture is sterility. Yeasts and bacteria grow much faster than mammalian cells, and so if just one bacterium gets into your cell culture it will soon outnumber the mammalian cells, and poison them with their metabolic wastes and endotoxins. So rigorous sterility is critical in cell culture.

Cells can be cultured in a variety of containers. The simplest is a Petri dish, called a tissue culture dish if made with special, cell-friendly plastics. Square-sectioned plastic bottles are also common. Larger cell culture is done in roller bottles, which are ordinary bottles which are rolled gently along their long axis to keep a gently movement of culture medium over the cells.

When the culture gets too crowded with cells, most cells stop dividing, but if they are diluted into new medium (a process called "splitting" the culture), then they will start growing again. For some animal cells this process can be carried on indefinitely. For others, after 40–60 doublings (i.e. when the original cell has divided and divided again, 40–60 times) the cells loose the capacity to divide at all, and become rather oddly shaped, sickly cells. This division limit is called the Hayflick limit. This process is called cellular senescence, and may be connected to the ageing process.

(*See also* Immortalisation).

Cell culture for production

This means producing materials using cultured mammalian cells (or occasionally the culture of isolated plant cells). Cell culture is a difficult process, requiring sterility and careful control of the conditions (*see* Cell culture). Large-scale cell culture has to take several additional factors into account.

Fermentor design. Confusingly, large-scale cell culture can take place in large vessels called fermentors or bioreactors. As mammalian cells are quite fragile, the usual stirring methods will destroy them, so gentle methods are used. For very large scales, airlift fermentation is established. Hollow fiber methods (see separate entry) are increasingly used not only for pilot scale but also for smaller production runs.

Small-scale systems. The favourite small-scale system is the roller bottle. This can be used in a research laboratory, and scaled up to a dozen litres simply by adding more bottles. Roller bottle culture is often used to "grow up" cells for use in a larger fermentor. Mammalian cells can usually only grow in

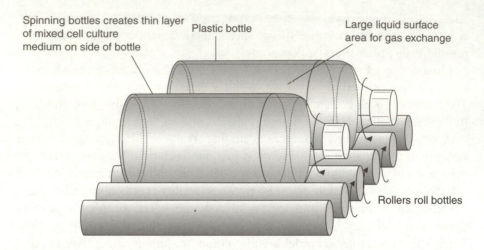

Spinning bottles creates thin layer of mixed cell culture medium on side of bottle

Plastic bottle

Large liquid surface area for gas exchange

Rollers roll bottles

relatively crowded conditions, so you cannot seed a 100 liter fermentor with a few cells and expect them to grow. You have to seed it with the output of a 10 liter fermentor: this might be a hundred 100 ml roller bottles.

Cell cycle

Cells grow in a carefully regulated cycle of events. In mammalian cells, the stages of the cycle are quite distinct. They are:

- G_1, when the cell is making all the components for the new cell except DNA
- S phase, when the DNA is replicated
- G_2, when the cell organises itself to divide
- M phase (or mitosis) when the cell actually divides in two.

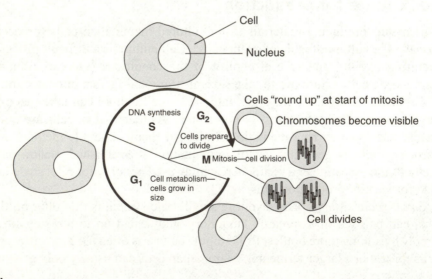

Cell

Nucleus

Cells "round up" at start of mitosis

Chromosomes become visible

DNA synthesis
S

G_2

Cells prepare to divide

M Mitosis—cell division

G_1 Cell metabolism—cells grow in size

Cell divides

During G_1, S, and G_2, together celled interphase, the DNA is contained in the nucleus, which looks like a relatively undifferentiated blob of DNA, RNA, and protein under the light microscope. During mitosis the DNA is organised into chromosomes. Cells in mitosis can be stained and examined under a microscope to count the number of chromosomes they have, a rough indication that they are the cells you thought they were: such a chromosome count is called a karyotype. A standard diagnostic technique called a metaphase spread fixes cells in metaphase so that their chromosomes are spread out nicely so they can be examined for abnormalities. *See* **Genes**.

Some cells do not divide at all. These are usually said to be in G_0, effectively a dead-end sideline off the main cycle at G_1.

A common laboratory trick is to arrange for all the cells in a culture to be at the same point in the cell cycle. There are a variety of methods for doing this "synchronization" step, at different phases. One method uses Fluorescent Activated Cell Sorter (FACS), which separates cells according to their **fluorescence** (see separate entry). If cells are labelled with a dye that fluoresces when it binds to DNA, you can tell whether a cell is in G_1, S, or G_2 by its relative DNA content.

There is a lot known about the control of the cell cycle, which is important for the development of drugs that target abnormal proliferation of cells, such as cancer. Some of these proteins are p53, the cyclins and cyclin-dependent kinases (CDKs), bcl, rb, and a host of others. These are of great interest in cancer research, where this control mechanism has gone wrong and the cells divide when they are not meant to.

Cell disruption

Many fermentation processes produce products which are inside cells. Examples are many proteins produced by genetic engineering, enzymes, and large molecules such as the biodegradable plastic poly-hydroxybutyrate. It is necessary to break the cells to get these products out. This process is called cell disruption.

It is easy to break up mammalian cells, which spontaneously break open (lyse) if you put them in solutions of low osmotic strength, for example, pure water. However, yeast and bacterial cells are specifically "designed" by evolution to be unbreakable. Thus, a lot of energy has to be put into breaking them, and there is the risk that that energy will also disrupt the product inside the cell. Plant cells in culture are not usually as tough as yeasts, but much stronger than mammalian cells. Methods used are:

Mechanical methods. The most obvious method is to break the cells mechanically. There are many ways of doing this:

- French Press, which forces the cells through a very small hole at high pressure. A large scale version of this is called a Manton Goulin homogenizer.

- Mills, in which the cells are shaken vigorously with abrasives, or metal balls (a ball mill), or rods (a cylinder mill).

- Blenders. Traditionally the laboratory uses a blender called the Waring blender (named after a 1930s New York dance band leader who invented or popularised it for making cocktails), but essentially this is a food processor with a powerful motor.

Autolysis. This simply alters the conditions so that the cell digests itself. This is the simplest possible method, but tends to be useless for protein products as the cell digests itself from the inside out, so breaking down the product before the cell wall.

Enzyme action. This is very effective—the cells are treated with an enzyme which dissolves some key component of their cell wall, which then simply falls to bits. Typical enzymes used are lysozyme (for bacteria), chitinase or glucanase (for yeasts), cellulase (for plant cells).

Detergents, alkali, osmotic shock (i.e. pure water), plasmolysis (treatment with high concentrations of salt), organic solvents. Any of these treatments will knock holes in the plasma membrane, the thin layer of lipids inside a cell wall which actually holds the cell's contents in (the cell wall, by contrast, is meant to keep the outside world out). These methods usually also include a DNAse treatment step, to break up the very long chains of DNA that are also released, making the solution a thick gluey mass.

Freeze–thaw. Freezing and thawing can break up any structure as ice crystals form inside the wet materials of which cells are made.

Cell fusion

The fusion of two cells together results in a new cell which has all the genetic material of the two original cells, and hence is a new type of cell. The ability to fuse different types of cell—from the same species or from different species—has been used widely in biotechnological research. Common methods used include:

- *Electroporation* (see separate entry).

- *Fusogen-mediated fusion*. Some chemicals, such as polyethylene glycol (PEG), can make cells fuse together: they are called fusogens. PEG is a polymer which binds into the lipid membrane of cells and causes it to merge with any other lipid membranes around.

- *Virus-mediated fusion*. Some viruses have lipid coats which fuse with the membrane of cells when the virus infects that cell. Thus, they also act as fusogens. Viral fusogens were discovered before PEG's fusogenic properties were, but PEG is preferred now because it is easier to get hold of and less potentially hazardous.

Cell fusion is used in a variety of techniques. Making monoclonal antibodies relies on making a fusion between lymphocytes and an immortalized cell line. Polyploid plants, plants with abnormally large numbers of chromosomes, can also be made by fusing cells from the same plant together.

Cell growth

The growth of isolated cells in culture follows a characteristic curve, shown in the figure. The phases of the curve are:

- *Lag phase*. This occurs when the cells are introduced to their new growth medium, and is the time taken for them to adapt to it. If their new environment is identical to their old medium, then the lag phase can disappear: however, even the mechanical shock of moving some cells around can cause a lag phase.

- *Log phase*. This is the main growth phase of the culture, when the cells are growing exponentially. When plotted on a logarithmic scale, the log phase shows as a straight line.

- *Transition*. This is the period (which can be minutes to days) between log phase and stationary phase.

- *Stationary phase*. Here the cells have stopped growing—they have reached the capacity of their growth system to sustain growth. In most bacterial systems, this is a balance between some cells dying and others continuing to grow. Mammalian cells usually stop dividing in stationary phase.

- *Death phase*. The total mass of cells remains the same, but increasingly few of those cells are alive in the sense that they could start a new growth

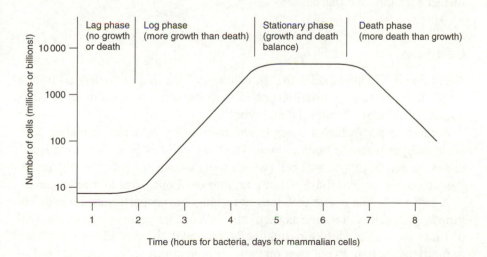

curve if they were given fresh growth medium. Death phase occurs because the cells have run out of nutrient, because they are too crowded (for mammalian cells in particular), or because their metabolism produces waste-products that build up and become toxic.

The length of the different phases varies enormously with different cell types. Many common laboratory strains of bacteria have a stationary phase lasting only a day or two before death phase starts. By contrast, mammalian nerve cells can last almost indefinitely in culture without dividing. A single mammalian cell isolated from skin or muscle and put into culture medium could take a week to divide—a single *E. coli* cell is unlikely to take more than 10 minutes to start growth.

The other key idea in cell growth studies is doubling time. This is the time that is needed for the cell population to double in number, and is equal to the time an "average" member of that population takes to go through one complete division cycle. "Growth Rate" is sometimes defined as 1/doubling-time. Doubling time depends on the growth conditions and on the organism being grown—some bacteria (e.g. *Clostridium perfringens*) can have a doubling time of 10 minutes in the right culture medium. Strictly speaking, the concept of doubling time only applies to organisms growing in log phase, that is, growing exponentially.

If fermentations are being used to make a product, the growth phases when the cells are being "bulked up" is called the tropophase: the phase when they are making the product we want is called the idiophase. In some cases these phases are quite distinct, in others they overlap.

This growth cycle is not the same as the overall life and senescence cycle of primary mammalian cells, which allow most mammalian cells to go through 40–60 cell cycles ("doublings") before they become incapable of further division. *See* Cell culture.

Cell line

A cell line is a clone of cells, that is, a "lineage" that has been derived from a single cell. The term is usually applied to mammalian cells cultured *in vitro*, outside their original mammalian body.

A cell line is capable of being grown indefinitely, which mammalian cells taken straight from the body are not. Thus, the cell has been "immortalized," that is, turned from a mortal cell (whose descendants are going to stop dividing after a few dozen divisions) into an immortal one. (*See* Immortalization.)

A stable cell line is one that does not change its properties as the cells are grown: each cell is the same as each other, even after years in culture. Not all cell line are stable. Unlike normal cells, mammalian cells which have been immortalized often do not pass on their chromosomes very faithfully, and so can loose genes which are not essential to the survival of the cell. These may include genes vital to the biotechnologist, such as the genes making an antibody in a hybridoma cell line. Strictly, before a cell clone is described as a cell line, its inventor has to demonstrate that it is stable in this sense. Some cell types are however called cell lines even if they are unstable—the most notorious is the HeLa cell "line," which varies substantially from laboratory to laboratory due to genetic instability.

Favoured cell lines for general biological research and for production of recombinant proteins in cell culture include Baby Hamster Kidney (BHK) cell lines, Chinese Hamster Ovary (CHO), etc. The names indicate where the cells to make these lines came from, and do not suggest that you have to cut up baby hamsters every time you need more cells.

See also entry on **Strain, Clone**.

Cell line and tissue rights

Ownership of a new cell resides in who made it. It does not reside in the source of material for the new entity. The Moore case in the United States (when John Moore claimed that a cell line used in cloning Interferon was derived from the hairy cell leukemia he was treated for in 1978, and hence was at least partly his) said that Moore had no rights to cell lines derived from his own cells.

Interestingly, if the Moore decision had gone against Sandoz and Genetic Institute (who now own the cell line), then many other people would have rights to a wide range of cells in research and industry. The descendants of Henrietta Lacks, originator of the HeLa cell line 40 years ago, would now have rights to a substantial part of all molecular biology and a mass of cells which probably exceeds her own weight in life.

In most countries, people do not have rights over organs removed in surgery either, providing that you survive. However, if you die you have more rights over your tissues and organs, at least in some countries. In the United Kingdom in 2001 Alder Hey hospital was castigated for keeping tissue from autopsies for future research, because the deceased (or their relatives) had a right to bury all of their bodies, not just the bits researchers did not want.

Centrifugation

This is a common biochemical purification technique. A sample is spun to generate centrifugal force, which makes cells, particles, even proteins settle to the bottom of a tube, so they can be easily separated from the liquid they were in. Centrifuges are widely used in research, in medical diagnostics (especially for separating blood into different cell types and plasma) and manufacturing. Research centrifuges in biology are usually categorized as

- *Bench-top*. Low-speed centrifuges, running at 100–5000 rpm, and generating forces of a few hundred to 20 000 gravities (one "gravity," or "g," is the pull on an object due to the Earth's gravitational field)

- *Microcentrifuge* (also microfuge). Eppendorf have cornered the market in these shoe-box sized machines that take 1.5 ml tubes and spin them at up to 15 000 rpm. Example use: separating bacteria for DNA preparation, separating affinity chromatography material from supernatant.

- *Mid-speed centrifuges*. (Also just "centrifuge.") Lab workhorses that are the size of a washing machine and spin at 1000–30 000 rpm. All sorts of uses.

- *High-speed centrifuge or ultra-centrifuge*. Spin at 10 000–100 000 rpm, generating forces of up to 200 000 gravities. Used for separating proteins, DNA and other macromolecules.

although there is a fair amount of overlap between these classes.

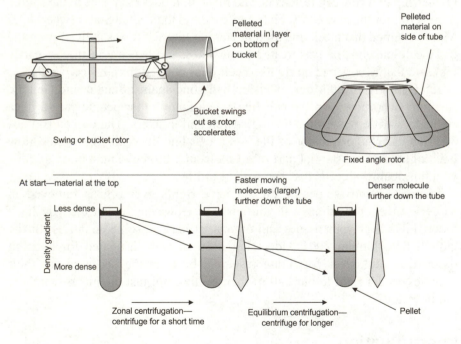

Key terms are

Zone vs. *equilibrium centrifugation*. Zone (or zonal) centrifugation separates essentially by size. Equilibrium centrifugation separates by some other parameter, such as density. This is related to:

Density gradients. Here the solution in the centrifuge tube is arranged so that it gets more dense as you go down the tube, by dissolving something in it: colloidal silica ("Percoll"), sucrose, or caesium chloride are common. When centrifuged to equilibrium, more dense parts of the sample will float further down the tube in more dense solution. This is also called "isopycnic" centrifugation.

Density gradient stabilization is also used in centrifugation: here a density gradient is created to stop the solution in the tube mixing, rather than to help separation.

Rotors. Most centrifuges consist of a drive unit (which powers it, controls rotation speed, etc), and a rotor in which the samples are placed and which goes round inside a bowl. In ultracentrifuges (centrifuges capable of tens to hundreds of thousands of gravities), the bowl is armoured steel to protect the operator should a rotor "fail" in a run. Legend has it that Svedberg, who developed ultracentrifugation for chemical and biochemical analysis, killed a couple of postdoctoral workers with bits of flying centrifuge when a rotor failed in an insufficiently armoured centrifuge.

Some rotors are special "zonal" or continuous rotors. Liquid is fed in the middle of them, and bacteria or other particulate matter is centrifuged out to the outside. These are of obvious use in separating microbial cells from their culture medium, but are an expensive way of doing this for large volumes. The hydroclone (whirlpool separator) is a version of continuous centrifugation.

Chaperones

Also known as chaperonins, these are a type of protein which help other proteins to fold up into their correct three-dimensional structure. Some proteins will fold up correctly on their own as soon as they are made in the cell, forming a working protein molecule. Most seem to do this very inefficiently, and to need other proteins to make them fold up properly.

It is speculated that this could be used to make production of proteins in bacteria more efficient, either by boosting the bacterium's own protein-folding machinery or by treating proteins with chaperones as part of the downstream processing. Both these are research ideas because the chaperone folding process is very complex.

Some chaperones are also known as heat shock proteins (HSPs), because they are made when a cell is stressed by sudden heat or cooling, as well as by some chemicals, or other insults. It is believed that the HSP chaperones are made to help to refold proteins or remove which have been damaged during the stress. Several groups are looking to harness this response so as to boost the cell's ability to repair itself after damage. Targets include reducing the damage caused by stroke, ischemic heart disease, and in Alzheimer's disease.

It has been found that the immune response to an antigen, such as a vaccine, is stronger if the antigen is attached to an HSP. Antigenics Inc is using this effect commercially in an autologous cancer vaccine program, to boost the body's response to a potential cancer vaccine by coupling it to an HSP. StressGen is aiming to increase the effectiveness of human papilloma virus (HPV, wart virus) by conjugating HPV proteins to HSPs.

HSPs can also be used as diagnostic markers for environmental stress. If an organism is making HSPs, then its chemical environment is poor.

Chemical library

A chemical library is an ordered collection of chemicals. Companies involved in discovering new chemicals usually start from such a library to see if they can find anything with the biological properties they want. This is then the basis for further development (*see* **HTS**).

Libraries come in two types. **Combinatorial chemistry** libraries are made all at once—see separate entry. Historical libraries (also called compound collections) are accumulated with time. They are usually more diverse (i.e. they contain chemicals with a wider range of structures, because they were made for many different reasons), but can be harder to manage and of variable quality.

Issues around chemical libraries are:

- *Library diversity*. Does the library contain a wide variety of chemical types, or is it like a book library containing only Westerns?

- *Quality*. Is it a collection of pure, stable chemicals, or a collection of tubes that had a chemical in once, and now have brown sludge in them?

- *Access*. How do you get the chemicals out again? This is both an issue of cataloging—can you find something again—and of physically collecting 50 000 compounds out of a library of 5 000 000 for a testing program.

- *Supply*. If five people have taken a sample from chemical X, have you actually got any left?

Related to chemical libraries is a chemical registration system—*see* **Chemoinformatics**.

Chemical space

Chemical space an odd concept that relates to how similar or different chemicals are. Each chemical can be described by some features that we consider important, such as the number of hydrogen atoms, its volume, the presence or absence of a particular group. We can show whether two chemicals are the same or different by using those "descriptors" to plot the two chemicals on a graph. If they are close together, then the chemicals are similar. If they are far apart, then they are different. If we imagine three such descriptors, then a collection of chemicals would be scattered over a three-dimensional volume, a "space."

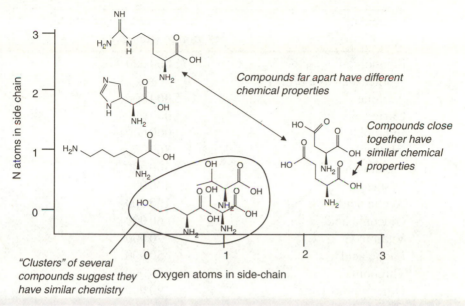

Simple, two-dimensional chemical space

Compounds far apart have different chemical properties

Compounds close together have similar chemical properties

N atoms in side chain

Oxygen atoms in side-chain

"Clusters" of several compounds suggest they have similar chemistry

The number of ways you can describe a chemical is enormous, and so the number of axes you need to plot out a "chemical space" is hundreds or thousands. Ordinary users reduce distances in such high dimensional spaces to a single measure of distance between two points. A common number is the Tanimoto distance (or coefficient).

The correct descriptors needed to describe chemical space is also hotly disputed.

Chemical space is important for designing chemical libraries. We want to get an example of as many types of chemical as possible into our collection, so we have the best chance to discover the one that has the properties we want, for example, that is an inhibitor of an enzyme, and hence may be a useful drug. This means that the chemicals we select should be scattered uniformly over chemical space, not all clustered together in one corner. We say that we want "to sample chemical space."

This is done by building a "virtual library" of chemicals—a computer-generated description of millions or tens of millions of possible chemicals. We then select a much smaller collection that is spread well over chemical space, and actually make those.

See also **Combinatorial chemistry.**

Chemicals produced by biotechnology

A number of chemicals are produced commercially by biotechnology in large amounts (apart from drugs and other specialist materials). Chemicals produced in large amounts by fermentation include:

Chemical	Amount produced worldwide per annum (tonnes)
Ethanol (other than as a drink)	75 000 000
Acetone	5 000 000
Citric acid	1 000 000
Butanol	1 000 000
Glutamate	800 000
Methionine	500 000
Lysine	400 000
Acetic acid	160 000
Gluconic acid	60 000
Vitamin C	70 000
Lactic acid	50 000
Threonine	30 000
Other amino acids	20 000
Other vitamins	16 000

The majority of these are for human or animal food.

Chemoinformatics

Chemoinformatics is the storage and retrieval of chemical information. It is not quite analogous to bioinformatics, as the technologies and issues are slightly different:

- *Compound registration and control.* A registration system is the software that catalogs the library, and tracks the properties and supply of all the chemicals in it. Ensuring that the right structure is entered and that the location of the compound (where the sample is physically kept) and other details are kept up to date is the role of the registration system.

- *Searching.* A chemical database needs to be searched fast not just for a specific chemical but for anything that "looks like it." In particular, you want software that will find any molecule with the same three-dimensional shape. Defining this and programming it so that it runs fast is quite tricky.

- *Associated data.* Most chemoinformatics systems are used in drug discovery, where the chemicals have been tested for their effect on biological assays. How is that data to be "linked" to the chemical, so that you can search it efficiently? Related to this is integration.

● *Integration*. How can many different pieces of software that relate to chemicals be integrated so that they work together without too much human intervention?

Chemoinformatics systems often include software for SAR and other property prediction (*see* **Rational drug design**).

Chemostat

A chemostat is a type of fermentor, a closed culture vessel in which new medium is continuously fed in and old medium and organisms continuously removed. It has a fixed rate of dilution, that is a fixed rate at which the new material is added and the old removed. This determines how fast the organism in the chemostat will grow. It contrasts to a fed batch fermentor, where new medium is added in a batch, fermented, and then removed.

An auxostat is a chemostat in which the dilution rate can vary. In an auxostat, the rate at which new material is added and old removed is determined by some feature of the culture. In a turbidostat the turbidity (cloudiness) of the culture is measured, and more nutrient added to keep it at a constant level. A pH auxostat will add acid or alkaline medium to keep the pH constant.

If the dilution rate is not high enough in a chemostat, the culture will grow at less than its maximum growth rate. If it is too high, the organisms will not be able to keep up with the addition of new medium and you will end up with an empty chemostat. An auxostat can be adjusted to keep up automatically with bacterial growth, and so maximize the growth rate. At such high growth rate bacteria which can grow fast are selected over ones which grow slowly. Thus, natural selection acts on the population, selecting fast-growing variants of the bacteria. Depending on what the auxostat is to be used for, this can be a good or a bad thing.

In practice, all large continuous industrial fermentation systems are auxostats rather than chemostats, as they have many feedback controls which allow the operator to adjust what materials the fermentation receives as it proceeds.

Chimera

A chimera is an animal which is a mix of several other animals. The chimera of mythology had a lion's head, a goat's body and a serpent's tail, and breathed fire. More prosaic and realistic chimeras can be made by a range of methods which mix cells from two sources to make an early embryo, which then develops into an animal which has cells derived from two sets of parents.

Chimeras in which the whole body is made from cells from two sources can and have been made by taking the cells from two very early embryos and mixing them together. This can be done at random, or the cells can be selected so that cells which are going to make specific areas of the body can come from one or other of the "parent" embryos. The techniques of *in vitro* embryology are then used to put the embryo back into a pseudopregnant mother (i.e. a mother animal which has undergone all the hormonal changes necessary to prepare her for pregnancy but who is not carrying any embryos). A sheep/goat chimera was made in this way in the late 1980s (it was called a "geep"), as was a cow/buffalo chimera. The former attracted such strong public disapproval that the latter was not publicized much (despite combining the resistance to Tetse fly of a buffalo with the milk-production ability of good dairy cattle), and further research on this line has been minimal.

Chimeras can also be made by mixing totipotent stem cells (ES or EC cells) with the early embryo. They can be incorporated into that embryo: the resulting mouse has cells derived from the EC cells in many tissues.

More limited chimeric animals are made by replacing some cells in an animal by those from another. In practice, this means the hematopoietic cells, the cells of the blood system. A common version is the SCID/Hu mouse. Here, a mouse with severe combined immune deficiency (SCID), caused either genetically or by a lethally high dose of radiation, is injected with human immune cells. The human immune cells in effect form a human immune system in the mouse, so the antibodies produced are the same as those a human would make.

An even more sophisticated version, the Trimeric mouse developed by XTL Therapeutics, uses a mouse body that has been lethally irradiated, bone marrow cells from a SCID mouse (which can make red blood cells), and then other tissue from humans, making an animal with tissues from three sources. This allows a researcher to put human liver into mice and test drugs for hepatitis C, something that cannot be done *in vitro* as hepatitis C virus will not grow in liver cells outside a body.

Chimeric/humanized antibodies

Monoclonal antibodies are mouse proteins, derived from mouse immune cells. When they are injected into a patient, they will be recognised as "foreign" by the patient's immune system, and the patient will quickly become immune to them. For long-term treatment, this means that after a few days or weeks the patient will have their own antibodies which bind to and neutralize the immunotherapeutic as soon as it is injected. This is known as the human anti-mouse antibody (HAMA) response.

To overcome this, scientists engineer the antibody so that it "looks" like a human antibody to the immune system. By replacing the constant regions of a mouse antibody with those of a human antibody, a protein which binds to an antigen like the original monoclonal antibody but which "looks" to the

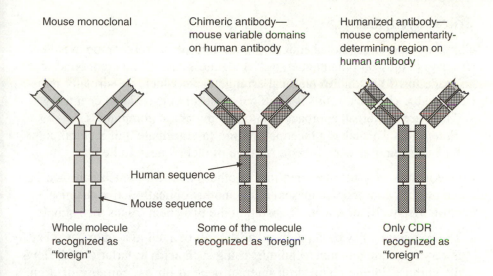

Mouse monoclonal

Chimeric antibody—mouse variable domains on human antibody

Humanized antibody—mouse complementarity-determining region on human antibody

Human sequence

Mouse sequence

Whole molecule recognized as "foreign"

Some of the molecule recognized as "foreign"

Only CDR recognized as "foreign"

human immune system like a human protein can be made. The fused protein is called a chimeric antibody.

More sophisticated engineering can be done to produce a "humanized" antibody. The ultimate engineering is to take only those parts of the antibody which determine the antibody's binding specificity (the complementarity determining regions—CDRs) and splice them into a totally human antibody. This is called CDR grafting.

An alternative version is called "antibody resurfacing" or "antibody veneering." This engineers all the surface residues on an antibody to "look like" the human residues, so that the human immune system does not recognize them as "foreign," but leaves the core residues and the binding site unaffected.

This has to be done on both of the antibody's protein chains. An advantage of Dabs and SCAs is that they are antibody-based proteins which contain only one chain, requiring less engineering to humanize.

The alternative approach is to make genuinely human antibodies. As human immune cells are not suitable for making monoclonals, this has to be done another way. This can be:

- by cloning the antibody genes from humans and expressing them in bacteria or cultured cells (*see* **Phage display**)

- by genetically engineering the human antibody genes into mice, so that their immune cells are genuine mouse cells but make human antibodies

- by grafting human immune cells into SCID/Hu mice (*see* **Chimera**).

Chiral synthesis

Chiral synthesis is the chemical production of chiral compounds (*see* Chirality) in only one enantiomer, or "handedness." As chiral compounds can be made in two (or more) physical arrangements which are virtually indistinguishable chemically, this is a difficult task for conventional chemistry.

There are two broad approaches to obtaining chiral compounds.

Resolution. This takes a racemic mixture (a "racemate") of a chiral compound, and removes one of them. A range of techniques can be used.

- One isomer may be converted into another chemical (which can then be removed by conventional means) using another optically active chemical or, most effectively, an enzyme. Separating the two chemicals is then simple.

- The compound may be crystallized as the salt of a chiral acid or base. This is very common research technique, using such acids as tartaric acid. How the chiral acid and chiral drug interact depend on the chirality of each, creating different solubilities of the two forms.

- Chiral chromatography. A racemic mixture of isomers is separated on a chromatographic column (the "stationary phase") that itself is chiral. Egg is made of a chiral material such as cellulose or protein. Because the difference between how each isomer moves down the column can be quite small, the chromatography column may have to be quite long, which is expensive at industrial scales. So adaptations such as SMB Chromatography are used on an industrial scale (*see* Chromatography).

Chemical synthesis. The alternative is to make only one isomer of the compound in the first place.

- Asymmetric catalysis. A catalyst that is itself chiral is used in a key step in the reaction. (Enzymes, of course, are one such catalyst—see below.)

- Chiral auxilaries. These are other components of the reaction that are themselves chiral, and "force" chirality on the molecule you are making. They are not catalysts, because they are used up in the reaction.

- Biotransformation. This is the synthesis of a compound using enzymes. As most enzymes product only one enantiomer as a product, they can be used to take symmetric (i.e. not chiral) starting products and produce pure enantiomers from them. This can be by isolated enzymes, or by whole bacteria (Bioconversion):the chiral drug Ephedrine has been traditionally produced by bioconversion.

- Fermentation methods. Chemicals made by fermentation or by plants will almost certainly be produced as one enantiomer. Many amino acids produced for animal feed supplements have been produced traditionally as single optical isomers by fermentation.

Often, chiral synthesis is not used to make the final chemical itself. Rather it is used to make a precursor to it which is easier to make using the available enzyme systems. This precursor can then be turned into the final chemical using conventional chemistry.

Chirality

Chirality is the chemical version of "handedness." Some molecules have distinct left- and right-hand forms, which, although containing the same atoms tied up in the same way, are not physically the same (just as your hands have the same number of fingers tied to the palm in the same way, but nevertheless are not the same). Such a chemical is called a chiral compounds, and like your hands they come in two mirror-image versions, called enantiomers (or optical isomers) of each other. Compounds which have two enantiomers are usually divided into L and D, or + and −, or S and R forms, so you have (L)-alanine or (+)-ephidrine. There are complicated rules about these nomenclatures for organic chemists.

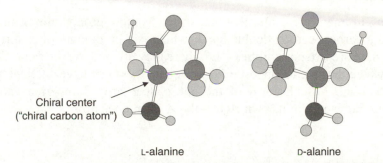

Chiral center
("chiral carbon atom")

L-alanine   D-alanine

Enantiomers are mirror image versions of the same molecule

Usually, the only difference between enantiomers of a compound is how they interact with other chiral materials, such as living things. All the chemistry of life is chiral, and so how other chemicals affect life depends on which enantiomer you have, just as it is easier to shake hands left-to-left or right-to-right, not left-to-right, because both hands are chiral, whereas it is as easy to pick up a light briefcase with left or right hands (because, although your hand is chiral, the briefcase handle is not).

Thus, different enantiomers of the same chemical can have different effects on living organisms. Because of this, any new drug or agrochemical that is chiral will only be approved for use as a single enantiomer. **Chiral synthesis** (see separate entry) is a major application of biotransformation and bioconversion technology for this reason.

For biopharmaceuticals, of course, chirality is not a worry—as biologically derived proteins, they all have the correct "handedness" anyway.

Related terms are

- *Geometric isomerism*. This is a general term for how molecules can have the same atoms linked together in the same way, but have those atoms arranged differently in space. Chirality is one example.

- *Chiral center*. This is the atom in a molecule that makes it chiral. This has to have four (or more) different groups attached to it, and so in drugs is almost always a carbon atom, hence the idea of the "chiral carbon" in a molecule. (Although this is just a carbon—its role in chirality is to do with the other groups around it.)

- *Diastereoisomer*. This relates to molecules that have more than one chiral carbon in them. The combinations become quite complicated here, and the resulting molecules are not always chemically identical. The two things to remember about diastereoisomers is that (i) the more chiral carbons they have, the harder they are to make by non-biological means, and (ii) like simple enantiomers, each diasteroisomer will interact with living things differently, so you have to make the right one.

- *Epimer*. Epimers are molecules that differ by how groups are arranged around a carbon–carbon double bond (which can be thought of as a rigid bar). There are two possibilities—E (where they are both on the same side) or Z (where they are on opposite sides). Again, these can have different biological effects. The E form of the anti-cancer drug Tamoxifen causes endometrial cancer in its own right—the Z form does not.

Chromosome

Genes are arranged in cells on long DNA molecules called chromosomes, that may contain a few dozen genes in a few tens of kilobases (one kilobase = 1000 bases) in the chromosome of a virus to tens of thousands of genes in hundreds of megabases (1 megabase = 1 000 000 bases) of DNA in the chromosomes of higher plants and animals.

In eukaryotes (living things with cell nuclei, as distinct from prokaryotes, like bacteria) the chromosomes can be seen under the light microscope when the cell divide. They can be chemically treated so that some regions take up dye more easily than others and so stain as darker "bands," reflecting the genetic structure of the DNA. These form rough chromosomal "addresses," which are often used to describe where a gene is on a chromosome. In addition, somewhere in a chromosome is a centromere, a waist-like constriction which is an essential part of the mechanism that makes sure each of two new cells gets one set of chromosomes each. The centromere divides the chromosome into two, usually unequal, parts. The short part is referred to as P, the long

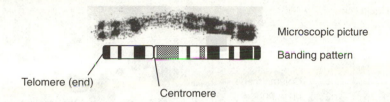

Microscopic picture

Banding pattern

Telomere (end)

Centromere

part Q, so 16p2.3 refers to chromosome 16, long "arm," band number 2, sub-band number 3.

Preparing cells so that the chromosomes can be seen and hence counted is called karyology, and the resulting count of chromosomes is called the cell's karyotype. Preparing the more complex chromosomal band analyses is covered in the broader term, cytogenetics.

All the genes (and hence, inevitably, all the chromosomes) in an organism is called its genome. The human genome is about 3 billion bases long. The record size is the lungfish, which has a 140 billion base genome, nearly all of which appears to be "junk" DNA.

Animals and plants usually have a double set of chromosomes—they are diploid (*see* Gene). There are a few exceptions: humans with three copies of Chromosome 21 survive, although they suffer from Downs Syndrome. Organisms with one extra chromosome, rather than a whole extra set are called Aneuploid. Other human aneuploidies are known, but are rare are produce extremely severe diseases. Cytogenetics is used to diagnose these conditions.

Chromatography

Many separation systems used in biochemistry, molecular biology, and biotechnological production are chromatography systems. Chromatography was originally developed as a way of separating pigments from plants by dissolving them and then "wicking" the solution through paper, an experiment that many schoolchildren do today.

The same basic ideas apply to all chromatographic separations: dissolve the mixture in some suitable solvent and then pass it slowly over some solid material. Depending on whether the molecules in the sample stick to the solid material (stationary or solid phase) or dissolve in the solvent (mobile phase), they either move over the solid or stay put. Most materials do a bit of both, and so move down the column of material slowly—the exact speed varies with each component of the sample, and so they are spread out.

There are a lot of variations of chromatography. The more common are:

- Gel chromatography/gel exclusion chromatography/size exclusion chromatography These sift by molecular size. The chromatographic material

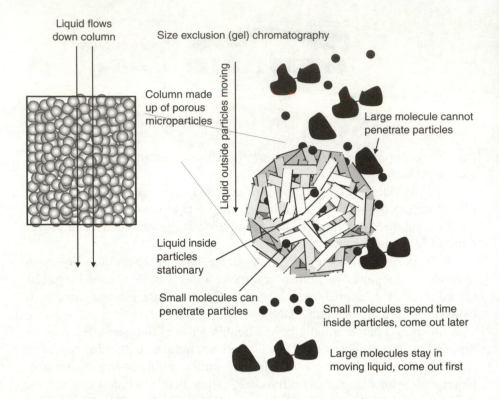

Liquid flows down column

Size exclusion (gel) chromatography

Column made up of porous microparticles

Liquid outside particles moving

Large molecule cannot penetrate particles

Liquid inside particles stationary

Small molecules can penetrate particles

Small molecules spend time inside particles, come out later

Large molecules stay in moving liquid, come out first

has small pores in it, into which small molecules can enter but from which large molecules are excluded. (Different materials have different pore sizes, so that the size limits "small" and "large" can be adjusted to suit whatever the scientist is trying to separate.) As the mixture of molecules passes down the column, small molecules diffuse into the pores, where liquid is stationary, and so spend some of their time standing still. Large molecules cannot enter the particles, and so spend all their time in the moving phase. Thus, large molecules move down the column faster than small ones.

- *Affinity chromatography*, where a specific molecule is linked to the chromatographic material and molecules separated by their ability to bind to it. See separate entry.

- *Hydrophobic chromatography*. This simply uses a hydrophobic material such as untreated silica as the stationary phase. Most HPLC systems are hydrophobic adsorption chromatography.

- *Gradient chromatography*. Here all the molecules in a sample are bound to the solid support material. The solution washing over the solid phase is then gradually changed to have more and more of some component, often salt, acid, or alkali. When the concentration reaches the point at which compound dissolves in the liquid rather than sticking to the column, it dissolves off and is washed out of the column. Because that point is reached at different concentrations, specific compounds will be eluted at different times.

Chromatography also varies according to the physical arrangement of the solid material (the stationary phase):

- *Column chromatography*. This is the most common by far—the solid phase is packed as small particles into a tube, and the liquid passed over it. Column chromatography methods capable of purifying kilograms of materials at a time have been developed. A variant is **high performance liquid chromatography** (see separate entry).

- *Paper chromatography*. This, essentially the same as the original method, uses a paper wick as the solid phase.

- *Counter-current chromatography*. Now not used very much at a laboratory scale, this uses two immiscible liquids (such as oil and water) that flow past each other to separate compounds. A compound that dissolves more in one than in the other will tend to get "pulled along" with the one it dissolves in, and so emerge purified from other material in that solvent.

- *Moving phase chromatography*. Here the solid phase is fed in one end of a tube (column), and fluid in the other. Sample is fed in the middle, and materials can come out of either end. This is efficient for industrial processes separating materials that would otherwise need very long columns to separate. You can achieve the same effect using a series of columns and pumps closed into a loop (simulated moving bed—SMB—chromatography), and this has been applied to separating enantiomers of a compound. The anesthetic Enflurane and the analgesic Tramadol have been purified into single isomers using SMB methods.

- *Thin layer chromatography (TLC)*. Here the stationary phase is a thin layer of treated silica coated onto a glass plate. This is widely used in research.

- *Gas chromatography (GC)*. This is quite commonly used to analyze the volatile metabolites produced during fermentations, and also to analyze other biomolecules which have been made more volatile by chemical modification. Here, the moving phase is a hot gas with the compounds to be analyzed vaporized into it. The volatiles partition between the gas phase and as liquid on the walls of the tube (hence this is sometimes called gas–liquid chromatography, GLC). The time that it takes them to get through the tube depends on how long they spend in the moving gas phase, and how long in the stationary liquid phase.

Last, there are the different materials that can make up the mobile and solid phase. In general, the mobile phase is water, or some watery solution—this is because nearly all the materials that biotechnology is interested in are soluble to various degrees in water, and some like proteins are not stable in any other solvent. The solid phase gives greater flexibility.

Cleaning-in-place (CIP)

Manufacturing equipment used in many biotechnological processes must be sterile and clean (i.e. not just contain sterile dirt). Before each "run," therefore, the system has to be cleaned and sterilized.

CIP is cleaning (and sterilization) of a manufacturing system without dismantling it, so that the parts are cleaned as a whole: it is also called "*in situ* sterilization*" or sterilization in place (SIP). This is a much easier operation to perform than cleaning and sterilizing all the components separately and then trying to assemble them under sterile conditions, or having separate cleaning and sterilizing operations. However, it requires some specialist techniques and equipment.

In particular, bioreactor machinery must be designed so that there are no "dead legs" (i.e. pipes which are blocked at one end), crevices or other areas where the cleaning fluid could not flow. It is also useful for the equipment to be so designed so that bits of it can be cleaned while the rest remains in operation.

CIP systems are a combination of the equipment used and the processes used in it. Thus, a full CIP system will include protocols for washing, usually with industrial-strength detergents, and rinsing out the wash agents.

Clean room

A clean room is a room which meets specific standards of cleanliness, especially with respect to what may go into and out of it and what the concentration of particles in the air might be. Clean rooms are central to pharmaceutical manufacturing, as it is by producing, formulating, and packaging drugs under rigorously sterile conditions that the sterility of the drug is assured. The same cleanliness conditions apply to a lesser extent to other biotechnological products, and also can apply to the research and development phase of recombinant DNA or plant or animal cloning work, where here the objective is to stop contamination getting into the experiments.

Clean rooms are classed in the United States by a number, which is the number of particles of more than half a micron diameter which are allowed per cubic foot of air. Thus, a class 100 clean room would have around 100 0.5 micrometer particles per cubic foot. At the moment, class 100 rooms are the highest category of cleanliness required by the pharmaceutical industry. Other countries have different rating systems (notably, most are based on SI units), but the level of air purity required is similar.

Clean rooms are kept clean by a variety of methods. Air going into the room passes through HEPA filters (high efficiency particulate air filters) to keep out even very small particles: ultra-clean rooms have several layers of filtering. Walls, floor, and ceiling are usually painted with materials designed not to stick any dust. People going into the clean room have to wear hats and overshoes, as hair and shoes are the most particle-laden parts of the worker, as well as the usual laboratory coat. For less rigorously clean areas there may be a "sticky mat" just inside the door, which pulls loose dirt off the bottom of the shoes of everyone who walks in.

To provide greater cleanliness within a clean room, laminar flow hood are provided, which keep a flow of filtered air over the experiment, separate from the air in the general laboratory.

Clean rooms use much the same air filtration technology are containment laboratories, but for a different purpose. Containment laboratories are mean to keep possibly hazardous material in the laboratory, rather than external contamination out. *See* **physical containment**

Clone

A clone is a collection of genetically identical individuals. They turn up in molecular biology and biotechnology in many contexts. "To clone" is to make a clone of something, to make several individuals that are genetically identical.

Clones of organisms. Clones of plants and some animals have been developed using many techniques. Members of a clone show less variability than collections of the same organisms bred by sexual methods, because their genes are identical. The trick is how you "clone" animals and plants, which usually reproduce sexually. *See* **Micropropagation** (for plants) and **Animal cloning**.

Gene clone. This means a collection of organisms (usually bacteria) which all contain the same piece of recombinant DNA. By extension, it means the piece of DNA they contain. *See entry* on **Recombinant DNA**.

Cell cloning. Some biotechnology methods produce a collection of single cells which are genetically different. If the resulting clone of cells is genetically stable it is called a cell line. Producing hybridomas is an example: the fusion step produces a large number of different fused cells. These variants are then "cloned," that is, separated out and individual cells grown up to produce a clone of cells.

Cloning cells and microorganisms is done by a combination of dilution and growth. The cells are spread out, so that they are spaced well apart. Then, they are grown up into individual colonies, each of which is therefore well separated, and the relatively enormous scientist can physically pick out one clone. The separation may be done by spreading cells onto an agar surface in a Petri dish—this is the usual approach for bacteria—or by diluting a suspension of cells into many small tubes—usually, the wells of a 96-well plate—at such an extreme dilution that most wells have nothing in them.

Then the few that do have something in them will, by statistical chance alone, have only one cell in, and so that cell will grow up into a clone. Because cells can stick together, this cycle has to be repeated several times to make sure that you have a true clone. This process is called dilution cloning or cloning by limiting dilution, and is the common way of cloning hybridomas to make monoclonal antibodies.

A novel approach is called gel microdrop cloning, where cells are encapsulated in a small droplet of a gel such as agarose. The cells cannot move between gels, so after some time for growth each drop contains a clone derived from one cell. This has been applied to fungal and mammalian cell cloning.

Co-enzyme

The term cofactor is used almost interchangeably with co-enzyme in most contexts. A co-enzyme is a non-peptide molecule which is needed by an enzyme to work, and is changed in the chemical reaction catalyzed by the enzyme. However, it is recycled after reaction so that it is ready to be used again.

The most biochemically common co-enzymes are a collection of molecules similar to nicotinamide adenine dinucleotide (NAD). These molecules shuttle hydrogen atoms around the cell. There are two flavors (NAD and NADP), and they come as hydrogenated (reduced) or non-hydrogenated (oxidized) molecules—NAD or NADP, oxidized; NADH or NADPH, reduced.

Many co-factors and co-enzymes are derived from vitamins. Thus, NAD is derived from nicotinic acid.

Some co-enzymes are tightly, even covalently linked to their enzymes—covalently linked ones are of called co-factors. Examples are FMN (flavine adenine mononulceoside) or FAD (flavine adenine dinucleotide) moieties that are needed by the common diagnostic enzyme glucose oxidase. If you remove the co-factor, the resulting factor-less polypeptide is called an apoenzyme.

Many enzymes which perform industrially important reactions cannot be used in the appropriate industrial process because the cofactors they need are too expensive and are gradually lost from the reaction as the enzyme is used. Biotechnology therefore seeks to either replace them with other molecules which are cheaper or more stable (biomimetics) or to recycle the cofactors more efficiently. Both are difficult to achieve economically on an industrial scale.

Combinatorial chemistry

Combinatorial chemistry ("combichem") is the production of a collection of variants on a known chemical in parallel. Such related variants that are all members of a chemical family are called congeners.

The chemist will start with a core or "scaffold," and then add a selection of side-groups or functional groups on to it. The result will be a series of new chemicals which are different but related.

Most combinatorial chemistry is performed using solid phase chemistry. Here a solid material (silica, or a plastic called a "resin") is used as a solid anchor on which the chemical is made. At each step liquid reagents are added to the solid, and then afterwards washed away again, leaving the new chemical attached to the solid. At the end, the resin is treated with a reagent that cuts the chemical off the resin, and the chemical is left pure in the solution. This gets round a major problem of traditional (liquid phase) chemistry of how you purify the material at each stage of the synthesis.

This very cool idea is, however, limited, because quite a lot of the reactions that chemists want to do would destroy the solid particles, or cut the growing chemical off them. So there is a trade-off of doing chemistry in the solid phase—easier, faster—and liquid phase—which is slower and harder but much more flexible.

Some combinatorial chemistry uses small microparticles in a "pool-and-split" scheme. At the end you have a huge number of small particles each with a unique chemical linked to them. These can then be tested separately for drug activity, or tested as a pool or set of pools. Once you find an "active" particle, you then have to find out what chemical is attached to it. A neat way of doing this is to physically tag the particles, and tags like coloured stripes, DNA molecules, even little radio transponders have been tested.

Other methods rely on using a flat surface as a solid phase. Affymetrix builds up its gene chips using combinatorial chemistry methods on a glass slide, which ends up with hundreds of thousands of DNA molecules in little squares, each square different.

The number of chemicals you could make using combinatorial chemistry is huge, but many of them are very similar. You therefore seek to design your chemicals in the computer to find the ones that are most diverse before you make them—the computer version is called a "virtual library." *See* **Chemical space**. This means that you make fewer chemicals, but they are more diverse. In the extreme this becomes parallel synthesis, just making a lot of unrelated chemicals at the same time. The simplest way of doing this is to build a robot that can do chemistry. Automated chemical synthesis machines can make up to a couple of hundred chemicals at once (when programmed properly), which eases the burden of repetitive labor.

In principle, the robotic synthesis systems could be miniaturized onto a MEMS system (*see* **Biochip**), so you can build a "chemistry chip" that would make 10 000 chemicals at once. Such ideas are not new, but are fiendishly difficult to get to work. Companies such as Orchid Biocomputer hoped to harness the microfabrication skills of the semi-conductor industry to making thousands of specific chemicals at once, but this turned out to be impractical.

A drawback of both parallel synthesis and combinatorial chemistry methods is that they tend to produce chemical libraries which ring "obvious" changes on a core structure. This is parodied by saying that they produce chemicals that are part of the well-known chemical series "methyl, ethyl, propyl, futile" (this is a chemistry joke). Thus, they are very good at finding a new chemical that is closely related to one you already know how to make among thousands of possibilities, but not very good at finding something radically new.

A variant on combinatorial chemistry is dynamic combinatorial library. This is a way (largely theoretical) of doing combinatorial chemistry in the presence of a molecular "target." Molecules that bind to the target are locked away, all other molecules are recycled by the chemistry, At the end you end up only with molecules that bind to the target.

This is a form of molecular imprinting—using chemistry to find a molecule that fits another. See separate entry.

Comparative genomics

This is the comparison of some genes or whole genomes of organisms that are related by descent from a common ancestor, to gain understanding of what the genes in one organism do by comparison with those of another. It is related to phylogeny (the study of how organisms are related), and specifically to phylogenetics (how genes are related by descent).

There are quite a lot of technical arguments about how you tell whether genes are really related by descent, and if so, how much. One of the basic tools of phylogenetic comparisons is the multiple sequence alignment—lining the DNA sequences up so that you can compare like positions with like. Often this is how the results are displayed, with the positions that are the same ("conserved") or different highlighted.

When two genes are very similar due to having a common ancestor, they are "homologous," and it is likely that they have similar or related functions. So finding a homologue to your newly discovered gene can give powerful clues to what the gene does. This is being used particularly in bacterial genomics, where comparison within and between bacterial genomes can identify "pathogenicity islands"—regions of the genome that have a lot of genes that are important for the organism to become a pathogen, for example, allowing it to invade the host or avoid the immune system.

If the order of a lot of genes in a chromosomal region is the same between two species, then those regions are said to show synteny. Gene order is quite highly conserved during evolution—even man and zebrafish show large clusters of genes in the same order. This again can give clues from one, known gene what another, unknown gene in a syntenic cluster might do.

Computational chemistry

This is a blanket term for using computers to predict or analyze the properties of molecules (as opposed to using computers simply to draw them, which is **molecular graphics**: see separate entry). Calculating the properties of molecules from first principles, which would be ideal, is impossible for practical purposes. Thus, computational chemistry uses the known properties of chemicals to calculate the properties of similar molecules partly from empirical rules ("heuristics"), partly from "rigorous" calculation. Methods used include "semi-empirical" calculations (a method that calculated quantum dynamics properties of the molecule, and is actually quite rigorous) and "molecular dynamics" methods, which treat atoms as hard spheres attached by springs with attractive and repulsive properties. Typical problems are "energy minimization," which calculates the total chemical energy of a molecular structure and seeks to adjust the structure to minimize it so that it accurately reflects what the molecule is like in real life.

One important application of computational chemistry in biotechnology is to predict how proteins will fold up just from knowing their **amino acid** sequence (see separate entry). Another is modeling how a small molecule will bind to a large one. The approach used is called "docking," because the binding site in the large protein molecule is regarded as a rigid "dock" into which the small molecule must fit.

Computational chemistry is quite different from computational biology— *see* **Systems biology**.

Concentration

Biological products are usually produced in rather low concentrations, by fermentation methods, enzyme reactions, or by extraction from animal or plant tissues. In order to keep the cost of purifying these materials down, it is useful to reduce the volume, that is, to increase the concentration, as early as possible in the downstream processing stages of a biotechnological process. Many concentration methods also purify the product to some extent as well. The very best processes concentrate and purify in one step, but that is rarely possible.

Concentration is inherently expensive. This is why extracting gold from seawater is not economic, whereas extracting salt is: salt is so much more concentrated than gold that it tips the balance. (*See* **Purification: large scale**).

Methods used in concentration are based on:

- *The size of the molecules.* In this category come various filter methods and reverse osmosis. In reverse osmosis, the sample is placed on one side of a semi-permeable membrane, and a very high pressure then pushes the water through the membrane, giving water on one side and much concentrated product on the other. Ultrafiltration is a similar technique. Here, the molecules are filtered through a membrane with pores of molecular size. Large molecules stay on the sample side, while water and small molecules including salt pass through.

- *The charge of the molecule.* This usually means ion exchange methods. Here, a polymer is made that has charged "side-groups" sticking off the main polymer chain. Molecules with the opposite charge to the one on the polymer will stick to the polymer, to be washed off later in a small volume.

- *The hydrophobicity or lipophilicity of the molecule.* This is also related to charge. As with charge, hydrophobic molecules can be extracted from water by absorption onto a hydrophobic solid like teflon or nylon, or by extraction into a solvent. Solid phase extraction is simple and convenient, but rarely has a very high capacity. Solvent extraction can be scaled up to enormous scales, but involves large volumes of often flammable liquids.

- *The volatility of the molecule.* If the product is volatile (boils easily and at a low temperature, like ether does and sand does not), it can be separated easily by fractionation. The classic example of this is separating ethanol from yeast fermentations to make distilled drinks.

If the product is not a molecule but rather is whole cells, then methods based on the relatively large size of the cells can be used. These include

- *Sedimentation.* This simple collects the cells by allowing them to fall out of the culture medium. It works well for large fungal cells: smaller cells have to be centrifuged to settle them out.

- *Flocculation* (making the cells clump together and then letting them settle out as a visible precipitate). This is used quite extensively in the sewage industry.

- *Flotation* (as cells can get stuck onto the walls of bubbles, and so be carried to the top of the liquid top be collected there as foam). This is a well-known technique from the mining industry.

Cows (and beef)

Agricultural biotechnology is concerned with cattle in several ways, other than the general programs mentioned elsewhere.

Cows and sheep both suffer spongiform encephalopathies, and "mad cow disease" has become a significant health concern in Europe (*see* **Transmissible encephalopathies**). Both can also be used in producing pharmaceutical proteins through genetic engineering (*see* **Pharming**).

Both beef and milk quality are the targets for cow breeding programs. A cow genome project is under way, aiming to identify linked markers that will allow breeding cattle to be selected very early in life for genes which only show themselves in adults. Improved milk quality for cheese production (primarily to improve the quality of the whey protein casein) and lower meat fat content are traits being addressed.

Mechanically-recovered meat (MRM), which is obtained from the carcass after the high-quality meat has been removed, is processed in a variety of ways to make "burgers," pet food, and other lower value beef products. Part of that conversion now involves biotechnological processes such as partial proteolysis, use of collagenase to break up connective tissue, and manipulation of the content of chemicals such as amino acids and vitamins. Other products are gelatin (used, *inter alia*, in confectionery and in drug capsules), and bone meal for fertilizer.

In the longer term is engineering the cow's physiology to be more efficient at converting food to cow. Cows convert cellulose to mammalian meat through rumination, the fermentation of their food in their four-chambered stomach: [the rumen (where anerobic bacteria break down the cellulose and release short chain fatty acids which the cow can metabolize), the reticulum, omasum, and absomasum]. Engineering the bacteria to make the process more efficient is a medium-term goal. The bacteria could also be engineered to provide other abilities, such as the ability to break down toxins in forage plants: the rumen bacterium *Butyrivibrio fibrisolvens* has been engineered to break down fluoroacetate, a poison in many tropical shrubs, and may enable cattle to eat these plants.

The abomasum of suckling cows is also a source of rennin, used to make cheese.

Crop plants

The large majority of commercial application of biotechnology to crop plants has been in the United States, where there is widespread acceptance of "GM crops" (*see* **GM crops**). About 52 million hectares of transgenic crops were planted legally in 2001 (slightly less than the area of Spain). (One hectare is nearly 2.5 acres.) This is still increasing: in 2000 it was 44 million hectares; in 1998, 30 million. Much of this is transgenic maize (corn): about 70% of this is planted in the United States, 20% in Argentina, 10% in Canada (Europe has effectively banned GM crops). Most of these are "input traits" crops—crops engineered to be easier or cheaper for the farmer, but not to have any specific benefit to the consumer.

The more commercially successful transgenic crop plants are:

- *Corn (maize)*. Twenty five percent of maize planted in the United States is transgenic, mostly herbicide tolerant strains. An exception is the Starlink corn that was found to have accidentally been used in human food preparation (*see* **Gene spread**).

- *Wheat*. Thirty percent of the wheat in the United States and 45% in Canada is transgenic. This is mostly for herbicide tolerance.

- *Soy*. Soy beans are a major crop in the United States. The first commercial plant transgenic was a soybean that is resistant to glyphosate, called Round-up Ready soy. Around 60% of soy planted in the United States in 2001 was transgenic crop.

- *Cotton*. Transgenic cotton that express the Bt CRY protein has been used widely. It is resistant to the Boll worm, which is a major problem for cotton farmers. Herbicide-resistant cotton is also being widely used in the United States. Around 65% of cotton planted in the United States in 2001 was transgenic, one-third insect resistant, half herbicide resistant, and the rest a combinations of traits.

- *Rice*. Although not yet a substantial product, "golden rice" has the potential to be a major crop in developing countries where traditional rice is a staple food. Rice is poor in iron and vitamin A. A team at Rockefeller Foundation spliced the genes for ferritin (an iron-binding protein) from *Phaseolus*, and phytase (an enzyme that breaks down iron-binding phytates) into the rice to increase available iron. They also introduced four genes from daffodils to turn a precursor metabolite into Vitamin A. The result was yellow and much more nutritious than normal rice.

- *Canola*. Eighty percent of canola grown in Canada in 2001 was transgenic. The large majority of this is herbicide resistant. GM Canola has caused a stir in the farming community because of claims that its pollen can spread up to 25 km, and hence that any GM canola crop can contaminate non-GM crops for tens of miles around. This needs to be confirmed by more rigorous tests.

A famous GM crop that is an example of a product that did not make it commercially is the Flavr Savr tomato. Flavr Savr was an example of an output trait crop—engineering properties that were (in theory) of value to the consumer. Flavr Savr was the flagship product of Calgene, one of the first food biotechnology companies. It was a tomato engineered to produce antisense RNA to the polygalacuronidase gene, which produces a protein involved in fruit softening during ripening. The idea was that Flavr Savr fruit would last longer on the shelf after picking. It was also speculated that they would be firmer *before* picking, and so could be allowed to ripen on the vine rather than being picked green when it is easier to handle. This turned out not to work very well, and that, the presence of a "marker gene" in the tomato, and the swing of public opinion against GM crops dealt a death blow of Flavr Savr and to Calgene. Research is going on to enhance tomatoes production of chalcone isomerase, which is involved in anti-oxidant production (as anti-oxidants are proposed to help prevent diseases of aging), and to block the genes that synthesize polyamines, which are one of the triggers of ripening.

Crop plants have also been used in a variety of other biotech applications, such as producing biopharmaceuticals, making vaccines, even making plastic (*see* **Biomaterials**).

Cross-flow filtration

This is a commonly used method of filtering thick fluids. If you try to filter (say) soup through a standard micropore filter in order to concentrate the particulate material, the pores will rapidly block up, and filtration brought to a halt. The build-up of material on a filter, blocking the pores or gaps, is called fouling, and is a major problem for any filtration process. Cross-flow filtration is one of the more effective ways of getting round fouling. It flows the liquid across the filter, allowing the carrier liquid to flow through. After it has passed over, the top (unfiltered) phase is more concentrated, and some of the fluid phase has passed through—meanwhile, the filter is not blocked up.

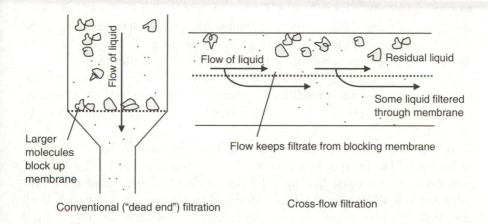

Conventional ("dead end") filtration   Cross-flow filtration

The alternative is "dead end filtration," which is when the filter is at the end of a tube or funnel and all the fluid being filtered has to pass through it. Most simple filtration (such as the filters used in coffee machines in the kitchen) are dead-end filters.

Other approaches to fouling include scraping the filter clean or replacing it periodically, or backflushing. In this last one, clean liquid is pumped back through the filter in the reverse direction, so that the accumulated material is pushed off the surface, and can then be cleared away. This is not as rough on the filter as scraping, and simpler to engineer than cross-flow systems. However it does not always work, if the fouling material "glues" itself to the filter.

Cross-flow filtration has been developed for the clarification of beer. It has the drawback that, for such large volume industrial processes, a huge area of filter is needed.

Cryopreservation

This is preservation of things by keeping them cold. There are several variations of relevance to biotechnology.

Freezing. The most obvious route. Just putting something in the fridge or freezer is fine for many biological materials. However, physically delicate materials, such as animal cells, are destroyed by the ice crystals that form inside them on freezing, leaving you with dead meat.

To prevent damage to cells on freezing, cells are often frozen in a mixture of a watery material (their usual growth medium) and another material called a cryoprotectant which mixes with water and stops it forming ice crystals. Glycerol is a favourite for bacteria, dimethyl sulfoxide (DMSO) for animal cells.

Bacterial cells preserved in this way can be kept in a conventional freezer, but animal cells need to be stored at liquid nitrogen temperatures if they are to survive more than a few weeks. Often they are kept in the flask, but above the nitrogen itself ("liquid nitrogen vapour phase" storage): this stops the tubes filling up with liquid nitrogen and then exploding when you warm them up. Even at these temperatures, mammalian cells eventually "go off."

Antifreeze proteins. There are proteins known which prevent ice crystals forming found in some arctic fish. In principle they could be used to replace the glycerol or DMSO (which are somewhat toxic), but this is rarely done in practice.

An odd application of cryopreservation was suggested by Robert Ettinger in his book "The Prospects of immortality," published in 1964. He suggested that dying people could be frozen until medicine had discovered a cure for whatever was killing them, when they could be thawed out and cured. Both ice crystal formation and the inevitable decay of mammalian cells even at liquid nitrogen temperatures means that this surely cannot work, but if you are dead anyway, why not try it?

A related preservation technique is **freeze-drying**. Here, material is frozen, and then dried while frozen. The resulting material is stable for a long time, even at room temperature. See separate entry.

Culture collections

These are central depositories where samples of microorganisms and cell lines are stored. Other name are strain depositories and type culture collections, the latter because they are where the "type specimens" (i.e. the definitive ones which describe that "type" of organism) are kept.

They have a triple function.

- They are a "bank" for your valuable microorganisms (against the risk that your laboratory burns down).

- They are a centre from which other people can get samples of your organism (if you want them to) without bothering you.

- They are somewhere where you can deposit an organism to prove you own it—a sort of biological patent office. (Some patenting systems insist that you deposit a sample of any organism mentioned in a patent which cannot be created easily by someone else at a recognized depository for future reference.)

The best known depository is the American Type Culture Collection (ATCC), which collects all types or microorganism and cell line. ATCC is also the World Health Organisation (WHO) international reference collection. There are a variety of other general depositories in other countries, some of which specialise in fungi, bacteria, or animal cells, speciality depositories for marine organisms, pathogens, mammalian cells etc., and industry-specific depositories.

Type culture collections became newsworthy in 2001 with the Anthrax attacks in the US—the most likely source of bioweapons bacteria for a terrorist is a culture collection. Substantial control always was exercised on who could get samples of potentially dangerous organisms from collections like ATCC, and this has been tightened up even more now.

Cytokines

Cytokines are proteins that signal between the cells of the immune system, and between those cells and other cells of the body. They are closely related to chemokines, materials which stimulate cell migration, and many cytokines act as chemokines. In structure, cytokines and their receptors are also similar to **growth factors** (see separate entry).

Cytokines are studied in mammals because they are important to many processes which involve cells moving about such as inflammation and development. Understanding them, and then isolating them and producing large amounts for therapeutic uses, has been a major research target of many "genetic engineering" and pharmaceutical companies. Nowadays the enthusiasm for cytokines as drugs in themselves has waned.

Cytokines attract the cells to sites of damage or infection where they can kill invading cells and, as a side-effect, produce inflammation, shock, and even death. So well understood are the immune system cytokines (compared to other cell mobility enhancers) that "cytokine" usually refers exclusively to cytokines which act on lymphocytes and macrophages. Cytokines are also involved in the body's control of how many blood cells are made in the bone marrow, and so are of general interest as potential stimulators of blood production (hematopoeisis).

Interleukins. There are over 20 known interleukins (IL-1 to IL-18). IL-2 has been used as a booster of the immune system in cancer and infectious disease therapy. IL-4 is linked to the allergic response (IgE-mediated immunity), and so agents which affect IL-4 response have potential for modulating allergies.

Colony Stimulating Factors (CSF). There are three varieties: G-CSF, M-CSF, and GM-CSF, which stimulate granulocytes, macrophages or both (respectively). They stimulate the differentiation of some types of white cell. Ten companies are trying CSFs as biopharmaceuticals.

Interferons (IFN). Well known as being one of the first proteins to be produced by the new biotechnology of the late 1970s, and touted as the wonder-cure for everything, these are actually three classes of these cytokines called interferons alpha, beta, and gamma. Interferon α (from Biogen) has been approved as a treatment of Hepatitis C. Interferon β has been approved for treating several cancers and for multiple sclerosis—it is produced by several companies including Ares-Serono and Biogen. The success rate for treating MS with IFN is not high, although it is higher than any other treatment. 2002 saw a viscous PR battle between Serono and Biogen about whose IFN was better—the statistics showed that there was virtually nothing in it.

Bovine interferon has also been shown to help improve the pregnancy rate in sheep, because it increases the process of "maternal recognition" by which the ewe's immune system learns that the developing fetus should not be rejected.

Tissue Necrosis Factor (TNF). Different sub-types have been suggested as an anti-cancer drug, and as the "toxin" part of an immunotoxin. It also triggers cell destruction in inflammatory disease, and anti-TNF antibodies such as Remicade have been shown to be effective at treating rheumatoid arthritis.

CD antigens. The "CD" letters turn up in many contexts, most notoriously CD4 as the protein that the AIDS virus uses to bind to its target cells. It stands for "cluster differentiation" antigen, and means a protein used to distinguish types of white blood cell. Many are actually interleukin receptors.

"Immunotherapeutics" is a now rather dated term for any biotherapeutic that acts on the immune system. As virtually all of them are cytokines, cytokine derivatives or anti-cytokine antibodies, saying that an immunotherapeutic is equivalent to a cytokine-based drug is not far off.

Dabs and other engineered antibodies

Dabs are single domain antibodies, recombinant antibodies in which there is only one protein chain derived from only one of the "domains" of the antibody structure. Greg Winter at Cambridge, United Kingdom, has shown that, for some antibodies, half of the antibody molecule will bind to its target antigen almost as well as the whole molecule. Usually, the binding site of an antibody consists of two protein chains (*see* **Antibody**).

Related ideas are single-chain antigen binding technology (SCA), patented by Genex, biosynthetic antibody binding sites (BABS) invented by Creative Biomolecules, and minimum recognition units (MRUs, or Complementarity Determining Regions—CDRs), which is a more general description of the smallest part of an antibody you need for it to bind to its target. SCAs and scFvs are terms for antibody binding domains in which the two binding regions from light and heavy chains are linked by a short peptide, so they can be produced from one gene.

The potential advantage of DABs, SCAs and the other variants mentioned above is that they are single protein chains. This makes them easier to engineer than regular antibodies (which have two chains), and practical to work on in **phage display** libraries (see separate entry).

Biospecific antibodies are antibodies with two different binding sites, so that each molecule has one binding site for each of two different antigens. They are made by a combination of protein engineering and protein chemistry. Their potential uses include ADEPT-type therapeutics, where the antibody binds both to the cell surface and to the therapeutic molecule (or even a cell type) that you want to target to the cell surface. Equivalent terms are hybrid hybridomas and diabodies.

A more extreme version is to fuse the constant part of an antibody with another binding molecule entirely, for example a cytokine or integrin receptor. The antibody part still activates the immune system, and the receptor moiety binds highly specifically to a target cell. Such constructs are called Immunoadhesins. This idea is related to that of an immunotoxin, where an antibody binding site is linked to a non-antibody "effector" molecule. *See* **Immunotoxins**.

Darwinian cloning

This means selecting a clone from a large number of essentially random starting points, rather than isolating a natural gene or making a carefully designed artificial one, to make something that nature has not seen before.

You start with random DNA, made by chemical means or by recombinant shuffling of natural DNA. From this mixture you select those molecules which look more like those you want than the rest. (How you select them

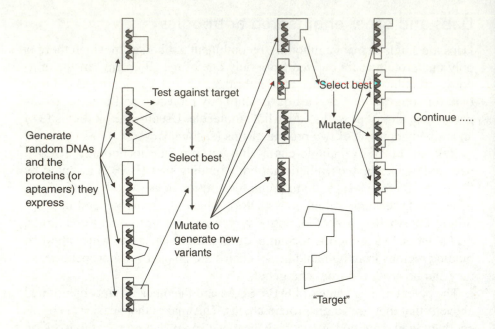

Generate random DNAs and the proteins (or aptamers) they express

Test against target

Select best

Mutate to generate new variants

Select best

Mutate

Continue .....

"Target"

depends on what sort of molecules you want.) You mutate these to generate a new set of variants, and reselect, make more variants and so on until you have the molecule you require. This is a process akin to natural selection, and so is sometimes called "directed evolution" or (incorrectly) "molecular evolution."

There are several classes of catalytic molecule suitable for this.

Catalytic antibodies. See separate entry. Natural antibodies are generated this way in the body: the cells of the immune system do the randomizing process, selecting antibodies that bind to foreign proteins.

Proteins. There are so many possible proteins that you need a way of focusing the randomness for this type of approach:

- **Phage display** *for specific bits of protein.* Only a small part of the protein is derived from the random DNA, the rest is kept the same (see separate entry)

- *Gene shuffling approaches.* Here different variations on gene from different members of a gene family, or parts of the same gene from different species, are shuffled together. This approach has had substantial success in making new members of extensive gene families, such as the polyketide synthases that make antibiotics (*see* **Antibiotics**), or the genes that make pigments in plants.

- *Incremental mutation and selection.* This is making a few random changes at each step. This can be done on small regions of a gene (by splicing error-filled oligonucleotides into them) or over the whole gene (by copying the gene with an error-prone PCR enzyme). This approach has been very successful in evolving luciferase and GFP that glow with different color light from the normal protein.

DNA and RNA. Aptamers are DNA or RNA molecules that bind tightly to another target (see separate entry) — they are particularly suited to Darwinian cloning approaches. The start-point is a random chain of bases, which is bound to the target molecule. Those that do not bind, or only bind weakly, are simply washed away and discarded. The few molecules (out of billions) that are left are then detached and amplified using PCR. These can then be altered by mutation, selected, amplified again, and so on. The same principle can be applied to selecting catalytic RNA, although this requires a transition state analog to bind rather than a simple molecule.

Genes. Of course, genes can be selected in this way—randomized DNA can be put into a cell and the cells selected for their ability to perform some task, or simply survive. This is a mainstay of strain improvement, and is not usually thought of as "directed evolution," but uses the same technology to essentially the same end.

The advantage to Darwinian approaches is that they can discover things that are quite unlike nature, and so offer new approaches to problems.

Deliberate release

This is putting something into the outside world ("the environment"), and in a biotechnology context means putting a genetically engineered organism into field trials. They are often called GMOs—genetically modified organism, or sometimes called GMMO—genetically modified microorganism. A wide range of such trials have been suggested, and some done, starting with the trial of a genetically engineered frost-proofing bacterial strain in California in 1986.

Deliberate release experiments are the natural follow-on from laboratory trials, and then, for organisms involved in agricultural applications, greenhouse trials. Laboratories can include a range of barriers to stop genetically engineered organisms getting out: negative pressure rooms, sterilization procedures, and genetically engineering the organisms so that they cannot survive in the outside world. But, of necessity, none of these can apply to release into the outside world. Here affected fields, animals, soil etc. is kept isolated from surrounding farms, and affected material is destroyed after the trial (except for some Australian pigs, which accidentally got to market as human food in 1988).

A wide range of social, scientific and political pressure groups support and oppose such trials. A major concern is whether the organisms concerned will breed with their wild cousins, allowing the engineered genes to spread (*see* Gene spread). It is rarely clear what the genes would do to the non-biotech ecosystem if this happened. *See also* GM.

Desulfurization

This is the removal of sulfur-containing materials, usually sulfides, from coal and oil. Sulfur residues in fuels end up as sulfur dioxide when the fuel is burned, causing acid rain. However, sulfur-containing fuels are often cheaper than "clean" fuels. As a rough guide, a "high sulfur" coal would contain 6% sulfur (mostly as pyrites), and cost $50–100/tonne less than a "low sulfur" coal containing 1% sulfur or less. Thus, there is an economic motive for removing the sulfur from oil and coal.

The same types of bacteria which are used in microbial mining can be used for desulfurization of coal. They oxidize sulfides (which are insoluble) into sulfates (which are soluble). The sulfates can then be washed away, with the bacteria. This does not work well on lump coal, as the bacteria cannot get inside the lumps fast enough to be economic, but can be effective for treating pulverized coal such as that used in electricity generating stations.

Crude oil also contains significant sulfur—from 0.1% (from the Far East) to up to 3% (for some Middle East oils). Usually, oil has its sulfur removed by Hydrodesulfurization, a physico-chemical technique, but work on using bacteria to remove the sulfur shows that it has some potential.

Developing world biotechnology

This is application of biotechnology to the developing world (there is an inherent bias in the assumption that Europe, the United States and Japan are "developed"). There are two, sometimes conflicting drives to do this.

On the altruistic side, biotechnological techniques are usually low energy and renewable, and so may be more suitable for countries without the infrastructure for the enormous energy and material consumption that is considered normal in the "developed" world. Thus, biotechnology may provide solutions to intractable problems that are cheap, appropriate and able to be implemented and maintained locally.

On the less altruistic side, there are far more of "them" than "us," and they are far more desperate for things ranging from AIDS drugs to clean water, so we can rapidly develop versions of "Western" technologies for use in the "Third World" that would not be economic at home. A substantial minority of politicians believe that if we do not genuinely help the rest of the planet, disaster in the form of global warming, complete deforestation, plague or global revolution will engulf us all.

In either case, appropriate and helpful technology seems to be a good idea, and so is applied with the usual mixture of motives. Application areas are:

Agriculture. Projects here include

• The "golden rice" strain, which accumulates vitamin A and iron, which conventional rice is very poor in. *See* **crop plants,**

- Pathogen-resistant fava beans which are a mainstay of the Egyptian diet and economy,

- Papaya, which is resistant to ringspot virus, which has in the past devastated crops in Hawaii, Brazil, and Jamaica.

- Cassava with enhanced nutritional content, which is a staple of many countries in sub-Saharan Africa, but is limited in some essential amino acids and, unless cooked properly, can have dangerously high levels of cyanide. Both properties could be altered with some fairly simple genetic engineering,

- *BT cotton*. Substantial acreage of cotton transgenic for Bt Cry proteins is planted in India, to reduce pest attack without substantial chemical use,

- *Drought-resistance and nitrogen fixation*. In the longer term, engineering plants to resist drought, to grow in brackish water (so estuary water or diluted seawater could be used for irrigation), and to fix nitrogen from the air (enhancing yields without nitrate fertilisers) are goals of plant biotechnology that could help developing countries.

Developing countries have been early adopters of some agricultural biotechnologies. China started planting GM crops for production in 1992, when the first was not planted in the United States until 1996 (and it is still not planted in Europe). Transgenics were seen as a way of reducing dependence on costly chemical weed- and pest-control agents. However, not all countries are so enthusiastic. In 2002, four African leaders said that any shipment of food aid that included GM grain must be milled into flour first, so that it could not be planted.

It is not clear if GM technologies are overall a good thing for the developing world. Their main effect is to increase productivity in the developed world, and so depress farm prices further in countries that already find it hard to export food.

Medicine. Most of the therapeutic products of biotechnology are far too expensive to be used outside the affluent West (some products are too expensive to be used in any country except the United States). However, vaccines are far cheaper, and their health benefits are enormous. Plant biotechnology that aims to produce edible vaccines is aimed at both developed and developing countries. Important disease targets are diarrhoea, cholera, and hepatitis B. If successful, they would be easy to make, easy to store, and would not require clean hypodermic needles to administer.

Diagnostics

This usually means chemically-based diagnostic tests for disease. Although it has its roots in traditional medicine (medical student lore has some ancient clinicians tasting urine to see if it was sweet, and therefore if the patient was diabetic), chemical diagnostics has come of age with the development of precise chemical methods. Diagnostics test for chemical changes in the body which are characteristic of specific diseases. Chemicals can be:

- simple molecules or ions, like sodium or potassium (as checks of kidney function), or glucose (an excess of which is definitive of diabetes)

- more complex molecules such as glucose or cholesterol. These are almost always measured using an enzyme assay, usually linked to a color change or fluorescent output.

- enzymes, such as creatine kinase (CK, characteristic of heart attacks) or aspartate aminotransferase (AAT, released by the damaged liver)

- specific DNA sequences, as a marker for some viruses and rare genetic diseases

- marker proteins, such as those made by viruses, bacteria, or cancers, or benign proteins such as the hCG hormone that marks pregnancy.

This last group now covers more than 40% of all diagnostic blood tests, and are performed by immunoassays.

Biotechnology helps develop diagnostics in two ways. It finds out more about living systems in health and disease, so we can discover the "markers" which a test subsequently detects. It can then develop reagents to do the tests, notably monoclonal antibodies. DNA probe-based tests are another rapidly growing area of diagnostics which is based on biotechnology.

Biotechnology can also take these techniques to other applications, for example veterinary or plant diagnostics. Usually, these industries do not have the money to develop new diagnostic technology from scratch. However, the technology that has already been developed for the lucrative human health-care market can be applied to less valuable testing of farm animals and plants.

The key measures of how useful a diagnostic is are its sensitivity and specificity. *See* assay.

Differential display

This is a group of techniques for showing the differences between the way that genes are working in two samples, for example between two related bacteria or between normal and cancerous tissue. There are lots of technical approaches.

The original differential display technology amplified fragments of cDNA from the mRNA of the target tissues, and then separated them on a poly-acrylamide gel to detect the different sizes of fragments. Each fragment came from just one RNA, so that if a band on a gel was darker or lighter in one sample as compared to the other, that meant that one RNA was present

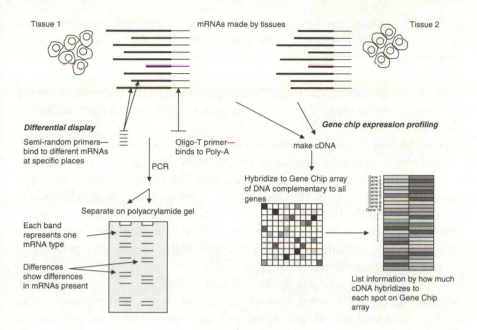

in higher or lower amounts. The band could then be cut out of the gel and the RNA identified.

A variant on this is to use RNA to select for RNAs that differ between tissues. In essence, you make cDNA from the RNA of one sample. You then hybridize the RNA from another sample to it. All the RNA from sample 2 that is matched with RNA in sample 1 sticks, forming double helices. Anything left is unique to sample 2. In practice, this is very hard to get to work reliably, and so is best used as a method of enriching your RNA for the RNA you want, which you then go on to analyse further with other techniques.

This approach is being replaced now by variety of genomics techniques—*see* **expression profiling**.

The protein equivalent of differential display and expression profiling is **proteomics**—see separate entry.

Digestor

This is a bioreactor that uses microbial growth to break down waste. The most common use is in disposal of sewage or agricultural waste.

Anerobic digestors operate in the absence of oxygen, and generate methane gas, which can itself be used as a fuel. Small anaerobic digestors for materials with high solids content such as farm waste can be very simple, basically just an air-tight bucket with a pipe on the top to collect the gas, and can generate useful amounts of methane. Typically 0.3–0.5 m³ of gas, which is 50–70% methane, can be generated from 1 kg of compacted material. Most of the rest of the gas is CO_2: the amount of CO_2 that comes out of the digestor depends on the amount of liquid there is to dissolve this quite soluble gas.

Digestors that can handle larger volumes of material, or can process material with more soluble organic matter in, such as sewage, need a more sophisticated design.

A variant on this is the anerobic contact digestor, where the living cells (active biomass) is separated from the material flowing out of the digestor and fed back into it to digest more substrate. The IRIS digestor, a commercial digestor system, is of this design, but the separation system is integral to the digestor itself. (IRIS stands for Institute de Recherche de l'Industrie.) Generally digestors for waste disposal operate at mesophilic temperatures (around 35 °C), although ones using thermophilic organisms operating at over 60 °C are also sometimes used.

Aerobic digestors, notably the activated sludge process in sewage disposal, destroys the organic content of waste more completely, and produces fewer noxious side-products. However, the growing microorganisms need lots of oxygen, which has to be pumped into the digestor at considerable cost in energy. *See* **sewage**.

DNA amplification

This is the use of enzymes to take a piece of DNA and multiply it in a test-tube into many thousands or millions of copies. It has been one of the central techniques in molecular biology since its invention in 1985.

Applications of DNA amplification include:

- detecting when a DNA molecule is present (by specifically amplifying it into enough DNA to detect chemically)

- making more of a very rare DNA molecule (egg from a rare cell or gene) so that the experimenter has enough to use in the laboratory

- making variants on DNA molecules, by amplifying only the ones the experimenter wants.

The best-known and by far the most commonly used in the **polymerase chain reaction (PCR)** system invented at Cetus (see separate entry). Others include:

- *Ligase Chain Reaction (LCR)*. Uses DNA ligase, a DNA-joining enzyme, to link two oligonucleotides together if a target DNA is present. The linked molecules can then be templates for more molecules to link together and so on. This technology is owned by Abbott Diagnostics, and was the first to be built into an automated test system reliable enough to be used in a routine medical diagnostic laboratory.

- *Nucleic acid sequence-based amplification* (NASBA, developed by Organon Teknika). This creates a new molecule of the DNA joined onto a promoter for RNA polymerase. RNA polymerase uses this to make many RNA copies of the new DNA. This is an *isothermal* technique—it all happens at one temperature (compare PCR). There are other transcription-based amplification strategies that vary from this idea.

The amplification systems above amplify the DNA itself. There are also systems that use DNA to amplify the signal that a DNA assay gives, without amplifying the DNA itself. The best known is the branched multimer amplification, developed by Chiron Inc. This builds a branched tree of separate nucleic acids on a single probe. At the end of each branch is a label enzyme, so that a single probe ends up "capturing" many enzyme molecules.

A probe amplification system is the Q-beta system of Gene-Trak. The probe DNA is connected to the RNA from a small virus—Q-beta—is duplicated by the RNA polymerase enzyme which the Q-beta virus carries. Add one molecule of Q-beta RNA to a tube of Q-beta replicase and the right chemicals, and the tube fills up with the hybrid Q-beta RNA.

As well as research use, all these systems have been developed to be used in medical diagnostics. All suffer from a greater or lesser degree to the problems of their extreme sensitivity to contamination (*see* PCR).

Most DNA amplification systems are qualitative—if there is target DNA there, they will amplify it. It is hard to turn them into quantitative assays, which could tell you how much DNA was there. The branched multimer technology is the only commercially used one which is relatively straightforward to quantify (*see* Gene quantification).

DNA fingerprinting

DNA (or genetic) fingerprinting, or profiling, is a way of making a unique pattern from the DNA of an individual, which can then be used to distinguish that individual from another. The technologies used can be based on DNA probes, PCR, or any other DNA detection technology.

To work effectively, the fingerprinting technology must be able to detect a piece of DNA that is highly likely to be different between any two people. The original DNA fingerprinting probes, discovered by Prof. Alec Jeffreys, detected sequences called variable number tandem repeats (VNTRs)—sequences where a simple short piece of DNA is repeated many times, the exact number varying with each person. In the 1990s, VNTR fingerprints were supplemented by SNP probes, and "single locus probes" which only detect a single VNTR, rather than a complex pattern. Each individual SNP or single locus probe is unlikely to be able to tell definitively whether two DNA samples are the same, but a combination can be as powerful as conventional DNA fingerprinting, and substantially easier to compare between experiments.

DNA fingerprinting has been used extensively as evidence in paternity, rape and murder cases to identify individuals. Until 1989 it was considered unassailable evidence, but since then several cases has questioned poorly collected or poorly analyzed DNA fingerprinting data. The O.J.Simpson trial in 1996 showed that the value of DNA fingerprint evidence rests more on getting the samples carefully and agreeing what they mean than on the molecular genetic technology of actually making the fingerprint.

There is substantial international disagreement about what crimes DNA fingerprinting can be used for, and whether the fingerprint data or the DNA itself can be kept afterward. The United Kingdom and France exemplify the extremes—in the United Kingdom, DNA data can be collected on suspects of many violent or serious crimes, and kept indefinitely if the suspect is convicted. In France, fingerprinting is only allowed for violent sex crimes, and nothing is kept even after conviction. There is also debate about what probes to use, as increasingly genetic mapping will allow us to say something about a suspect's possible height, colour, weight, etc. These are social and legal matters, not ones that need a scientific solution.

DNA fingerprinting has many other, less high profile applications. These include its use in animal forensics (such as racehorse and pedigree dog tracing), identification of meat sources, fibre testing (to show, e.g. that a cashmere sweater is actually made from cashmere, and not sheep).

DNA fingerprinting can be extended to non-natural DNAs, which are made as very specific "tags" for materials to prevent counterfeiting. There are 10 trillion possible sequences of a 20-base DNA, which is a substantial number of "codes" that can be used to tag something. Tiny amounts of DNA can be detected by PCR, and if the DNA sequence is chosen so that it is different from any naturally occurring DNA then its presence is a definitively unique "marker" for the original, authentic product.

DNA probes

As well as being used as the genetic material to "program" cells to do things, DNA can be used as a reagent in its own right. One common way that DNA is used is as a DNA probe, also called a hybridization probe (another is as a PCR primer—*see* **PCR**).

Natural DNA usually exists as a double helix, because the order of the bases in each of the two strands is complementary: A is next to T, G to C. If such a double helix is separated, it will spontaneously tend to come together again. However, two strands that are not complementary in this way will not come together: no helix will form.

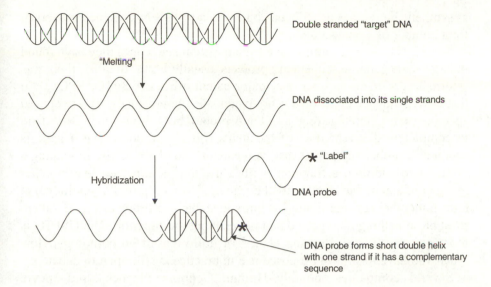

Double stranded "target" DNA

"Melting"

DNA dissociated into its single strands

"Label"

Hybridization

DNA probe

DNA probe forms short double helix with one strand if it has a complementary sequence

Thus, we can use a single strand of DNA with a known sequence to detect whether one with a complementary sequence is present in a mix—if it is, our "probe" will form a double helix with it. This process of getting a DNA probe to bind to a target sequence is called hybridization, and can be used to detect DNA or RNA. The same chemistry is behind the use of PCR primers—they bind to their target DNA, initiating the amplification process—because they are complementary to that target.

Probes can be natural DNA or RNA, cloned DNA, or a synthetic oligonucleotide. The last is preferred in most cases on technical grounds.

Applications of both probe technologies and PCR in medical diagnostics are:

- *Testing for viruses*. Because viruses cannot be seen down a conventional microscope and are very hard to grow, chemical tests are used for them. The most sensitive of these are DNA-based tests

- *Human genetics*. A wide range of human genetics applications use DNA probes and PCR. *See* SNPs

- *Cancer typing*. As cancer is a disease caused by mutated genes, it makes sense to identify the type of cancer a patient has using a DNA-based test.

DNA (gene) patents

As biotechnology companies want to protect their intellectual property with patents, and as some of the leading edge of biology is concerned with gene sequencing, it is inevitable that researchers would try to patent DNA sequences that they had discovered. (Patenting invented DNA sequences, such as synthetic genes, has never been controversial.) This has been controversial for a number of reasons.

Since 1991, several groups have tried to patent sequences they have found in large-scale genome sequencing projects, usually ESTs. In 2001, there was a formal agreement between the major patent offices (USPTO, EPO, JPO) that DNA sequence on its own is not patentable. A sequence with a proposed specific use is patentable (e.g. as a DNA probe), but this use does not extend to completely different uses of the gene. (So if you show that a gene is useful as a probe, and patent that, you cannot also claim its use in making a protein.) In addition, a fragment with a use that is expected [e.g. another member of a gene family identified because of its similarity ("homology") to known members of that family] is not patentable in Europe or Japan—there must be something unexpected and novel about the function. The USPTO is more generous about allowing claims of utility based on homology. How much you have to *show* a proposed use in practice is still open to debate.

Several companies, notably Human Genome Sciences and Incyte Pharmaceuticals, based their business in part on the patent-protectable value of the sequences they discover. The US pharmaceutical company Merck deliberately put all the DNA sequence they discovered into the public domain as a "spoiling action" to this strategy (because once it is described in public, it is no longer novel). Both Incyte and HGS have moved on from a purely patent-based business model in the late 1990s.

There is also a technical issue of whether a DNA sequence patent covers only the sequence claimed, or all "reasonable" variants of it. In the case of BRCA2, one company (Myriad Genetics) has patented the gene, another has patented a polymorphism of the gene. Which is "right," and which patent "dominates" the other?

There is a side issue that the public is adamantly opposed to the idea that a gene patent allows a company to "own life," or least a part of life. Patent lawyers argue that this is based on a misunderstanding of what a patent is: it does not give someone "ownership" of the patented invention, but rather allows them a monopoly to pursue practical applications of it for a couple of decades. Scientists argue that DNA is not life, anyway. In reality, practical solutions will be found for all this, and in 20 years' time the whole debate will look rather like angels and pins.

DNA sequencing

Determining the sequence of the bases in DNA (DNA sequencing) is one of the mainstays of gene cloning technology. At root, this is a chemical procedure. By far the most common chemistry used the Sanger technique (di-deoxy method, chain termination method). This uses enzymes to make a new DNA chain on the target you want to sequence, using the "dideoxy" reagents to stop the chain randomly as it grows. The result of a series of reactions are analyzed to show what length the DNA molecules are. As each molecule will end in a known base (depending on which di-deoxy reagent you put in that tube), you can tell from the pattern what the order of bases was in the original DNA.

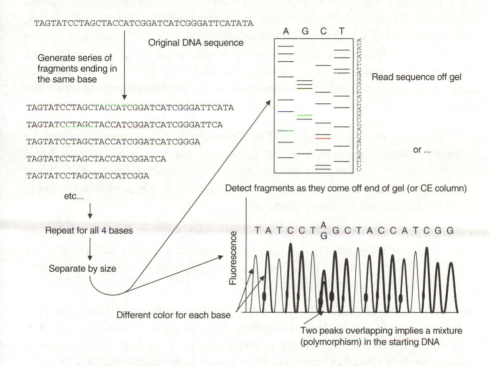

Measuring the length of the DNA molecules is done by electrophoresis, traditionally gel electrophoresis but now often capillary electrophoresis. Capillary gel electrophoresis has a higher capacity, and most genome scale sequencing laboratories uses this.

Machines for both methods are made by Applied Biosystems Inc (ABI), which is the leader in providing instrumentation to genome centres. These machines all use fluorescence to detect the DNA that has been labelled with a fluorescent "tag" as part of the DNA sequencing reaction. ABI machine uses four different coloured dyes, so you mix the four different base reactions in one lane on the gel, resulting in greater efficiency. Sequence data is sometimes shown as traces of the fluorescence intensity detected by the machine, as shown.

DNA sequencing is starting to be used as a medical diagnostic technique as well as a research tool. An increasing number of gene mutations are associated with disease, or with the patient's response to drugs. Because of the number of mutations involved, it is impractical to design a DNA probe to test that will cover them all. So clinicians would like to determine the whole gene sequence in their patients. At the moment this is still a very costly and complex process. Gene chip tests, that test tens of thousands of gene probes at once, might be another way to achieve the same goal.

The rate at which DNA can be sequenced has doubled once every 18 months for the last 20 years—what once took years now literally takes seconds. Today, DNA sequencing can determine the sequence of tens of thousands of bases an hour per DNA sequencing machine. There are a variety of super high throughput technologies being developed which may be able to increase this already phenomenal rate to millions of bases per hour, and continue this trend in the next decade. These include:

- *Sequencing by hybridization.* (*See* **Gene chip**). Not a good method for sequencing DNA one its own, but a powerful tool for re-sequencing, that is checking the sequence of a gene when you are already fairly sure of what it is. In principle, you could use this to detect every single sequence variant in a human genome in about 60 minutes.

- *Single molecule sequencing.* Several companies, such as Solexa, are seeking to carry out DNA sequencing on single molecules. If this can be achieved, then a human genome could be broken into million of large molecules, which are then all sequenced at once.

- *Nanopore techniques.* DNA is driven through a nanometer-diameter pore by an electric field (electrophoresis). As each base comes through, it is detected by its electric field, by spectroscopy (either of the original base or of a labelled base that you have enzymatically built into the DNA), or by enzymatically cutting it off and analyzing it. Detection sensitivity is the big problem with these systems.

- *SEM, STM, AFM etc.* Many microscope techniques can, in theory, "see" the sequence of DNA. If this can be realized, then the sequence of a genome could simply be read off under the appropriate microscope. So far, this is a long way from reality.

DNA structure

DNA is a polymer, a very long molecule made of repeating units linked together. In DNA, there is a "backbone" that holds the molecule together, on which other units called bases as linked like side-branches on a trunk. Information is stored in the order of bases on that backbone. There are 4 bases—adenine, guanine, cytidine and thymidine. In the very similar RNA molecule, thymidine is replaced by uracil.

How to draw DNA:

Molecular structure | Outline of double helix | Helix "straightened" | As sequence of bases | As genetic structure

5' end

base

Sugar

Phosphate

3' end

atgatcgatcgccatcggtacctg

base pair

"Backbone"
Arrows show "direction" of strand (5' to 3')

Missing base pair

Shows which bases are paired to which in non-standard structures

Symbols represent genetic structure, reflecting sequence

The three-dimensional structure of DNA is, famously, a double helix. Each "strand" in the double helix runs in one direction (called 5'–3': "five prime to three prime"), determined by how the sugars and phosphates are linked up. Two strands wrap around each other so that they run in opposite directions: the 5' end of one is next to the 3' end of the other. This arrangement is called antiparallel. (Dual carriageway roads are antiparallel—the road signs on each side point in opposite directions, as do the cars in most circumstances.)

The double helical structure of DNA was discovered by Watson, Crick and co-workers in the early 1960s. It is a stunning insight into how DNA, the fundamental molecule of genetics, works. However, most molecular geneticists today completely ignore it, and treat DNA as if it were a long piece of string that they cut up and tie together. Only rarely does its double helical nature matter at all. At a biotechnology conference, the only double helix you are likely to see is on the conference logo.

One example of when it does matter is when the DNA is as a closed loop. Because the strands of DNA are wrapped round each other, a closed loop is permanently knotted, and you cannot separate the strands from each other without breaking them. All natural closed loops of DNA are like this. To uncoil them for transcription and replication therefore requires unknotting enzymes called topoisomerases, which are important drug targets for anti-cancer and anti-bacterial therapy. Finding that the DNA is a knotted closed

131

loop also is a quality control measure for DNA made for gene therapy. Technically, such DNA is called supercoiled.

Downstream processing

This is a general term for all the things which happen in a biotechnological process after the biology, be it an enzyme reaction, fermentation of a micro-organism or growth of a plant. It is particularly relevant to fermentation processes, which produce a large amount of a dilute mixture of substrates, products and microorganisms. These must be separated, the product concentrated and purified, and converted into a product which is useful.

There are three general steps to downstream processing:

- Separation

- Concentration

- Purification.

These all have separate entries. The first step separates the crude product from the other ingredients, the second removes most of the water (and hence is often called dewatering), the third takes the concentrated product and purifies it. The orders can be different, but generally fall into this scheme.

In addition, if a product is made inside a bacterial cell you must break the cell open to get at it (*see* Cell disruption).

The key issue in all of this is "recovery" or "yield"—how much of the product in the original mix is still there in the final product. "Loss" is how much of the product is *not* there in the final product. Sometimes recoveries can be 10% or less, if most of the product is washed away with the impurities. Whether this is acceptable or not depends on how expensive the process is to run and how valuable the end product is. Recoveries or yields can also be worked out for each step in the process—the overall yield is the product of the individual yields. Thus, if step 1 has a 30% yield, step 2 an 80% yield and step 3 has a 40% loss, then the overall yield $= 0.3 \times 0.8 \times (1 - 0.4) = 0.144$, or nearly 15%. In this case, increasing the yield of step 2 to 100% is not going to make much difference, but increasing step 1's yield by only 10% more would raise the overall yield to nearly 20%, so that is the step that needs work.

Drug development pathway

A substantial amount of biotechnology, and the majority of the more public aspect of biotechnology start-up companies, is concerned with developing new drugs. As a consequence, some of the jargon of drug development and licensing creeps into biotechnology discussions. This entry summarizes the key points in the pathway a new drug candidate has to go through. *See also* drug discovery, for how you find the drug in the first place.

Pre-clinical research. This means all the research that goes on before you try the compound out on people, but is often taken to mean animal studies of the drug. As well as research, this covers the formal testing for toxicology and ADME that have to be done under GLP before you are allowed by law to give a new chemical to a human. (*see* ADME)

Phase I trials. These are the first trials in which a drug candidate is given to people. The purpose of the trial is to establish pharmacokinetics of the drug, and to confirm that there are no adverse effects in man at a dose likely to be effective (having already established that there are none in animals). The people involved are often normal, healthy volunteers (often students). Most phase I trials start out with a very small dose and works up (dose escalation). Usually a relatively small number of people—as many as 100 or as few as 10—are involved.

Once a drug has got to phase I trial it is said to be "in the clinic." This is a big milestone for a small company to achieve.

After phase I the developer applies for an investigational new drug application (IND, in the United States), or the equivalent in other countries (e.g. the clinical trial exemption certificate—CTX—in Britain). This is the regulatory hurdle necessary to go on to phase II trials, and at this point the developer must show that they have complied with a wide range of GLP practices in their pre-clinical and phase I trials (*see* entry on GLP/GMP). For medical devices such as prostheses (whose development path is essentially the same), the IND is replaced by the 510(k) application in the United States.

Phase II trials. This is the first time the drug is tried on ill people. This trial is usually done at one hospital centre on a small number of patients, and looks for any evidence that the drug actually has a medical effect on the disease it is meant to treat. The drug is said to be being developed for one "indication," that is one collection of symptoms or one disease type. The object of this and subsequent trials is to show that the drug has an effect for this indication.

Some drugs go straight into man in a phase II trial, skipping the phase I stage, if they are likely to be dangerous to healthy people. Most cancer drugs are like this—conventional cancer drugs are extremely toxic to normal cells, and so would make healthy people very ill. So it is not ethical to try them on healthy volunteers. The other category of drug that goes straight to a phase II can be some types of new, aggressive therapy, such as gene therapy, where the risks are completely unknown.

Phase II trials are sometimes called "proof of efficacy," because this is the *first time* after years and tens of millions of dollars that you have some evidence that the drug will actually work on people.

Phase III trials. This is where very big money is spent on drug development. The object of this phase is to see whether the drug is worth launching, because it is better than existing therapies, does not have severe side-effects and so on. This requires hundreds or thousands of patients (all of which have to be followed in detail), usually in at least six hospital centres. The trial is done double blind, so that neither the people giving the drugs nor the people analyzing the results know who has received the drug and who has received a placebo until the study is over (when it is "unblinded," i.e. the codes as to who received what are made available to the investigators). It can be a crossover trial: halfway through, those who had been getting placebo get given drug and those who had been getting drug get placebo. (This helps to avoid problems due to the differences that people show in their response to drugs.)

At the end of phase III the drug is submitted for a new drug application (NDA, in the United States) or a product licence application (PLA in Europe). (For a medical device, the equivalent is the pre-marketing approval—PMA) If this is approved, then the drug can be sold. Approval can take a long time—therapeutic proteins that were approved between 1995 and 2000 took between 5 months (for Herceptin) and 4 years (for Gonal-F, a fertility hormone from Serono) just to be approved by the FDA.

Phase IV trials. Post-marketing surveillance or pharmacovigilance then takes over. Pharmacovigilance looks for rare adverse reactions (adverse drug reactions—ADR), to look for opportunities to decrease the dose (because the initial estimates of the best dose are often rather high), and to extend the range of indications for which the drug may be used. Extension of indications can come about because of "off label" use, that is use of the drug by physicians for indications other than those for which the drug is licensed. There is nothing to stop people doing this, providing that they are very careful to emphasize to their patients that they are effectively doing an experiment on them. Successful experiments lead to new ideas for the use of the drug, and hence new clinical trials to see if the new indication is a suitable one for this drug.

Much of the legwork of clinical development can be contracted out to contract research organizations (CROs) (also sometimes called clinical research organizations).

All this takes years, and has a very high failure rate (*see* **Drug discovery**). Because of this, almost none of the biomedical technology described in this book will not result in improvements in human medicine before 2010, and most will not result in any new medicines at all.

Drug discovery

Over half the biotechnology industry (by capital value, if not activity) is concerned with discovering new drugs. The reason is that the potential value of a new drug is huge, there is a never-ending demand for them, and once you have a patent on a new drug it is effectively yours for 20 years (although if it takes 18 years to develop, that is not as valuable as it sounds).

There is a "standard" path to discovering new drugs, which is now driven by the molecular biology revolution (sometimes called the "genes to drugs paradigm"). That is:

- *Discover a target.* Much of the biomedical technology described in this book is concerned with this. *See* **targets for drug discovery**

- *Discover a chemical.* Once you have a target, discover a chemical that interferes with it. This can be by structure-based drug design, or a combination of high-throughput screening (HTS) and a method for making many chemicals, such as combinatorial chemistry.

- *Development.* Once you have a chemical that you think might be a drug, there is a well-worn path to finding out whether it is, a path called drug development, that is a series of heavily regulated tests that start with testing in animals and go on to test in increasing numbers of people. *See* **Drug development**. (This is actually 90% of the work.)

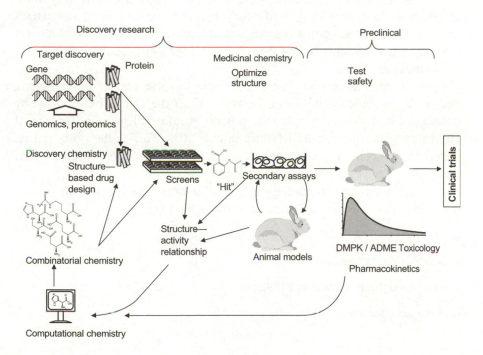

DRUG DISCOVERY

The problem with this is that it is very failure-prone, very slow, and staggeringly expensive. Typical costs, durations and chances of success are:

Stage	Typical duration (years)	Typical cost (millions of dollars)	Typical chance of success
Research—discover a target	1.2	14.5	70%
Research—discover a chemical	2	17	75%
Preclinical (animal safety tests)	0.4	7	65%
Phase I trials (including approvals, re-trials)	1.7	3.5	80%
Phase II trials	3.5	120	65%
Phase III trials	2	200	70%
Approval	2	40	70%
Overall	12.8	400	8.5%

The values—especially the success rate—depend a lot on how optimistic you are about how well people's new technology will actually work. I have taken a fairly optimistic view here. Many in the industry would say that the chance of succeeding in clinical trials alone was no better than 1 in 10.

This is sometimes called the "attrition rate" problem. It is why most investors in biotechnology, while they may be dazzled by a company's science, consider the company's drugs to almost worthless until they have at least been into humans: before then, there is a 90% chance they simply will fail to make it.

Many biotech companies claim to be able to solve part of this problem, and no doubt some of them can. However, if *all* the claims being made by companies in 2000 were right, the process would still take an average of 10 years and have an overall failure rate of 80%. This is the other reason that most biotechnology companies are in drug discovery—even if their competitors are successful, the market will still exist.

As well as general technology entries, this book has specific entries for some drug discovery topics relevant to biotechnology:

- Genomics
- Combinatorial Chemistry
- High Throughput Screening (HTS)
- Pharmacogenomics

Eggs

Hens eggs are being investigated by several groups as a novel production technology. Eggs have been used for years to make influenza and MMR vaccines (as the disabled virus grows in eggs). This works because eggs are a sterile environment with a rich medium in for cell growth and many rapidly dividing cells in the early bird embryo, but no immune system.

This is being harnessed by companies such as TransXenogen to make recombinant proteins, initially antibodies. Transgenic chickens with appropriate promoters on the transgene secrete large amount of recombinant protein into the egg. The advantages are that this is a vertebrate protein source (hence the glycosylation patterns should be similar to man), eggs are a validated production technique (from vaccine production), and egg production is a huge and highly efficient industry, producing about 260 eggs per person per year in the USA, so this can be scaled up very easily and (in theory) cheaply. Chickens also do not carry BSE or viral contaminants that might affect antibodies produced from mammals or mammalian cells.

Electrochemical sensors

These are types of biosensor in which a biological process is harnessed to an electrical sensor system, making a sensor. The most commonly discussed type of electrochemical sensor is the enzyme electrode (see separate entry). Other types couple a biological event to an electrical one via a range of mechanisms. Among the more common are:

Oxygen electrode-based sensors. These are based on a well-known electrochemical cell called an "oxygen electrode" (or Clark electrode) which generates a small current when oxygen is present. This can be turned into other sorts of sensor by coating it with a biological material which generates or (more usually) absorbs oxygen. When the biological coating is active, the amount of oxygen next to the electrode falls and the signal from the electrode changes. Typical coatings might be an oxidase enzyme (which consumes molecular oxygen to oxidize a particular substrate) or a whole cell (which consumes oxygen when presented with a range of substrates). This latter type of biosensor—a microbial- or cell-based biosensor—can be used to detect poisons, as poisons damage the cells and hence reduce the rate at which they consume oxygen.

pH electrode-based sensors. Again, a standard electrochemical pH electrode (one whose current changes as pH changes) is coated with a biological material. Many biological processes raise or lower pH, and so can be detected by a pH electrode. Examples could include hydrolysis of an ester to acid and alcohol, or the metabolism of neutral pH substrates by a bacterium.

Electron microscopy

This is a microscope that uses electrons instead of light. A microscope magnifies the image formed when light falls on something. However, a "light microscope" of this sort is inherently limited—it cannot resolve (make out features in) something that is smaller than the wavelength of light, which for blue light is about 400 nm. So if we want to get accurate microscope pictures of cellular organelles, proteins, or single atoms, we cannot use light.

The electron microscope (EM) uses electrons instead. These can be "shone" onto a sample, and then the electrons that go through (transmission microscopy) or that are reflected (scanning electron microscopy—SEM) are focused to make an image. You use magnets and electrically charged metal plates to focus electrons, and they are to be detected on a fluorescent screen or CCD camera. A vacuum is created inside the EM, as electrons are absorbed by air—thus, living samples cannot be directly viewed by EM, as they would dry out, or explode, when you took the air away.

EM has been used very widely to uncover the structure of cells—chances are that any image that shows a cross section of a cell with lots of vesicles, mitochondria, looking like an aerial picture of a major road junction in black and white, is an EM picture. EM can also be used for protein structure determination, through a technique called electron tomography, where sample is "lit up" with very low energy electrons from several angles, and the result built up into a computer. This has relatively low resolution (2 nm, compare to 0.2 nm for a good X-ray structure), but is useful for big protein complexes, very "floppy" proteins, or membrane proteins, all of which are very hard to crystallize.

If the limit on light microscopes is the wavelength of light, another approach is to use light of shorter wavelength. UV microscopes (down to 250 nm) are known, although they are rather exotic as the lenses have to be made of quartz crystals, not regular glass. X-ray microscopes have been proposed, but have not proved practical yet, mainly because making and focusing X-rays is incredibly hard.

Electrophoresis

Electrophoresis is separating things by moving them in an electric field. Nearly always this is separating molecules, although in principle organelles or whole cells could be electrophoresed as well.

In gel electrophoresis, separation is done in a gel, a polymer matrix. Samples are put at one end of a slab of polymer gel (any jelly-like material). An electric field across the gel pulls the molecules through it—smaller molecules can pass through the gel more easily and so move towards the other end faster. Thus, molecules are separated mainly according to size.

A large number of materials are used to make the gel, but by far the most common are agarose (for DNA and RNA) and polyacrylamide (for DNA in DNA sequencing and for proteins). Polyacrylamide gels are often called PAGE gels (polyacrylamide gel electrophoresis). Various chemicals can be included in the gel to help the separation, such as the detergent sodium dodecyl sulphate (SDS) in protein gels which unfolds all the proteins, and urea in DNA sequencing gels which does the same to DNA.

A recent variation in DNA gels is pulsed field gel electrophoresis (PFGE) and its variants, where two sets of electrodes are used—this allows very large molecules of DNA to be separated, although not all that well. Several terms such as orthogonal field gel electrophoresis and field inversion gel elec- trophoresis (FIGE) refer to how the different fields are arranged.

Variants on gel electrophoresis are isoelectric focussing (IEF) gels, which separate macromolecules on the basis of their isoelectric point (roughly, the number of different charged groups they have) rather than on size. The gel has a gradient of pH inside it, and the molecules migrate to the point where the pH of the gel is the same as the molecule's isoelectric point—at that pH they have no charge, so they stop moving and form a narrow band. O'Farrel gels (also called 2D gels, or 2D-PAGE) run an isoelectric focussing gel down one side of a gel slab, and then do a standard PAGE at right angles along the length: this produces a two-dimensional pattern of protein spots which is as characteristic of a mixture of proteins as a fingerprint. Nearly all proteome analysis (the systematic analysis of all the proteins in a sample) is done using 2D-PAGE.

Electrophoresis without gels has been used. It is called free zone elec- trophoresis, and uses a flow of water, or sometimes a density gradient column, as a medium for electrophoresis (such density gradients are discussed further in the entry on **centrifugation**). However, the stirring effects in these systems can be substantial.

See also **Capillary electrophoresis.**

Electroporation

This is opening cells by exposing them to a strong electric field. This can be used to get large molecules into the cells, or to fuse cells together.

Transformation of cells—getting DNA into them—can be achieved simply by exposing the cells to a suitable electric field while they are in a solution of the DNA. The electric field seems to modify the lipid membrane that surrounds the cells, and greatly increase the rate at which pinocytosis, a normal mech- anism by which cells take up chemicals from solution, takes DNA into the cell. It is not widely used for bacterial cells, where other methods have been devel- oped which are quite reliable. However, electroporation is discussed fairly extensively when talking about getting DNA into plant protoplasts and animal cells, and to a lesser extent into fungal cells. Amaxa specialize in technology to do this.

Fusing cells. This was the first application of electroporation. Protoplasts of plant cells or whole animal cells, can be made to fuse by putting them next to each other and exposing them to a strong electric field, a process called electrofusion. Typically, the cells are exposed to a low electric field, which induces a dipole in the cells and causes them to line up. They are then exposed to a quick pulse of very high voltage to make their membranes "leaky" and so to fuse. There seems to be no limit to the types of cells which may be fused together using this technology. Applications in plant genetics include making hybrid plants and making polyploid plants.

There has been some work on using electroporation to deliver drugs, specifically DNA and peptides, directly into the skin cells. This certainly works to some degree: it is possible that the matrix around the skin cells protect them from the damage that electroporation usually does to cells in culture.

Embryo technology

Embryo technology is a generic name for any manipulation of mammalian embryos.

Embryo technology encompasses:

- *Cloning*. This making a clone of an organism, a collection of genetically identical individuals. Cloning mammals is very difficult (see separate entry). Cold-blooded animals easier to clone, and cloning as part of fish breeding is well established (*see* Fish farming.)

- *In Vitro Fertilization*. A widely used technique for animals and people, this means fertilizing the egg with sperm outside the body. Usually the fertilized egg is cultured outside the body for a few days before reimplantation into a female to make sure that fertilization has occurred. A related technique is GIFT, which injects sperm directly into the fallopian tubes, and so is a half-way house to the fully external fertilization of IVF.

- *Gamete and embryo storage*. This is the storage of eggs, sperm or fertilized embryos outside "their" original source (an animal or person). Almost invariably this means freezing them at liquid nitrogen temperatures. There is also intense ethical debate about this.

- *Stem cell technology*. Most stem cell work has to use tissues from embryos as a source of cells, although the technology is not aimed at embryos directly. See separate entry.

Some of the other key debate points about embryo technology are:

DNA-based genetic diagnostics. Because DNA probes can detect "defective" genes whether they are actually doing anything or not, they can and have been used to detect whether a fertilized egg, an embryo or a fetus is carrying an undesired gene. If it is, then it can be aborted before the gene has a chance to express itself. This is tied up with the abortion debate.

When is a fetus...? There is heated and unresolved debate about when a fertilized egg becomes recognized as a human, with the rights to protection that a human has. In the United Kingdom, following the Warnock Report, the embryo has no rights until day 14 (before that it is called a "pre-embryo"). After 14 days it becomes an embryo, and starts to acquire some rights as a human being. Sometime between then and around 15 weeks the embryo is renamed as fetus. The fetus is not usually considered capable of independent life until 24 weeks gestation (and even then only with heroic medical intervention, and a high risk of "congenital" malformation).

See also Yuk factor.

Embryogenesis (in plant cell culture)

Embryogenesis is encouraging plant tissues to form new plants *in vitro*, without first forming a seed, and hence usually from cultured plant cells or a small piece of plant tissue. The generation of embryos takes two stages: initiation and maturation. The former needs a high level of the group of plant hormones called auxins: the later need a lower level. Other chemicals have to be at suitable levels, too. Thus, the procedure is usually to take a piece of plant tissue and put it on a high-auxin medium, where it grows into a mass of undifferentiated cells called a "callus." This is then transferred to a maturation medium, where the callus starts to develop initial organs, ultimately growing a root and a shoot.

In plant culture circles, embryogenesis is used to describe the generation of new plants from bits from old plants. If you generate a plant from a single cell, it is organogenesis, although the techniques have many similarities. Embryogenesis is critical for plant cloning and micropropagation technologies.

Encapsulation

This is any method which gets something, usually an enzyme or bacterium, into a small package or capsule while it is still working (or alive). The capsule can be any size, but usually is no bigger than a few millimeters across. If it is too small to see fairly easily with the naked eye, the process is called microencapsulation.

Encapsulation is one method of immobilizing cells for use in a bioreactor. Encapsulating agents can be anything which will form a shell around something, but usually are polymers, especially polysaccharides such as alginate or agar, because they are inert, allow nutrients and oxygen to diffuse into and

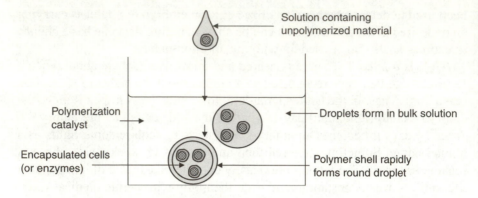

out of the capsule readily, and are easy to convert from a gel (solid) to a sol (liquid) or solution form by altering the temperature or the concentration of ions such as calcium. Proteins such as collagen (gelatin) are also used. The usual procedure is to make droplets of liquid polymer with the enzyme or cell in them, and then drop them into a solution that causes the polymer to solidify, because it is cooler, or contains specific ions. This easily forms spherical blobs of polymer with the stuff you want on the inside. The size of sphere just depends on whether you use big drops or a fine spray.

Cell encapsulation is also used to try to develop **artificial tissues** (see separate entry).

Enzymes may also be encapsulated, although they are more usually immobilized on the surface of polymer particles (*see* **Enzyme Immobilization**).

Environmental biotechnology

Environmental biotechnology is a general term covering any biotechnological product or process which can be considered to be helpful to "the environment." Usually, this means control, reduction or disposal of waste, removal of chemical pollutants, or reduction in power use, especially in industry. Because of the high political profile of "the environment," a number of diverse biotechnology activities have been included under "environmental biotechnology."

Biotechnology is well-placed to address some ecological and environmental concerns. As opposed to traditional heavy industry, biotechnology is likely to use potentially renewable resources, inherently low power processes, materials which are unlikely to be dangerous, and to produce products which can be justifiably labelled as "natural."

The most commonly discussed topics in environmental biotechnology are:

- **Bioremediation**: cleaning up contaminated soil using biological processes (see separate entry).

- *Soil amelioration*: improving soil quality through manipulation of its microflora (see separate entry).

- Developing biodegradable replacements for plastics, and particularly developing biotechnological ways of making them (*see* **Biodegradable plastics**).

- Waste disposal: developing bacterial methods for disposing of waste, or at least of disposing of the biodegradable part of it more rapidly. (*see* **Waste disposal**).

- Creating alternative energy sources: specifically biofuels, biogas and solar energy methods. These are much more speculative, and rarely are economic alternatives to oil, coal or nuclear power at the moment (although it is emerging that nuclear power is actually horrendously expensive, when the cost of decommissioning the power stations is taken into account, and so biofuels are almost certainly cheaper than nuclear fuels). They are covered in separate entries.

- Addressing global warming issues, especially by reducing emissions of "greenhouse gases" from waste disposal processes and from farming, or (and this is very long-term) enhancing organisms' ability to capture CO_2 from the air and so reversing CO_2 build-up.

A major issue for all these and for other applications of biotechnology to environmental problems is a lack of legislative commitment. Industry does not want to invest in expensive "solutions" to things that are not actually required, but equally does not want to build technology that next year's legislation will ban. So the temptation is not to invest. This affects all environmental technology, but particularly more research-intensive areas such as biotechnology, where substantial investment in expensive research is needed to find and test possible solutions at an industrial scale.

Paradoxically, in the past this has meant that industry has actually lead the way in environmental standards. CFCs were effectively banned by industry (who saw what the scientists were saying and hence how the legislation would eventually go) before laws about them crept into the statute books. Industrial work on CO_2 emissions is starting to go the same way.

Biotechnology is also providing tools for more conventional environmental agendas, such as improved tests for pollutants using immunoassays or biosensors.

Enzymes

The core of traditional biotechnology, and a key feature of the new biotechnology of gene cloning, is the use of enzymes. For practical purposes these can be considered to be catalytic proteins, although recent work has shown that RNA can act as an enzyme.

Commercially useful enzymes come from a huge variety of organisms, from viruses to cows. In general, they may be extracted from some organism which already produces the enzyme, made from an organism that does not normally make that enzyme but which is induced to do so, or made from an organism which has been genetically engineered to produce the enzyme.

Enzymes are so widespread in biotechnology that they crop up in many entries in this book. Specific classes of enzymes covered are amylases and glucose isomerase, kinases, lipases, and proteases. Also, entries refer to enzyme production, kinetics, and the use of enzymes in biotransformation.

Some classes commonly used in biotechnology and chemistry applications that do not have their own entries here include

- amidase (make or break peptide bonds—*see* **peptides**)
- dehydrogenases (remove pairs of hydrogen atoms from molecules)
- oxidases and peroxidases (oxidize molecules, often by adding an oxygen atom to them)
- aldolases and transketolases (catalyze specific types of rearrangements within or between molecules)
- glycosidase and glycosyltransferases (add or remove sugar molecules from other molecules—important for glycobiology)
- sulfotransferases (add or subtract sulfate from other molecules)
- hydrolases (break molecules down using water).

Enzymes also crop up in biotransformation, protein engineering, enzyme production by fermentation and expression, as well as *en passant* in many other entries.

The value of some of the major industrial uses of enzymes is:

Industrial Enzyme	Market value ($millions) (estimated 2000 values)
Detergents (proteases and lipases)	500
Dairy industry (mostly rennin)	170
Starch processing	340
Other food industry enzymes	160
Textile processing	140
Leather processing	40
Biotransformation	55

Enzyme Commission (EC) number

All enzymes are given a systematic name and number which identifies them in technical literature. (They may also have a "common" name, like trypsin or rennin, that is easier to remember for the non-specialist.) These names are given by the Enzyme Commission. The name and number are systematic descriptions of what the enzyme does. The number is a four-digit number. The first digit classifies the enzyme into one of six broad groups:

Number	Class
1	Oxido-reductases (transfer H atoms or electrons)
2	Transferases (transfer of small groups between molecules)
3	Hydrolases
4	Lyases (addition to double bonds)
5	Isomerases
6	Ligases (formation of bonds between C and another atom, using ATP as an energy source)

Each of the groups is subdivided into sub-groups, each subgroup into sub-sub groups, and the last number is specific for the enzyme. The systematic name describes the reaction catalyzed. Thus, creatine kinase is EC 2.7.3.2 (2 because it transfers a group from ATP to creatine, 2.7 because the group is phosphate, 2.7.3 means the sub-group that transfers the phosphate to a nitrogen atom, and the last 2 is because this particular enzyme was the second classified in this group). Note that the dots are essential, as some classes of enzyme have more than ten members. The systematic name is ATP : creatine phosphotransferase—the enzyme that transfers a phosphate group from ATP to creatine.

Enzyme electrode

A type of biosensor, in which an enzyme is immobilized onto the surface of an electrode. When the enzyme catalyzes its reaction electrons are transferred from the reactant to the electrode, and so a current is generated. (This is distinct from other types of electrochemical biosensors, where the enzyme generates a distinct chemical product, for example an acid, which is then detected by a separate electrode system.)

There are two types of enzyme electrodes:

- *Ampometric*. Here, we measure the current generated by the reaction, and the electrode is kept as near zero voltage as is practical. When the enzyme catalyzes its reaction, electrons flow into the electrode, and so a current flows.

- *Potentiometric*. Here, we measure the voltage generated by the reaction, and the electrode is held at a voltage which counteracts the voltage created by the enzyme's tendency to "push" electrons into it. This may be done by actively adjusting the voltage or by not connecting the electrode to anything (as is the case in ISFET devices). The device's output is the voltage necessary to prevent any current flow through the electrode.

Usually enzymes transfer their electrons inefficiently to the electrode, so a mediator compound is coated onto the electrode to help the transfer. Favored mediators are ferrocenes (a specific class of iron-containing organic compounds), because they can easily carry a single electron across the voltage difference that is generated when an enzyme carries out an oxidation or reduction. A range of other organic chemicals have been considered, and the "organic metals," that is organic compounds which conduct electricity, hold promise as electrode materials. Ionomers are also used. These are polymers which are not charged (and so stick to the electrode), but which have a charged group as a side chain.

The enzyme has to be immobilized on the electrode in some way. Common methods include:

- *Physical adsorption.* The enzyme is encouraged to simply stick to the enzyme surface. Many proteins will stick quite avidly to some surfaces, held there by small patches of electrostatic charge or because they sit in a hydrophobic "pocket." This approach is simple, but the enzyme can simply leach off again unless it is held very tightly (which they usually are not).

- *Chemical cross-linking.* The enzyme is chemically linked onto the electrode surface. Rarely do the chemistries of the enzyme and the electrode match to allow this route.

- *Immobilization in a gel.* The enzyme is mixed with a polymer such as agarose or polyacrylamide, and then chemically cross-linked to the gel to form a solid capsule around the electrode.

- *Capture behind a membrane.* Here the electrode is inside a small sack which is permeable to the analyte but not to the enzyme. The enzyme is inside the sack.

A vast number of enzyme electrode systems have been developed in laboratories, and the early 1980s saw a boom in interest in their application. However, they nearly all proved to be hopelessly impractical for commercialization. The major exception was the glucose biosensor for diabetic monitoring: a few other medical sensors are now being commercialized.

Enzyme immobilization

This is linking an enzyme to a solid support material, typically polymeric beads in a column or bed. Typical immobilizing materials could be silica or carbohydrate polymers such as sepharose, although Altus Biologics has cross-linked enzymes in pure crystals to stabilize them (to form cross-linked enzyme crystals—CLECs).

Immobilization has two general advantages.

It enables you to recover the enzyme quickly and easily from the reaction mixture, by separating the particles out. This is useful both for purifying the reaction mixture and for recycling valuable enzyme. The alternative is some sort of chromatography (slow and expensive) or an extraction technique with a solvent (which is fast and cheap, but often destroys the expensive enzyme).

Many enzymes are more stable when bound to a surface, which appears to "lock" their conformation into an active form, so slowing down the inevitable reduction in enzyme activity that occurs with use. (Sensitive enzymes can be inactivated in a few minutes of use. Robust industrial ones usually take hours or days.)

Immobilization is widely used in industrial biotransformation processes for these reasons. The synthesis of the anti-cancer drug Camptosar, for example, uses a lipase to resolve a chiral intermediate (*see* **Chiral synthesis**): the process is only economic if the lipase is immobilized, to give it added stability and efficacy.

Enzymes are also commonly immobilized in biosensor and bioassay applications. Here, the main reason is to hold them in one place, so they do not wash off the assay strip.

Enzyme mechanisms

As the use of enzymes is one of the most commercially important areas of biotechnology, understanding how they work is an important part of the research underpinning the technology. Indeed, one of the reasons that enzymes are so widely used is that their mechanism of action has been studied for most of a century, and the science of enzymology is correspondingly mature (as opposed to, say, the relatively new science of molecular genetics).

A chemical reaction does not usually happen all at once. The reacting molecules have to find each other, their component parts at least partly break open, and then rearrange to make some new molecules. The first part depends on how many molecules there are—the more concentrated a solution is, the more likely the molecules are to bump into each other. Once they have met up, the molecules need energy to reform into new molecules. This is called the activation energy of the reaction, and is often thought of as the energy necessary to make a molecule that is a half-way house between the starting materials and the products. This is called the "transition state." Of course, it is not a stable molecule in its own right, so you cannot isolate it. However, chemists can deduce what the transition state must be by knowledge of how chemicals react.

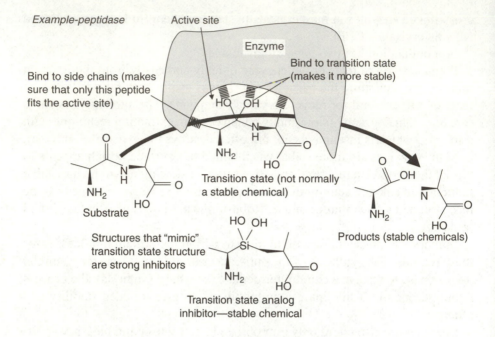

Example-peptidase

Active site

Enzyme

Bind to transition state
(makes it more stable)

Bind to side chains (makes
sure that only this peptide
fits the active site)

Transition state (not normally
a stable chemical)

Substrate

Products (stable chemicals)

Structures that "mimic"
transition state structure
are strong inhibitors

Transition state analog
inhibitor—stable chemical

Enzymes stabilize this transition state, by binding tightly to it, and so reducing the energy needed to form it. The enzyme may also enable a new reaction mechanism, for example providing a metal ion to join in the transition state and stabilize it.

You can inhibit enzymes by making molecules that "look like" the transition state—the enzyme binds to them very tightly, but they do not go on to form product. The enzyme is therefore "blocked." *See* **Catalytic antibodies** for an application of this idea.

Specific aspects of how enzymes may be improved for a particular application are dealt with elsewhere. In general, modification aims to improve the enzymes specificity (i.e. what reactions it catalyzes), its kinetics (how well it catalyzes them) or its stability (how long it lasts). However, there are several lines of research which are relatively new technologies to use in enzymology:

Chemical modification. Changing an amino acid in the protein into another by chemically reacting it. This usually results in a change in the enzyme's activity, almost always "for the worse," that is it reduces the enzyme's catalytic effect, specificity, or both. Sometimes the change can result in a commercially more useful enzyme.

Site directed mutagenesis. The directed change of an amino acid into another amino acid by genetic modification, to achieve one of the goals above. See separate entry.

See also **Kinetics**.

Enzyme production by fermentation

Industrial enzymes may be made by extraction from a naturally occurring source, often part of a large animal or plant, or by production from a micro-organism in fermentation. The former requires less equipment, but is more prone to interruption of supply by seasonal variation, the vagaries of climate, international trade, and (in extreme cases) war. Fermentation has the potential to provide a more uniform and reliable source of material.

The majority of industrial enzymes are commodity products. A substantial part of the cost of their production is the raw materials and power necessary to produce them. Thus, a successful fermentation process must use low cost feed materials, an organism which does not require excessive heating or cooling, and one which makes a lot of the enzyme.

Typical feedstocks are hydrolyzed starch, molasses, whey, or cereals as a source of carbon, and soy flour, fish meal, blood, and cotton-seed meal for nitrogen. For high value enzymes (for example for use as drugs), some of these are inappropriate as they contain insoluble dirt which will have to be rigorously removed from the final product. As well as the usual control of fermentation conditions, foaming has to be prevented in some enzyme man-ufacturing processes, because some enzymes are denatured on surfaces or may concentrate at them and be washed away. In addition, inducers must be present and repressors removed from the fermentation (*see* **Induction**).

Many industrial enzymes are sold as fairly crude preparations, with a mix-ture of proteins in them. These have been prepared by separating the cells from the fermentation broth, and then partially purifying the protein from the liquor by precipitation, ultrafiltration or similar technique.

Epigenesis

This is inheritance of biological traits outside the normal genetic mec-hanisms of DNA and genes. It is usually applied to biochemical aspects of a cell's metabolism. The first epigenetic traits known at a cellular level were the patterns of cilia (short "tentacles") on the surface of pond-dwelling protozoa, which are passed down from mother to daughter organism even though they are not coded in any DNA: the pattern in one cell is copied onto its two daughter cells.

The best characterized epigenetic phenomenon in higher organisms is DNA methylation. A small fraction of the cytosine bases in DNA are altered by adding a methyl group onto them, using an enzyme called DNA methylase. This affects whether the genes containing those methyl cytosines are active or not. When DNA is made during cell growth, the enzyme has to "decide" whether a particular cytosine should be methylated or not. It does so by methylating all the cytosines in the double helix which are next to another methylated cytosine. Thus, the pattern of methylation is copied from mother to daughter DNA. (This only works because the methylated cytosines are always in CG pairs, called CpG dinucleotides.) Thus, although the pattern of methylation is not "coded" in the DNA, it is nevertheless inherited from one cell to the next. The scientist can use this methylation as a method of finding which genes are active and which are not in a cell. (*See* **Gene control**.)

Another example of epigenesis is the pattern of chromosome activity in women, who turn off one of their X chromosomes early in fetal life: which one their cells chose appears to be random, but all their descendent cells stick with the decision.

Recently, RNAi has risen as a topic in discussions of epigenetic control of the genes. See separate entry.

Essential nutrients

There are a range of nutrients which mammals need in their diet: these are called essential nutrients. As well as minerals, mammals need varying combinations of essential amino acids, fats and vitamins. The particular combination depends on the species.

Essential amino acids—There are 20 amino acids needed to make proteins, but our bodies can only make 12 of them: the other 8 have to come from food. The essential amino acids for man are leucine, isoleucine, valine, threonine, methionine, phenylalanine, tryptophan, and lysine. For children, histidine is also considered to be an essential amino acid, because their ability to make it is rather limited. Food protein with a good supply of all these is called "first class protein." Many single crops are second class protein—they are notable deficient in some essential amino acids. One aim of crop genetic engineering is to boost the supply of specific amino acids in common crops like maize and cassava to make them into sources of first class protein.

Many microorganisms can make all the amino acids, and so can be used to produce them for use as food supplements. Other mammals, such as farm animals and pets, share much the same list, and much of the food supplement amino acids made are for farm animals rather than people.

Fatty acids—People also need particular fatty acids, primarily to make fatty acid esters involved in cellular signalling. These are polyunsaturated fatty acids, fatty acids with several double carbon-carbon bonds in their backbone chains. Babies and very young children also need DHA (docosahexaenoic acid), a major component of the myelin sheaths of brain cells: DHA is produced by fermentation from microalgae by Martek Biosciences and Merck.

Vitamins—The third category of essential food additive is vitamins, which are a wide range of chemicals which we happen to need for our metabolism and not be able to make. Some vitamin requirements are common among the mammals, some are rare—thus vitamin C is only an essential dietary requirement of man, chimp, gorilla, fruit bat and guinea pig, as far as is known. The vitamins sold as "health supplements" are usually produced by fermentation. Thus, vitamin B12 is made by *Pseudomonas denitrificans* in a complicated series of fermentations: over $70 million worth of Vitamin B12 is made this way per year. Vitamin B2 (riboflavin) is produced by *Candida* fermentation, and biotin from *Serratia marcescens*. Quite often these are very specialist strains, selected for their ability to produce large amount of a vitamin that they would usually only make in tiny amounts. *See* **Strain development**.

Essential oils have nothing to do with diet. These are oils which have are the "essence" of a flavour or aroma. Examples are mint oils (from mint leaves, containing a high concentration of menthol), lemon oil (from lemon rind, containing limonene) and so on. While people are quite used to using vitamins that are made synthetically, the term "essential oils" means the oil extracted from a plant, and conspicuously labelled "natural".

Expression compartment (inclusion bodies)

Getting a protein made in a recombinant cell is relatively straightforward, as a wide range of expression vectors exist to be used to clone the relevant gene. However, often the protein is produced in a form which is not useful to the genetic engineer. This is often a feature of where in the cell the protein is made. Common sites of synthesis are:

- *Inclusion bodies*. Inclusion bodies are condensed particles of protein which are formed inside bacteria and (to a lesser extent) eukaryotic cells when the cells are forced to make large amounts of protein. Inclusion bodies are the result of "off pathway" aggregation or folding—proteins failing to follow the usual folding pathway, and forming inactive aggregates.

  Inclusion bodies were the bane of early recombinant DNA production methods, but the skills of manipulating bacterial physiology (i.e. how they grow) to avoid inclusion bodies are now much better developed. In addition, denatured proteins can be refolded by dissolving them in a detergent or "chaotropic agent" solution, and then gradually removing the detergent by dialysis to refold the proper structure. Some companies such as Scil Proteins have specialized in recovering functional protein from inclusion bodies.

- *Periplasmic space*. This is the space between the cell membrane and the outer cell wall in bacteria. Many proteins which are secreted (*see* **"Secretion"**) end up here. This has the advantage that it gets them out of the cytoplasm, but does not release them free into the medium (so they can be harvested simply by collecting the cells). However the periplasmic space has its own set of digestive enzymes which can break down proteins.

- *Secretion.* Some proteins can be exported from the cell entirely, to remain free in solution in the culture medium. This makes it much harder to collect them, but if they can be exported successfully they rarely precipitate as inclusion bodies. There may, however, still be proteases in the solution which can break them down.

- *Cytoplasmic expression.* This is where the protein remains if none of the others apply. Usually only a fairly small amount of protein can be accumulated in the cytoplasm before the cell starts to break it down as an unwanted material.

Expression library

An expression library is a gene library that actually "expresses" the genes that have been cloned (*see* **Expression system**). Usually this means making the proteins from them. It can also mean that those proteins are allowed to perform whatever function we want, usually an enzymatic reaction. There are quite a variety of expression library technologies.

The most commonly used is to make a phage or bacterial library in which the protein is made ("expressed") on the outside of the phage or bug. The library is then screened for the protein by testing each clone with an antibody, or with a reagent that detects the enzyme or other activity that we are looking for. *Phage display* libraries are of this sort—see separate entry. Expression libraries in mammalian cells are usually screened using antibody labelling and/or fluorescence activated cell sorting (FACS).

More complicated is when we make an expression library, and then select the clone we want by some specific function, thus identifying the gene that performs that function. This is useful if we do not know the protein involved, or do not have a reagent to identify it directly. Expression libraries made from the DNA extracted from soil bacteria have been screened in this way for bacteria making new antibiotics. Mammalian cells can be selected for their expression of hormones that enable them to keep growing.

The genes for expression libraries are usually naturally occurring DNA, in which we want to find a gene for further use or study. However it can also be synthetic DNA, in which we want to discover a new gene that performs some useful function. *See* **Darwinian cloning**.

Expression profiling

Expression profiling is the generation of a profile of which genes are making how much of what RNA and protein in a sample at any one time. Usually it is RNA that is measured, as this is easier to do, and so the term "expression profiling" usually refers to RNA profiling—finding how much of all the different RNAs there are in a cell or tissue. It is therefore also called transcript profiling (as RNA is said to be transcribed from DNA—the genes: *see* **Gene structure**). Profiling all the proteins is usually called proteomics.

There are many technologies used for expression profiling. They fall into four classes.

- *Differential display*. This is based on making cDNA copies of genes and then either use them as probes to select RNA from other samples or "display" them on a gel to show differences between samples. See separate entry.

- *Hybridization based methods*. Here an array of DNA probes is used, one for each RNA, and the amount of RNA bound to each is measured. This used to be done with "dot blots" or northern blots (*see* **Blots**), but is now done with gene arrays or gene chips. This is not very accurate, so each experiment must be repeated several times to make sure that the differences seen are not errors. Not that other methods are all that much better, but the crisp graphics of gene chip arrays suggests a level of accuracy that is not realistic.

- *Cloning methods*. You can clone and sequence every cDNA (or EST) in a cell. This is usually done to discover new genes, but also gives information about the number of copies of each mRNA in the cell. However, it is impractical even today to sequence a million cDNAs per experiment.

- *Sequence based methods*. A more practical approach using sequence is to determine the sequence of a very short section of each RNA and use that as a "tag" to identify which RNA it was from a genomic database. Sequential analysis of gene expression (SAGE), a technique commercialized by Genzyme, is the longest established of these techniques, using a combination of restriction enzymes and sequencing to generate a series of short "tag" sequences that you can then use to look up what RNA it came from in a database. Sydney Brenner invented an enormously parallel, bead-based technique, where each RNA is assigned to a microbead and a short sequence reaction done on them under a microscope. In both cases, each sequence corresponds to one RNA molecule in the original sample, so the more times a particular sequence turns up the more common that RNA was in the cell.

Traditionally, the amount of a specific RNA in a cell is determined by a Northern blot (*see* **Blots**). The same information can be gained by comparing databases of sequence information, SAGE data or other sequence-based results: this is often called a digital northern.

An issue with all these techniques is the sensitivity (can it detect the very rare RNAs that might represent control genes), dynamic range (if it can detect one molecule per cell, will it be swamped by another RNA present at

10 000 molecules per cell). The dynamic range is sometimes described in terms of "copies" or "copies per cell," meaning individual RNA molecules per cell. It is not uncommon to have *less* than one copy per cell—this means that, in a sample, only a minority of the cells had any molecules of that RNA in at all.

Applications of expression profiling include:

- *Use in drug target identification and validation.* The idea is to find the differences in gene activity between normal and diseased tissue, and then target those differing genes with a drug (*see* **Targets for drug discovery**). This is related to:

- *Use in functional genomics.* If two genes always increase or decrease their expression together, we might suspect that their function is related. This therefore can be a tool in "functional genomics"–trying to make sense of the function of genetic data.

- *Use in diagnostic marker development.* The same approach as above, but you then use the RNA or the protein as a test for that disease, ideally coming up with one that is more sensitive or more specific than anything existing, or one which can detect the disease earlier. Often the diagnostic is not a single gene, but a combination of the rise and fall of several. This is related to:

- *Use in systems biology.* This is a new trend. Biology is not just a series of isolated genes, but is made up of networks of genes. By identifying how *all* the genes in a cell go up or down, it is hoped that those networks can be identified.

- *Chemical genomics.* Rather than waiting for disease to alter the gene expression profile, we can deliberately probe the cells with a chemical and see what happens to the expression profile of the cell as a whole. This can be both a probe for the biology of the cell as a system, and also a check on what the chemical (maybe a candidate drug or potential toxin) is actually doing. This is also called "chemogenomics."

Expression profiling is usually done on tissue samples or cultured cells. However, it can be performed on individual cells, or even on regions of larger cells. Thus the RNA present in the ends of single nerve cells has been analyzed by sampling their cytoplasm with a very fine micropipette and carrying out the cDNA chemistry in picoliters of liquid. Single cells can be dissected out of a tissue sample using laser capture microdissection (LCM), a method that uses a laser to dissect out small clusters or cells or individual cell on a slide. Usually you need to dissect out a lot of cells from many different samples to get enough material to work with, so this is painstaking and extremely skilful work.

Expression systems

Expression systems are combinations of host cell and vector which provide a "genetic context" which makes the gene function in the host cell. Usually, this means make a protein at high levels. They are therefore a way of making a recombinant protein, either a replica of the natural protein made from a recombinant gene, or a protein that is itself altered.

Normally a cloned gene will be inert as far as the cell is concerned: it will not perform the function that it usually performs when put into the new host cell, because the various signals that tell the cell where the gene is and what to do with it are missing (*see* **Gene control**). To make the gene function, the cloning vector must include appropriate promoter sequences, and sometimes other control elements. It sometimes also needs to have a short region of coding DNA, that allows protein synthesis to start properly. This produces a protein that is mostly the gene we want, but with a few "foreign" amino acids stuck on the front—amino terminus—end (a fusion protein).

The proteins made in such systems are called foreign or "heterologous" proteins, because they are not native to the cell in which they are being made. Thus, this is "heterologous gene expression".

The majority of heterologous expression systems are single cells, usually bacterial or yeast cells although mammalian cells are of increasing importance. Expression in whole organisms, such as animals (*see* **Pharming**) or plants (*see* **Protein production in plants**) is also possible.

Because making lots of foreign proteins is often lethal to the host cell, there are several variations on the theme of the expression vector which allow the level of the protein made from a cloned gene to be increased. In general, these allow the cells to grow with little of the protein in, and then expression of the gene is turned on, either directly (by inducing the gene activity with a suitable chemical) or by amplifying the number of copies of the expression vector using some genetic tricks.

Secretion vectors are vectors which allow the protein product of the cloned gene to be secreted from the cell. This can be very helpful for purification, as all the other proteins in the host cell are removed with the cell itself, but does not always work because the target protein is broken down in solution, is not stable, or is incapable of being secreted effectively. *See* separate entry on **Secretion**.

Obtaining the best performance from an expression system requires considerable knowledge of how the host cell's internal machinery (its physiology) works. But there are no general rules about which expression system is best to make which protein. Thus there are a variety of systems available among microorganisms, *E. coli, Bacillus subtilis* and "yeast" (*Saccharomyces cereviseae* and *Saccharomyces carlbergensis*) are commonly used. Others include *Hansensula polymorpha* (a yeast, commercialized by Rhein Biotech) and *Aspergillus* (a mold).

See also **Expression library, Expression compartment.**

Extremophiles

These are organisms which live in "extreme" conditions, that is conditions substantially different from Western laboratories. They include

- *Thermophiles*—organisms that live at high temperatures. See separate entry.

- Psychrophiles—organisms that live at very cold temperatures. Despite living at very cold temperatures, they move, grow and reproduce at rates similar to warmer organisms, so enzymes must be as efficient at these low temperatures as ours are at 37 °C. They can be useful sources of enzymes that are active under cold conditions: most enzymes act only very slowly at temperatures much below the normal living temperature of the organism they came from. They can also be sources of heat-unstable enzymes, enzymes which are destroyed by moderate heating to 40 °C. This can be useful if you want to "turn off" an enzyme without affecting other components of a complicated process.

  Another interesting target for the biotechnology of psychrophiles is fish that can live in arctic seas at temperatures which should freeze them solid. They are sources of "antifreeze proteins," proteins which prevent their blood freezing.

- *Halophiles*—organisms that can live in concentrated salt solutions. Apart from their basic scientific interest, halophilic bacteria are interesting because of their metabolic curiosities, including the purple membrane protein of *Halobacterium halobium* (*see* **Molecular computing**). Biotechnology is also interested in engineering plants to be halophiles, so that they can be watered with sea water.

- *Piezophiles (also known as barophiles)*. Organisms that thrive under high pressure. In fact, piezophiles are an extreme response to pressure, the other being piezosensitive organisms that cannot stand high pressures. Piezophiles usually come from deep in the sea: the pressure at, for example, 3 800 m depth (the average depth of the ocean floor) is 38 Megapascals (MPa), or 375 times normal atmospheric pressure, and organisms must adapt to this.

- *Acidophiles and alkophiles*. Organisms that can live in extremely acid or alkaline conditions respectively. Acidophiles particularly are of industrial importance in microbial mining (*see* **Leaching**).

Enzymes from extremophiles are potentially valuable in their own right because of their stability in industrial processes, many of which are run at high pressure or temperatures to make chemical reactions proceed faster. They also have value for what they teach us about enzyme stability, and hence how we might make other enzymes more stable. Most normal proteins denature at extreme pressures (greater than 300 MPa), so, like the enzymes from thermophiles, those from piezophiles must be unusually stable. Such enzymes from extremophiles are sometimes called "extremozymes."

FACS

FACS (pronounced "fax") stands for fluorescence activated cell sorter, and is an instrument that can measure the fluorescence on individual cells, and then sort them into different sample tubes according to the result. Cells are suspended in individual droplets, squirted past a laser that excites any fluorescent dye in them, past a detector that measures the fluorescence, and then into one of several receivers into which they are steered using electric fields. The detectors can detect one or two different colors of fluorescence (the latter being called two-color FACS). Usually one of the colors is provided by a fluorescently labelled antibody. Sometimes the other is a DNA-binding dye, to measure where the cells are in the cell cycle.

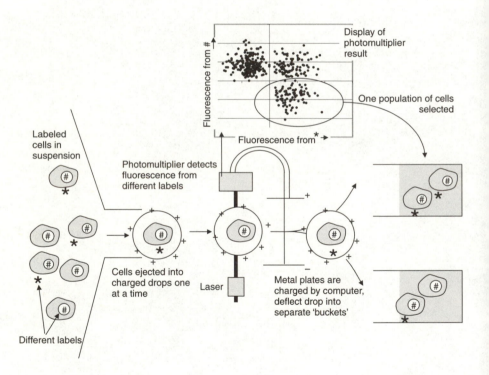

FACS machines are widely used in immunology, where they are used to identify and select populations of cells bearing specific combinations of cell surface markers. They can also be used to identify cells that are expressing specific proteins that we have cloned into them, or functions that we have added to those cells (such as an enzyme activity). Thus, it can be used as a selection technique for an expression library.

Fermentation

A central discipline of traditional biotechnology, and a key part of all biotechnology today, is fermentation. This is the growth of microorganisms, and encompasses a wide range of technologies which are dealt with in different entries in this book. Fermentation includes:

- *Bioreactor design*—the design of the container in which fermentation is to take place.

- *Substrates*—what the microorganisms are to grow on.

- *Growth rates and conditions*—under which the organism will grow and produce what we want (which are often mutually exclusive).

- *Supports*—whether the organism is to be on solid supports or in suspension.

- *Fermentation processes*—how you arrange to do the fermentation.

There is also a wide range of options in the **downstream processing** (see separate entry).

Fermentation processes

Strictly speaking, fermentation is microorganism metabolism under anerobic conditions on a carbon substrate. However, it has been extended to mean growing microbes in liquid under any conditions.

There are three general ways in which fermentations are done, each with a variety of associated terms. In all cases, there are some common terms in bacterial growth, such as the bacterial doubling time (the time needed to double the number of bacteria there): these are discussed further in the entry on cell growth.

Common terms. For all bioreactor processes, the first thing that happens is that the bioreactor is sterilized. This can be done with steam, chemicals, washing or some combination of them. Often fermentors have a "disinfection cycle," which is an automatic series of operations that cleans and disinfects the machine ready for the next use. The fermentation is then started with an inoculum, a small, actively growing sample of the organism to be cultured. Fermentation then proceeds according to one of the schemes below.

- *Batch fermentation*. Here the reactor is filled with a sterile nutrient substrate and inoculated with the microorganism. The culture is allowed to grow until no more of the product is being made, when the reactor is "harvested" and cleaned out for another run. The culture goes through lag, log, stationary, and death phases (*see* **Cell growth**). Depending on what the product is, the "useful" part of the growth cycle can be any one of these four stages, although it is usually the growth or stationary phases.

- *Fed batch fermentation.* Here the batch culture is fed a batch of nutrients before it gets to the stationary phase, so that it never runs out of nutrients. At the same time some of the fermentation is removed and taken off for processing, so the microorganisms do not run out of space or poison themselves.

- *Continuous culture.* This is the logical extension of Fed Batch fermentation. The fermentor is fed continuously with nutrient and the culture medium removed continuously. This has some advantages over Fed Batch systems in that the culture conditions are always the same, but also is harder to control. This is essentially a large-scale chemostat.

- *Cascade fermentation.* Here the fermenting "liquor" is passed through a series of ferments, so that more and more of the product builds up each time. Each step can then be optimised for a specific condition. A typical example would be in brewing, where the beer would be fermented in several stages to increase the alcohol content, each stage using a yeast adapted to working in alcohol of that concentration. At the end of each stage the yeast is separated from the beer and used again in that stage, while the beer goes on to the next one.

Fermentations may also be classified according to when the product is made:

- Type I fermentation—product is made from primary metabolism.

- Type II fermentation—product is made from secondary metabolism at the same time as primary metabolism is going on (i.e. when the cells are growing).

- Type III—product is made by secondary metabolism at a different time from the primary metabolism (i.e. during stationary or death phases of the culture).

Lastly, fermentations may be classified according to how the culture is kept "clean."

- *Aseptic/sterile fermentation*—all other organisms are excluded by the biotechnologist. This is by far the most common case.

- *Consortium fermentations.* Here a group of organisms are growing together, rather than just one organism. For this to work, the organisms must be dependent on one another, as otherwise one will outgrow the others and dominate the culture.

- *Protected fermentations.* Here the culture is not aseptic, but is performed under conditions under which only one type of organism will grow. Thus, fermentations at extremely high temperatures, extremes of pH or with very hard-to-metabolise substrates will tend only to be able to support the organism the biotechnologist is after, thus removing the problem of having to keep contaminants out.

Control is critical for all fermentations: usually, basic measurement of pH, CO_2, O_2 gives a guide to how the fermentation is proceeding. Other specific substrates or products may be measured "on-line" as well, as well as samples being taken off for analysis at intervals using more complex tests. *See* Process control.

Fermentation substrates

Many materials are used as food for growing microorganisms. These are referred to as the substrate. The substrate and the trace materials needed, together with chemicals added to make the fermentation easier (such as anti-foam agents to stop froth forming) make the culture medium. (Similarly, when mammalian cells are grown outside the body, the liquid that they are fed with is called the culture medium, although it has a very different composition from what you would grow bacteria in.)

The substrates can be divided into those providing each of the different essentials for life: a source of carbon, nitrogen, and (in the case of aerobic fermentation) oxygen. Usually carbon substrates cost the most, because you need most of them. Among common carbon substrates are:

1. *Molasses*. A side product of sugar refining that contains most of the material from sugar beet or sugar cane which is not sugar, molasses is one of the cheapest substrates available. There are several varieties with different properties, components, and costs:
   - Beet molasses—the molasses left from processing beet.
   - Black strap molasses—from an early stage in cane sugar processing.
   - Cane refinery molasses—from a later (more sugar rich) stage in processing.
   - High test molasses—actually, concentrated cane juice treated with invertase.
2. *Malt extract*. Made from malted barley by soaking it in water. As in beer brewing, this partly breaks down the starch in the barley to make it easier for microorganisms to ferment.
3. *Starch and dextrins*. Polysaccharides often made from cheap crops like potatoes.
4. *Cellulose*. The world produces about 100 billion tonnes of cellulose a year, so it is a potential raw material for large-scale fermentation. But only a few organisms can degrade it.
5. *Sulphite liquor*. A by-product of paper pulp production, which contains much of the fermentable sugars from wood without substantial cellulose.

6. *Whey*. A side-product of dairy processes, it is cheap but expensive to store or transport.
7. *Methanol*. A very cheap chemical from the oil industry, but only a restricted range of organisms can grow on methanol. Similarly ethanol ("alcohol") can be used, but is more usually the product than the substrate.
8. *Oil, gas*. Some organisms can use natural gas or some components of crude or refined oil as carbon substrates. However, their commercial use depends critically on the price of oil. (*See* **SCP**.)

Nitrogen substrates include:

1. *Ammonia*. A very smelly gas produced as a bulk commodity for the chemical industry. Most organisms can use ammonia. Sometimes it is converted into ammonium salts or into urea for ease of handling.
2. *Corn steep liquor*. The liquid generated in the early stages of wet-milling maize to produce starch. Maize grains are submerged in water to soften them before milling. Low molecular weight sugars and peptides accumulate in the water.
3. *Soy protein*. The protein left over when you have taken the oil out of soybeans.
4. *Yeast extracts*. Made from waste yeast from industrial fermentations, they have everything necessary for microbial growth.
5. *Peptones, casein hydrolysates*. These are partially digested meat or milk proteins respectively. The proteins used are usually waste material from the food industry—nevertheless, this can still be an expensive source of nitrogen.

Fingerprinting

In biotechnological terms, fingerprinting means making a characteristic chemical profile of something, to identify it. There are many variants.

- *DNA fingerprinting*. Generating a profile of the DNA variants of an individual. See separate entry. Note that DNA fingerprinting can be applied to animals, plants or bacteria as well as its widely known application to potential rapists and murderers.

- *Protein fingerprinting*. This generates a pattern of the proteins in a cell or organism, which provide a completely characteristic "fingerprint" of that cell at that time. The technique most usually used is two-dimensional gel electrophoresis. This gives a complicated pattern of spots, which can be compared between cells, although the pattern you get will depend on the cell's metabolic state as well as its origins. This is in fact a different name for proteomics.

- *Peptide fingerprinting*. This can be one of two things. The first is to use peptides (i.e. the degradation products of proteins) to characterize a cell or organism, as above. The second is to characterize a single protein using its component peptides. A protein is digested into short peptides using a protease enzyme: the pattern of sizes of the peptides is characteristic of the protein. Then you can search all the proteins in a genomic database and work out which protein they must have come from, without having to do any protein sequencing.

- *Chemical fingerprinting*. Here the low molecular weight chemicals in a cell are analyzed, usually by GC or HPLC but increasingly by mass spectrometry, and the resulting pattern used to identify a cell or microorganism. As some metabolites are very characteristic of groups of organisms (penicillins of some moulds, different chlorophyll types of different algae for example), this can be a useful way of telling which organism is which.

Footprinting is a distantly related idea. This is primarily applied to DNA, although the same logic can be applied to other molecules. It is a method of finding how two molecules stick together. In case of DNA, a protein is bound to a labelled piece of DNA, and then the DNA is broken down, by enzymes or by chemical attack. This produces a "ladder" of fragments of all sizes. Where the DNA is protected by the bound protein it degraded less, and so the "ladder" appears fainter. Footprinting is a common technique for homing in on where the proteins which regulate gene activity actually bind to the DNA.

Fish farming

Fish farming is raising fish in tanks or holding nets, as opposed to hunting them at sea. Biotechnology techniques are developing more effective technologies to improve yield, and particularly to improve the fish's genetics. The activity falls into three areas:

Sex manipulation. Fish's sex can be manipulated quite readily by exposing them to hormones—many fish will produce genetically male but physically female fish if they are exposed to estrogen during development. This makes it much easier to develop inbred-lines of fish. The genetically male females produce fertile eggs, some of which contain Y chromosomes instead of X ones—when fertilised by Y-chromosome sperm, they produce "supermale" YY fish. (This has been done for tilapia, a popular fish in Asian fish farms, and the channel catfish in the USA.) When crossed with normal females the YY males produce only male fish, which grow faster and put more metabolic energy into meat (as opposed to eggs).

163

Transgenic fish. Fish cloning is well established, so a single "desirable" fish can be cloned, some of its descendants sex-reversed, and a whole new strain developed. Generating transgenic fish is relatively easy (compared to transgenic mammals). However, very few fish genes have been identified that carry "desirable" traits, so this is research-stage work. A/F Protein's transgenic salmon containing a growth hormone gene that gets it to adult size in half the time (18 months rather than 3–4 years), 80% of the feed of normal salmon. The final adult size of these salmon is no larger than normal, although others may be.

There is concern about what these salmon would do to wild salmon stock if they got out. Conceivably, they would compete for resources successfully with wild salmon (being larger then their wild cousins of the same age), and also possibly be more attractive to mates. Aqua Bounty Farms in US produce growth hormone transgenic salmon, and have not been allowed to sell them in the United States (as of mid 2002) because of this concern, although they will only sell sterile females. This is more of a concern for fish farming than for animal farming because the fish are kept in cages in the sea near their natural habitat.

Vaccines and diagnostics. In fish farms, controlling diseases is a substantial problem. There is a growing need for diagnostics to tell which sort of fish disease is present, and vaccines to protect the fish from them. These are most easily derived through cloning the pathogens involved.

Food. Fish food is a complicated mix of nutrients that must not only be formulated for the dietary needs of the fish (which vary as the fish grow, sometimes quite dramatically) but which must also sink at the right rate and not dissolve in water. Several dietary ingredients are made using biotechnology, including essential amino acids, proteins, and other trace nutrients. Biotechnology also makes the pink pigment astaxanthin, which can be extracted from crab shells or from algae, and is included in salmon and trout feed to make the flesh pink. (The same material is fed to flamingos for the same reason.)

Flavor chemicals

Flavor and fragrance chemicals are usually complex mixtures of very complicated molecules, and so very difficult to reproduce with conventional chemistry. They arrive in food either as part of the ingredients or as chemicals created during cooking. Biotechnology can help both processes. (Most flavor chemicals are actually fragrances—we taste most of our food with our noses.)

Flavor chemicals in ingredients are usually secondary metabolites from plants (especially spice plants) or bulk ingredients like salt and sugar. Often, plants are unreliable industrial sources of chemicals, so biotechnology has sought to grow their tissues in vitro to make chemicals. *See* separate entries on **plant cell culture** and **hairy root culture**.

Enzymes are used extensively to help to mature the flavor ingredients of food during processing. Examples of the use of enzymes in flavor development include:

- enzymes to generate artificial cheese flavor by simulating what happens in maturing cheese, and using endopeptidases (proteases) to break down bitter-tasting peptides in low-fat foods (fat usually dissolves these hydrophobic peptides and so disguises their taste)

- using a variety of oxidases to generate synthetic flavors to replace products such as cheese

- making synthetic peptides which mimic the flavor precursors in chocolate and cocoa, so allowing cheaper cocoa to be used in high-quality products.

The same technology could also be used to make fragrance chemicals for the cosmetics industry. Biotechnological methods are often needed because fragrance chemicals are often complex, chiral molecules: the characteristic scents of oranges and lemons, for example, are caused by the different stereoisomers of the same compound—limonene. However, this application has not taken off. Cheaper fragrances can be made successfully from conventional chemicals. Top-of-the-range perfumes are still made by traditional methods, and the perfumes gain an appeal through their use of "natural" and "traditional" recipes of this sort.

Fluidized bed

Many types of bioreactor use particles to support the enzyme molecules or cells. These particles form a "bed." There are a range of ways of making sure that the reactor contents flow smoothly and uniformly past such a bed.

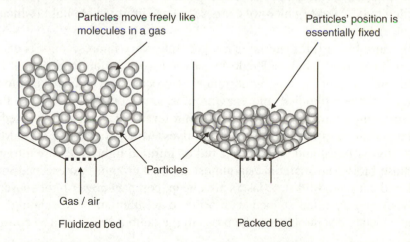

Particles move freely like molecules in a gas

Particles' position is essentially fixed

Particles

Gas / air

Fluidized bed

Packed bed

Fluidized beds. Here the flow of material through a bed of particles is enough to shake them loose from their neighbours, but not enough to lift them. The bed of particles then behaves like a fluid, not a solid, and mixes very well with the material flowing through it. This technology is more widely used in the industrial processing of gases (such as burning coal dust in power stations), but can be used with liquids.

Entrained bed reactors. Here the flow is higher, and the particles are kept in suspension by the fluid flow. Some cell culture systems use variants on this approach, with the cells growing on the surface of polymer particles that are kept suspended in a flow of growth medium. This is not the same as keeping the particles in suspension by stirring.

Packed bed reactors. Here the particles are packed into a bed, and remain essentially solid. This is the most common type of bed reactor, and is widely used in chromatography—a chromatographic column is in fact a columnar packed bed reactor. Fluid can flow unevenly through such a bed of solid particles, and so variants have been developed to encourage more even flow, such as the tapered bed reactor (with a conical base filled with particulate solids) and the disc bed reactor, where a disc of solid material has liquid fed in from the centre and flows out for collection at the outside.

Fluidized bed reactors have been applied to beds of particles with biofilms on the particles and with enzymes chemically linked to their surfaces, as well as particles of conventional catalyst particles in a wide variety of chemistry.

Fluorescence

Fluorescence is a very widely used way of detecting things in biology and biotechnology. A fluorescent chemical absorbs light of one wavelength ("exciting" light) and re-emits it at once at another wavelength ("emitted" light). (A rarer phenomenon, phosphorescence, is when it re-emits it again later after the exciting light has been turned off.) Traditional flurophores are excited by UV light (which we cannot see) and emit visible light (which we can see).

Because biological material is not generally very fluorescent, it is easy to see a fluorescent label molecule introduced to it. Fluorescent labels have been used to label a very wide range of biochemical reagents, including DNA probes, antibodies, and enzyme products. Common labels are fluorescein (which glows greenish-yellow) and texas red. Increasingly used are red and near infrared (NIR) fluorescent dyes, which are excited by red light (e.g. from a laser) and emit deeper red or infrared light. They are attractive because biological material has almost no background fluorescence in the red, and because solid state lasers are cheap, compact power light sources.

DNA is very often stained with a UV dye, often an "intercalating" dye (one which slides in-between the bases in the double helix) such as ethidium bromide. When it is bound to DNA in this way, it glows much more brightly than on its own. If an agarose gel is soaked in a dilute ethidium bromide solution, the DNA shows up under UV light as bright bands on a dark background.

There are a range of techniques used in fluorescence, which enhance its power. Two of these—fluorescence transfer and time resolved fluorescence methods, are discussed in the entry on FRET. Another is the use of confocal microscopy to home in on the fluorescence being emitted from a particular part of a cell (*see* **Microscopy**). Machines to measure fluorescence, such as the molecular dynamics fluorescence imaging plate reader, [abbreviated to FLIPR ("flipper")], are common pieces of laboratory equipment.

An increasingly popular fluorescent label is green fluorescent protein (GFP), a protein which has a very strong green fluorescence. As it is a protein, it can be made inside cells, and so you can engineer cells to make the GFP as a marker of gene activity. GFPs have been engineered to emit a range of colors between yellow and blue-green, so in principle you could have two or even three GFPs distinguishable within one cell. Other proteins with altered colours and faster rates of action are being investigated, such as red fluorescent proteins from corals.

Food

Biotechnology can be used to enhance existing foods, improve the manufacturing of standard products (*see* **food processing**), or generate new ones. In the former category, biotechnology has been used to remove lactose from milk (for lactose-intolerant individuals), make rennin-free cheese (for vegetarians and people allergic to rennin). A very wide range of science and technol-ogy is also used more mundanely to monitor, control and improve quality in all food products, and to understand how food can be kept fresh and palatable for the consumer. Among the other enhancements to existing products are:

Genetically engineered plants. So far, with the exception of the Flav Savr tomato, these have been engineered to help the farmer, not the cook, but other applications such as plants with altered oil compositions are in development or in the field. *See* **Agricultural biotechnology**.

Food components. Biotechnology produces a wide range of chemical components of food, such as vitamins, some coloring agents, modified starches, fats and lipids, HFCS, etc. The most extreme example of this is SCP, producing bulk protein for food use using fermentation. Biotechnology can also contribute to making **probiotics** (see separate entry).

There are few completely biotechnological foods. An example is Quorn, a fungal protein product made from the fungus *Fusarium graminearum*, grown by fermentation and subsequently processed to look and "feel" rather like meat.

Flavor is added using synthetic flavor chemicals. Other organisms, including algae, are also grown as food products for processing in a similar way, but not as widely in the West. (In Japan, seaweed is a traditional staple food anyway.) In fact, many traditional foodstuffs such as Tofu, cheese, soy sauce, and all alcoholic drinks are made by fermentation processes, but these have evolved over hundreds of years, rather than being the result of biological knowledge, and so are not really biotechnology.

"Food" in this regard can have a pretty wide meaning: European regulation says that food is anything that is regularly consumed, including chewing gum and water.

Food processing using enzymes

One of the major users of enzymes is in the food industry. Enzymes are used to control food texture, flavor, appearance, and to a certain extent, nutritional value. This use requires fairly pure enzymes (as they, or their cooked remains, are going to be eaten), and enzymes that are quite specific, so that all the other complex components that make up food are not affected.

As with pharmaceutical products, in the US the FDA provides a rigorous regulatory gateway to using new enzymes in food, especially genetically engineered enzymes, and approval for a food material in the United States is generally taken as being a clear signal to European authorities that the new ingredient is safe. A much wider range of "novel" food ingredients is approved for use in the Far East, including Japan, than is found in food in the "West."

Major applications are:

- Amylases are used to break down complex polysaccharides, which form viscous solutions or solid gels and do not have much flavor, to simpler sugars which form more fluid solutions and taste sweet. The same enzymes can be used to modify the properties of polysaccharides so that they have the gelling properties that the food technologist wants.

- Proteases are used widely, especially to tenderize meat proteins by using collagenase to break down collagen, the major protein in connective tissue such as "gristle" in meat. This allows cheaper meat to be made more edible (although it does not make it look or "feel" like steak, so you cannot usually use collagenase to "upgrade" meat.) Proteases are also used in to clarify beers and condition dough for bread making.

One of the most widely used proteases is Rennin (chymosin). Genetically engineered Rennin was the first enzyme produced by recombinant DNA to be approved for food use: it was cloned by Collaborative Research and marketed by Dow Chemicals.

- Pectinases for fruit processing, both in producing jams and preserves and also to digest pith between orange or grapefruit segments leaving segments themselves intact (for canning etc). (The same enzyme can be used to remove the bark from trees.)

- Flavor modification. *See* **Flavor chemicals**.

These enzymes are often added to food during processing, so the amount of enzyme added and the stage of the process at which it acts can be controlled. These are "exogenous" enzymes. Food also contains endogenous enzymes, enzymes which are naturally present in the food materials. These are also responsible for the changes in texture, flavor, and appearance of the food as it is processed, but are harder to control. Thus, allinase helps to develop the characteristic odor of onions, but also can create a bitter flavor in the same food.

Formulation

Formulation in pharmaceutical terms is how you make a drug into a medicine. Most drugs are given in small amounts and taste horrible. So they are made into something that is easier to handle, has an exact dose in every portion, and in which the taste of the drug is masked. This process is called formulation. It can be formulated into a tablet, a capsule, a cream, a liquid solution that can be drunk or injected, an aerosol spray (as for asthma drugs), or more complicated systems. How this is done is the science of pharmaceutics.

Formulation has to combine the drug ("active pharmaceutical ingredient"—API) with other materials to make the medicine. These are called excipients (to provide some function, like mask the drugs taste or dissolve fast), and bulking agents (that just make the tablet big enough to handle easily). Tablets then usually also have lubricants to make it possible to get them out of the tablet press, and a coating to make them easier to swallow, provide a surface to print the company's name on and so on. "Syrups" also have flavor agents added.

Some new technologies for helping formulation and drug delivery are:

- *Cyclodextrins*. These are cyclic carbohydrate molecules that form a pocket in the centre. Very hydrophobic drugs can sometimes be made more soluble by complexing them with cyclodextrins.

- Nanoparticle technology, where drugs are ground up into extremely small particles so that they dissolve more readily because of a greatly increased surface area (1 g of sugar ground to particles 40 nm in diameter would have a surface area of about 100 m).

- Ion exchange resins to mask taste. Many drugs have very bitter tastes, which are inherent to the drug and so you cannot remove the bitter ingredient.

However, ion exchange resins will bind the drug so that it does not touch the surface of the tongue, but is carried into the stomach before it is released to have its effect.

Coupled to formulation is the issue of stability. A tablet has to be able to be kept on someone's shelf for at least months, usually years, be able to be accidentally heated up in the sun or frozen, and still be safe and active. So all formulations have to undergo stability testing before they can be used, a process that usually takes months. In some cases this is "accelerated testing," where the tablet is heated up or exposed to high pressure water vapour to simulate what will happen at lower temperatures over longer times. But this is not always possible.

Formulation is a particular problem for proteins, as they have to be stabilized so that they can be stored and transported, and then dissolved prior to their use ("reconstituted," usually in a saline solution). Proteins drugs are always injected—the ability to deliver protein drugs as a pill is one of the "holy grail's" of drug delivery research (*see* **Protein drug delivery**).

Freeze Drying

This is a common technique, also called lyophilization, for preserving biomolecules and microorganisms. The sample is frozen, often in a solution containing another material such as lactose or trehalose (an excipient) which acts to stabilize it (and is called the excipient). It is then put into a chamber attached to a vacuum pump and, while the sample is still frozen, the chamber is evacuated. The ice sublimes under vacuum (i.e. turns directly into vapor without melting), and the water vapor is removed. After a while all the water in the sample has been removed, and what is left is a dry powder or pellet of material.

Commercial freeze-drying apparatus can control the temperature and pressure of the vacuum chamber very accurately, and can heat up the samples to be freeze-dried during later stages to drive off the last remaining water. However, simply connecting a frozen sample up to a vacuum pump often suffices for research freeze-drying applications.

Freeze-drying is the standard way of preserving microorganisms for long periods of time. It is also a favourite way of formulating biopharmaceuticals, as these protein drugs are often not very stable in watery solution. A good freeze-dried preparation of a protein is a very light fluffy material which, when water or buffer is added, dissolves almost instantly.

Freeze-drying can also be applied to mammalian cells if they are well enough protected to withstand the disruption. Trehalose as an excipient is unusually good at this, and is a natural way that some plants can be dried completely and still "come back to life" in a matter of hours when wetted.

Even (dead) pets have been preserved in (where else?) the United States by freeze-drying, a sort of high-tech mummification.

FRET

FRET stands for fluorescence resonant energy transfer. This is a fluorescence technique that allows you to tell when one molecule is near to another.

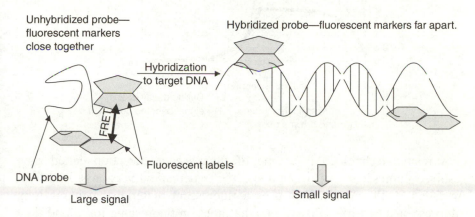

Unhybridized probe— fluorescent markers close together

Hybridized probe—fluorescent markers far apart.

Hybridization to target DNA

FRET

DNA probe

Fluorescent labels

Large signal

Small signal

A molecule that fluoresces does so in two stages—it absorbs light (often UV light is used), to become "excited," and then loses the energy as light of a longer wavelength. However, it can also lose its excess energy by transferring it to another molecule if that molecule is close enough. Then that other molecule can get rid of the energy as heat, or as light. If it is heat, then the fluorescence is said to be "quenched"—light output that would have happened before now does not happen. If the other molecule gives out light (almost always at a different wavelength from the fluorescence emission of the first molecule), then the fluorescence is "transferred" to that second molecule.

Both can be used to see when two molecules have been brought very close together: in the first case, the light output stops, in the second it changes color. The molecules need to be within a few nanometers for this transfer to work.

There are several other technologies that detect whether two molecules are close together, which rely on something being given out by one label, and absorbed by another to generate light. Scintillation proximity assay (SPA) generates alpha particles, which can only travel a short way through water before being absorbed. AlphaScreen generates singlet oxygen (an excited state of the oxygen molecule) on the surface of a bead, which can only travel a few nanometers in water before it reacts with it and becomes inert again. Both of these are used in assays that test whether one molecule has bound to another.

171

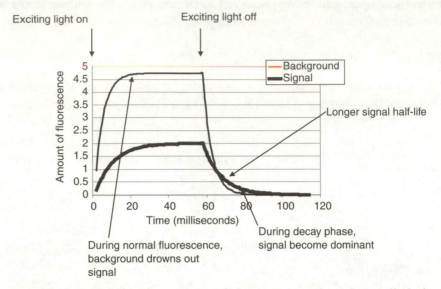

Exciting light on   Exciting light off

Longer signal half-life

During normal fluorescence,
background drowns out
signal

During decay phase,
signal become dominant

A related technology is a type of fluorescence detection called time resolved fluorescence, of which a specific trademarked version is DELFIA, marketed by Pharmacia. This uses a fluorescent material that has a long "fluorescence half-life," that is, one that goes on fluorescing for a long time after you have turned off the exciting light source. Then you look at the fluorescence after the exciting light has been turned off, to distinguish very faint fluorescence from any other light around.

Functional genomics

The "functional genomics" movement aims to add high-throughput biology to the genomics programme of high-throughput gene identification and sequencing, to identify what all these genes do. In pharmaceutical discovery, this is linked to "target validation"—finding out whether a gene is a good drug discovery target (*see* **Targets for drug discovery**).

The approach is to link gene identification and sequencing with a tool or set of tools which give clues about the gene's function. The tools used include:

- *Proteome technology*. Looking at all the proteins produced in a cell or tissue. See separate entry.

- *Animal genetics*. Cloning genes from animals which have known effects, and then seeing if related genes can be identified in the human genome. The mouse is the favoured organism here, although Zebrafish, *Drosophila*, *Caenorhabditis*, yeast, and other organisms can also provide insights.

- *Knockout mutants*. If you do not know what a gene is doing, knock it out in an animal (again, usually a mouse) and see what happens. Increasingly, the answers are very unexpected. Related to this is "knock-in"—making a transgenic organism with an added gene to see what it does. See separate entry.

- *Cell biology*. As a last resort, scientists are going back to cell biology and biochemistry and dissecting what the genes are doing one at a time in the lab. This is a slow process—it can take a team of 20 scientists a decade to find what a new gene may be doing in even one cell type, and there are 100 cell types and 30 000 genes in man.

Fusion biopharmaceuticals

Several biopharmaceutical proteins have been developed which are fusion proteins—that is, they are the product of two genes which have been fused together so that the proteins that they code for are joined end-to-end (*see* **Fusion protein**). The advantages of such proteins as drugs can be:

- They have two complementary or synergistic activities in one molecule. Thus, when the molecule binds to a cell, it does two things at once. To get the same effect with two molecules could need much more of both of them, to increase the chance that both would bind at once to one cell.

- The adverse effects or poor stability of one molecule is offset by the properties of the other.

- One molecule acts as a "targeting" mechanism to bring the other to the site where it is meant to act (*see* **Immunotoxins**).

  Examples of such fusion peptides are the CD4–lgG combined molecule which Genentech has developed as a potential AIDS treatment, and the Immunex GM–CSF–IL-3 fusion. The CD4–IgG blocks binding of the AIDS virus to cells, and is much more stable in the blood than the free CD4 molecule itself. GM–CSF and IL3 have synergistic effects at stimulating bone marrow to produce white blood cells, so linking the two together produces a potentially more powerful compound than the two molecules separately. Immunotoxins can also be fusion pharmaceuticals.

Fusion protein

A fusion protein is a protein in which part of the amino acids come from one protein sequence and part from another. "Biotechnology" is a fusion word, with the "bio" of "biology" fused onto "technology". Fusion proteins are produced by splicing the gene for one protein next to or into the gene for another: the genetic apparatus reads the gene fusion as a single gene, and so produces a fusion protein.

FUSION PROTEIN

Fusion proteins are used in a number of biotechnological applications.

- To add an **affinity tag** to a protein (see separate entry).

- To produce a peptide as part of a larger protein, which is then cut up after it has been made by cloning (*see* **Peptide synthesis**).

- To produce a protein with combined characteristics of two natural proteins (i.e. a chimeric antibody).

- To produce a protein where two different activities are physically linked (e.g. enzymes for substrate channelling or as a fusion biopharmaceutical).

In practice many proteins are expressed as fusion proteins during research. It is easier to splice the gene for a potentially interesting protein into the middle of another gene than get it positioned exactly correctly behind a promoter sequence so as to express it as a protein with no additional amino acids.

Gas transfer

One of the most important characteristics of tank fermentation systems (*see* **Tank bioreactors**) is the rate at which gas can be transferred from gas (the gas phase) into solution. Often, the rate at which the organisms in the fermentor can metabolize is limited by how fast they can be provided with oxygen or have carbon dioxide, ammonia, or other "waste" gases removed. Oxygen is poorly soluble in water, and so the liquid itself holds very little, which the organisms in a dense culture can use up in a few seconds. Thus, they must be constantly supplied with oxygen gas, either as pure oxygen (efficient but expensive) or as air. Many fermentor design features are aimed at optimizing this transfer rate.

Smaller bubbles of gas have a larger surface area per unit volume and greater internal pressure than large bubbles, and so gas diffuses out of them faster. Therefore, the smaller the bubbles you can make, the faster oxygen diffusion will occur. However, creating small bubbles requires power, may cause disruption of the organism growing in the liquid, and may fill the reactor vessel up with viscous foam. Anti-foaming agents can help this latter problem (which is also a problem when the organisms produce a lot of carbon dioxide gas).

There are wide variety of methods for making small bubbles, keeping them suspended in the fermentation broth, and encouraging their content gasses to dissolve. Among them are the deep jet fermentor, which injects the gas very fast into the bulk of the liquid, and the pressure cycle reactor. In the latter, both gas and liqor are recycled around the reactor loop at high hydrostatic pressure. At the base of the reactor the pressure is much higher, encouraging the gas to dissolve. The constant recycling ensures that bubbles stay in the liquid phase, and do not separate as froth.

Other methods ensuring good gas transfer rely on increasing the surface of liquid in contact with gas. Bubbling the gas through the liquid spreads the gas out—other methods spread the liquid out, for example, in a thin sheet (in an oxidation pond), or in a thin permeable tube, as in a hollow fiber bioreactor.

Gene

A gene is a section of DNA that codes for a defined biochemical function, usually the production of a protein. Because the information in DNA is "coded" in the sequence of bases (*see* **DNA structure**), a gene can therefore be thought of as a stretch of DNA base sequence.

Molecular biologists use base and base-pair fairly indiscriminately to mean the length of a piece of DNA or RNA, as RNA is copied off DNA base-for-base in the process of transcription. Thus, it is often said that such-and-such a gene is so many bases (or kilobases or megabases) long.

Many organisms have two copies of each gene—they are called diploid. Animals, and higher plants, are diploid. Organisms with only one chromosome set per cell are called haploid, and include yeasts and bacteria. Some species have many gene sets: these are called triploid (for three), tetraploid (for four), and so on, or more generally "polyploid." Polyploid organisms are usually sterile: triploid fish, for example, are deliberately created for use where you do not want them to breed. This has implications for genetics, but does not affect how we understand genes' structure very much.

Separate entries in this book deal with:

- Genetic code—how genes code for proteins, and also what function different bits of genes have

- Gene control—how genes are controlled

- Polymorphism—how genes vary between people, and what effects that has.

Gene chip

A gene chip is a type of DNA array, that is built into a very small area in a "biochip" type device. The term actually covers many different types of device, but they all have many DNA probes packaged into a small device of the same general size as a large semiconductor "chip." "Gene array" and "DNA array" are effectively synonyms.

Gene chips have many DNA molecules arrayed in a small space. There are two approaches to achieving this: pre-make your DNA and put it on a surface, or synthesize the DNA *in situ*. Affymetrix dominates the commercial DNA chip field with a synthesize-*in-situ* technology using light to target the DNA to tiny spots on a glass chip. Large collections of specific DNA probes can also be made and used on microparticles, an approach commercialized by Luminex, Lynx Therapeutics, and Smartbead Ltd. Part of the reason for bead approaches is to enhance the performance of gene chips, but part is to side-step the patent problems in the field (*see* **Patents**).

More sophisticated systems have added features that allow the user to move sample around the chip. Nanogen has developed chips where each spot where DNA can bind is also an electrode. DNA is steered onto the spot or driven off it by electric fields. Agilent has built ones where chemical reactions, including PCR, can be performed, and the result then fed onto a small number of DNA probes to get a readout.

Gene chips are being used widely for genome-scale analysis of genes and their activity. Many DNA tests actually need you to conduct dozens or even hundreds of hybridization reactions to get all the information you need. The biochip idea allows you to do all those reactions on one small sample (because the chip is very small) and one chip (because it has all the probes on it).

Applications include:

- *Expression profiling*. This is measuring all the RNA in a cell. See separate entry.

- Checking a DNA sequence, to see if there are mutations in it. This is being used in research in cancer and AIDS biology. It is thought that one day the same chips could be used in medical diagnosis, to see exactly what the mutations in a cancer or some other disease are (*see* **Pharmacogenomics**).

- *SNP analysis*. Looking for markers for genetic analysis (*see* **Polymorphism**).

If you use enough DNA probes on a biochip, you can actually sequence the DNA with them, a technology called sequencing by hybridization (SBH). However, you need millions to sequence a few thousand bases of DNA, which is scarcely practical. The less ambitious aim of resequencing genes to check for changes is a more practical application.

Gene control

This is the genetic control of how a gene acts. Genes are pieces of information coded in DNA—on their own they do nothing. There must be a control system to make sure that their information is used at the correct time.

Prokaryotes (bacteria) and eukaryotes (yeasts, plants, animals) do this differently. In both types of organism there is region which says "gene starts here" on the DNA. This is called a promoter region, which contains DNA sequences which control *when* the gene is active. The promoter can respond to an inducer or a repressor protein, proteins that bind to the DNA in the promoter and turn the gene on or off, respectively. These proteins, in turn, might respond to many things, such as a chemical in the environment, temperature, or other proteins (*see* **Inducer**).

Other parts of the promoter element can include elements that perform other control functions, including (in eukaryotes) response to hormones.

Eukaryotes also have enhancers, pieces of DNA which alter the activity of a gene but which are not next to the promoter.

In eukaryotes, promoters are often rich in the two-base sequence CG, in which the C can be altered by turning cytosine into methyl-cytosine (methylation). For historical reasons, CG is usually called CpG (the P is the phosphate bond between them). Regions of DNA that have a lot of CpG are called "CpG Islands," and looking for them is a good way to get clues as to where eukaryotic genes start.

In bacteria, genes that are regulated together (i.e. that are turned on at the same time and by the same stimulus) can be arranged in a tight cluster called an operon.

Gene family

Many genes are not unique, but are closely related to some other gene or genes. Such a collection of related genes is called a gene family. The genes can be anything from 30% the same (i.e. one in three of the bases are the same between the two—only a few more than the one in four that would be expected by chance) to 95% the same. Other DNA sequences which are not genes (i.e. they do not have a recognizable function) can also come in families of similar sequences.

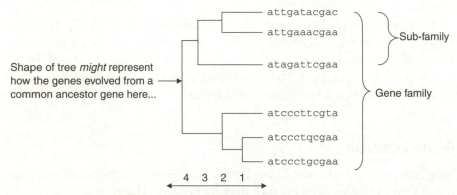

Shape of tree *might* represent how the genes evolved from a common ancestor gene here...

attgatacgac
attgaaacgaa
} Sub-family
atagattcgaa
} Gene family
atcccttcgta
atccctqcgaa
atccctgcgaa

4 3 2 1

Distance along tree from sequence to branch point tells the difference between the sequences

There can be any number of members of a gene family, from three or four to thousands, with varying degrees of similarity between them. Gene families can be clustered into superfamilies, with even more members which are even less similar to each other. An example is the immunoglobulin superfamily, which include a huge number of proteins with very different structures and functions. However, they all contain a core of sequence that makes a protein called the "immunoglobulin fold," which is like one domain of an immunoglobulin structure (*see* **Antibody structure**). On the other hand, some proteins can come in very similar forms (usually called isoforms or isotypes), which make a very closely related gene family, so close that they are really naturally occurring variants on a single protein.

A gene is usually identified as a member of a gene family because its sequence is similar to known members. The 3D structure of gene family members is often much more similar than their sequence—nature has allowed the amino acid sequence to change over time, but has kept the structure the same. A better way to detect gene family members is therefore to look at their 3D structure. As this structure is related to function, showing a gene is a member of a known gene family can give clues about its function (*see also* **Comparative genomics**).

Other sequences are found in many slightly varying copies in the genome. These include:

- *Pseudogenes.* These are sequences that have nearly the same DNA sequence as a real gene, but do not work as genes because some critical bases are mutated. This has been considered as "junk DNA" or "selfish DNA": however, the discovery that RNAi is quite a common effect has revived the idea that these genes do something useful.

- *Repetitive sequences.* These are sequences repeated many time, with no apparent relationship to a working gene at all. The most common are called short and long interspersed nuclear elements (SINEs and LINEs, respectively). In man, the most common SINE is the Alu sequence, which is present in 300 000 copies per genome, a huge amount of DNA and no one has any real idea whether it has a function at all, never mind what it is.

The "average sequence" of a gene family is called its consensus sequence. It is a sequence which, at each base, resembles the family members the most. A consensus sequence is an abstraction which represents the family as a whole— it may not represent any one member of it. Gene families are often shows as family trees (strictly, phylogenetic trees) with the consensus at the "root."

Gene library

A gene library is a collection of gene clones, which contains all the DNA that is present in some source, but split up and joined onto suitable vector DNAs. It is also sometimes called a gene bank. If the original source of the DNA was the original DNA from a living organism, then the library seeks to include clones of all that DNA: it is called a genomic gene library, because it contains all the DNA from that organism's genome ("genome" simply means all the genes, or DNA, in an individual). If the DNA is from some other source such as copy-DNA (cDNA) made by enzymatic copying of RNA, then the maker of the library seeks to include representative clones from all that source, and, in this case, would be called a cDNA library.

Gene libraries are not organized like book libraries, and can only claim to be complete because the number of clones in them is sufficiently large for us to be very confident that all the clones we want to be there are there, that is, that there is a very small chance that anything has been left out.

The number of clones needed to get a complete gene library depends on how big the cloned pieces of DNA are. Thus, if you are using a Lambda phage vector to make a genomic gene library from human DNA, you need about 500 000 clones to be 90% confident that you have not left any out. However, cosmid cloning vectors can hold substantially more DNA—you would only need 200 000 of them. YAC vectors hold 10 times as much DNA again, so you only need 10 000 of them. This is why people use YAC vectors to make genomic gene libraries, as screening 10 000 clones for the one you want is almost invariably easier than screening 500 000.

Gene quantification

This involves measuring the amount of the DNA for a particular gene in a sample. This is used directly to look for mutation in the genes, especially associated with diseases such as cancer, but also to look for the amount of cDNA copied off RNA in a sample. So the same technology can be used as an approach to **expression profiling** (see separate entry).

The most reliable methods are based on hybridization. This can be based on one of the various blots, or on gene chips. The Invader Assay is a sophisticated version of this, marketed by Third Wave Technologies—used the activity of a nuclease enzyme that recognizes DNA "branches" to release a label when the probe is bound correctly, but not when it is unbound. The amount of probe released is proportional to the amount of target DNA.

An alternative is to perform PCR on a sample. Normally, you run PCR until all the reagents are used up, so the same amount of DNA is made each time. However, the rate at which it appears depends on how much you had to start with, and either this rate or the time it takes to reach a certain amount of DNA can be used to estimate the amount of target DNA there was there to start with. This is very prone to error, and so it is usual for an experiment to have several "standards" included as reference points.

A clever combination of hybridization and PCR called MAPH has been developed by researchers at Nottingham, UK, which binds the PCR probes to target DNA and then washed them off and counted them using PCR. Because the probes are synthetic, their behavior is more reliable than trying to amplify the DNA directly.

Gene silencing

Gene silencing is the "turning off" of a gene. This usually refers to a gene you have put into a cell or organism, with an appropriate promoter so that you expect it to be active. Sometimes, however, it is not—the gene is said to have been silenced.

This can happen at two levels.

For a gene to be active in a eukaryotic cell, the chromatin—the protein package that surrounds it in the nucleus—has to open up to allow the enzymes that make RNA access to the genes. If the chromatin remains closed (called "heterochromatin"), then the gene will be very unlikely to be active no matter what promoter or other control regions it has. This is important for the scientist making transfected cell lines or transgenic animals: if the gene they put in integrates into the genome in a region of heterochromatin, then it will be silenced, and hence not do what we want it to do.

This is known as a *cis* effect—it is an effect of the chromosome next to the gene, not to another genetic element a way away.

The other silencing effects work in *trans*, that is, that genes on one chromosome work to silence genes on another. This is often also called post-transcriptional gene silencing, because the most common mechanisms operate to stop the RNA being translated or to break it down faster once it is made. Common mechanisms for PTGS are:

- *Antisense*. This was the first RNA level *trans*-silencing effect observed. It has been developed as a method of turning genes off for some time, with variable results. See separate entry.

- *RNAi*. This is RNA interference. How it works is still slightly mysterious, but is it known to be double-stranded RNA that has a similar sequence to the target gene causes the RNA transcribed off that gene to be broken down. It works on introns as well as exons. This is probably related to the cell's defence against viruses, many of which make double stranded RNA in their life cycle. RNAi is very potent—only a few molecules of dsRNA are needed to trigger the breakdown of all the RNA transcribed off a gene. It is also fast. The RNA is broken down almost as soon as it is transcribed, and never gets out of the nucleus. So there is a lot of excitement about using this as a method of gene control, both in transgenic organisms and in therapy, for applications similar to those tested for antisense technology.

- *Co-suppression*. This is a similar effect to RNAi, but apparently works through a different mechanism. It was found in the 1980s that sometimes transgenic plants did not only not express the transgene (which could be due to silencing "in *cis*"), but also turned off similar genes in the plant as well. This effect has now been found to occur in most eukaryotic cells, and is called co-suppression. The transgene needs to be expressed, but can be expressed as the sense gene or as an antisense, so this again is something that works at the RNA level.

Gene spread (GM crops)

A major concern with genetically modified (GM) crops is that their use in the field will mean that the modifying genes will escape and spread into the wider ecology. This is quite probably true: the issue is how much this will happen, and how much it matters.

There are two versions of gene spread.

The first is the problem of crops remaining from one year to contaminate the next year's crop. These are called volunteer crops. They might become weeds in their own right, and are then an especial problem if they have been engineered to be resistant to herbicides. Round-up ready canola plants becoming a significant weed pest in Ontario. (This is a problem that is general for new farm crops, not just GM ones. Unmodified oilseed rape—canola—is now a frequent garden weed in some parts of England where the crop is intensively farmed.)

The second is that pollen from the GM crop drifts away from the field, and could fertilize nearby weed plants, creating hybrids that are still weeds but carry the GM trait. This is known as gene flow, and is a serious concern. Canola pollen is generally believed to be able to drift 10 m from a field, but has (controversially) been claimed to be detectable 25 km away. Maize pollen drifts 50 m, wheat 400 m, and this latter is especially serious as wheat is a "domesticated" relative of many wild species of grass that are ubiquitous in temperate countries.

There were claims in the late 1990s that so much GM pollen spread from fields of GM crops that the pollen itself was starting to kill off butterflies. Researchers from Guelph and Cornell claimed that the pollen from BT corn contained Bt toxin, and so killed insects on plants it fell on, specifically the monarch butterfly. This generated huge media interest, and claims that this proved the environmental dangers of GM crops. In fact, that year the amount of Bt corn planted increased 40% on the year before, and the Monarch butterfly population (as monitored by amateur butterfly enthusiasts) rose 30%, but this was largely ignored. (However, introduction of Bt corn has not resulted in any decline in insecticide use, which was one of the benefits that was meant to come out of this technology.)

Such spread of genes between crops and weeds occurs anyway. The problem with transgenic plants is that they contain genes for resistance to antibiotics or herbicides that might pose problems if found outside the field in which the drop was planted.

What is not clear is the extent to which this means that GM genes are entering the wider ecology. Claims in 2000 that wild Mexican corn plants were being found in many sites that were contaminated by Bt transgenes made a very substantial press impact, but in 2002 other scientists trying to repeat the experiments showed that there was a basic flaw in the method, and what the original study was picking up was not a transgene at all.

By far the biggest "accidental release" was the discovery of 11 000 hectares planted illegally with transgenic cotton in Gujarat state, India in 2001. Ironically, it was only discovered because boll-worm was devastating the non-engineered crops in neighboring fields, but not the engineered ones. Navbharat Seeds claimed that their variety NB-151 was naturally derived, but gene tests showed that in fact it was derived (without permission) from a Monsanto transgene carrying a Bt insecticidal protein gene.

The accidental appearance of Starlink corn in taco shells in September 2000 has been blamed on gene drift, but is more likely to be due to a mix-up of batches of corn. Starlink corn, a transgenic maize containing the Bacillus thuringiensis Cry9C gene to give it protection from insect attack, was found in taco shells by the FDA. Subsequently 1 in 10 US Department of Agriculture tests showed that supposedly non-Starlink corn had some traces of Starlink genes in it. (Starlink was not approved for human consumption at the time because the EPA could not be convinced that the Cry9C protein would never be allergenic to people.) All Starlink corn was withdrawn, at a cost of around $1 billion, and general corn exports from the United States dropped.

As well as scientific concern about what any potential gene spread might do to the world's plants, commercial concerns are pushing to reduce the spread of genes. Organic farmers are claiming compensation from agro-chemical companies and the US and Canadian Governments for the contamination of their crops by GM pollen, which renders it "non-organic." This is because the rules of organic farming associations forbid any transgenic crops at all. In a curious turnaround to this, Canadian farmer Percy Schmeiser claimed that the Starlink corn he was growing was all the result of contamination, and not because he planted Aventis' seeds. The US patent office took a dim view of this argument (*see* **Patents**).

Gene therapy

Gene therapy is changing the genetic makeup of a human. Gene therapies had been administered to more than 4000 patients by the start of 2001: however, nearly all of these were phase I trials, testing to see if the approach was safe (or at least, less dangerous than the disease they were suffering from), and whether gene transfer took place (i.e. whether the gene was getting into the patient, and functioning there at a molecular level). Thus, there is not a proportionate number of "cures" from gene therapy yet.

There are two approaches, germ-line gene therapy and somatic cell gene therapy. The former changes the "germ cells," the cells which make sperm or ova. This has a permanent effect on all the individuals which are the descendants of whoever had the therapy. Somatic therapy alters the somatic cells— all the non-germ cells in the body. Changing them does not affect the germ cells, but does affect the engineered person.

Gene therapy of the germ cells of animals is called **transgenic** technology (see separate entry). Germ line gene therapy of humans is considered unethical, and has not (legally) been tried.

Somatic cell gene therapy can be aimed to correct a genetic or a non-genetic defect. Current therapeutic targets include both categories.

The main issue with somatic gene therapy is getting the DNA into all the cells that you want to treat. Standard transfection methods get DNA into 1 in 1000 or 1 in 10 000 cells, which is rarely useful for a patient. So the major area of work is to develop new **vectors** for gene therapy (see separate entry)

An alternative approach is to take the patients' cells out of the patient, treat them, select the ones where this has worked, and then put them back. This approach was used in a lot of early gene therapy trials, which focussed on blood cells which are easy to remove and re-inject. Targets for this approach include gene therapy of the bone marrow, where it has been used to treat severe combined immunodeficiency disease (SCID, a very rare genetic disease caused by a deficiency in the enzyme adenosine deaminase (ADA)). W. French Anderson and Michael Blaese used a gene therapy treatment for SCID on a 4-year-old girl in late 1991, with reasonable success. Other targets include several cancer treatments, including introducing white blood cells engineered to produce more tumor necrosis factor (TNF) or interleukin (IL-4) into cancer patients where they are hoped to be able to assist in destroy the cancer.

Gene therapy was originally thought to be useful for correcting the gene defects in inherited diseases such as cystic fibrosis or muscular dystrophy, where a single gene was lacking in patients. However, because they are inherited, every cell in the body is affected, and it is not practical to get a gene into every cell. Cystic fibrosis (CF) may be an exception, as it causes imbalances in the way chloride ions are transported across cells. If some of the cells lining the lung can be corrected, it is hoped that the proper balance will be restored and the disease greatly improved.

Alternative approaches to replacing the gene in a cell is to use trans splicing (*see* Genetic code) to try to correct the RNA after the gene has been transcribed. Intronn Inc. is trying to apply this technology to CF gene therapy.

After a wave of hype in the early 1990s, it is now accepted that gene therapy is extremely difficult to get to work, for a variety of technical reasons. In 2002, it is starting to make a comeback, as part of the usual cycle of hype.

See also Transfection, Genoceuticals, Gene therapy—regulation.

Gene therapy—regulation

Applying gene transfer techniques to humans, usually called gene therapy, has been a major problem for legislators and regulators as well as for scientists. Today, regulation is through a combination of approaches.

- Most work is funded, at least in part, by government grant-giving bodies. These can impose rules on who can do what.

- Any clinical trial is governed by regulatory bodies such as the FDA in the United States.

- Most clinical trials are also governed by local ethical oversight committees.

On September 17, 1999, Jesse Gelsinger, an 18-year-old Arizonan being treated by gene therapy for ornithine transcarbamylase deficiency, died of sudden, complete organ collapse. He was in an escalating dose Phase I trial: the other patient in his high-dose group had no adverse effects. The program was being run by The University of Pennsylvania.

This was the first *reported* death due to a gene therapy therapeutic (others had died, but of acute illness). A problem here was the lack of coordination between the FDA and the NIH, which was funding the trials: the FDA has to be notified of "adverse effects" by law, but does not publish them. The researchers said that they believed that their requirement to notify the NIH as well (who do publish deaths during trials) was fulfilled by notifying the FDA. In the publicity that followed, six other deaths were uncovered in trials on using gene therapy to rebuild hearts after heart attacks.

Genetic code and protein synthesis

The genetic code is the code used by living cells to turn the information in DNA into the information needed to make protein. Originally, it meant how a code of bases in DNA could be used to code for a sequence of amino acids in proteins, but the meaning has been extended to mean anything to do with how genes code their function.

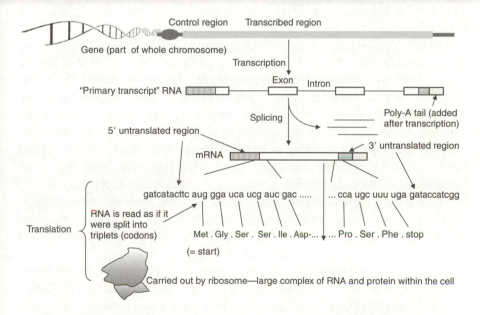

How this works is not essential to an understanding of much of biotechnology—the genetic machinery can be treated as a "black box" for even quite advanced discussions.

The information in DNA is held in the sequence of the four bases of DNA (adenine, guanine, cytosine, and thymidine). This information is transcribed into base sequence in RNA, and then translated into amino acid sequence in protein, the latter occurring on the ribosomes. RNA is a very similar chemical to DNA—this "transcription" is therefore like copying instructions from parchment onto paper.

The RNA is made starting at the 5′ end (*see* **DNA structure**), and is translated starting from that end too: protein is made starting from the amino end (N-terminus). Because there are four bases but 20 amino acids, you need three bases to code for one amino acid. Thus, the RNA is "read" in triplets, called codons. The part of the RNA that is "read" in this way is called the reading frame. A specific start codon ("initiation" codon) starts the reading frame, one of three special termination codons finish it.

Because the RNA is read in threes, the loss of a single base would then throw out all the cell's reading of the subsequent triplets. Such a mutation is called a nonsense mutation, because it makes a nonsense of the rest of the protein.

Although most of the code is shared between all living things, there are some differences: for example, mitochondria, which have some of their own DNA, do not have quite the same genetic code as the cells in which they reside.

As well as the RNA bases that code for protein, RNAs have pieces of extra RNA of the front and back, called "untranslated regions" (UTRs). The 5′UTR ("Five Prime UTR") is on the "front" end, the 3′ UTR on the back.

The DNA sequence in the chromosomes need not be the same as the RNA sequence that is eventually translated into protein. The RNA that is copied from the DNA is called the primary transcript. This can be edited, most usually when segments called introns (which occur in most eukaryotic genes, also sometimes called intervening sequences) are removed, a process called splicing, rather like cutting the ads out of a taped TV movie. The parts that are kept and spliced together are called exons. The final RNA is called mRNA (m for messenger, because it carries the sequence "message" from the nucleus to the cell cytoplasm, where it is translated into protein).

There are a couple of known cases where splicing involves joining bits of different RNA molecules together, known as trans-splicing. This turning out to be more common that once believed. It has also been researched as a method of correcting gene defects in gene therapy.

These processes are called post-transcriptional processing or modification, because they happen after the primary transcript has been read off the DNA.

These complications have two implications for biotechnology. First, it is often not possible to express a eukaryotic gene in a prokaryote. Even if prokaryotic promoter sequences are spliced on, the prokaryote will be incapable of carrying out eukaryotic splicing to the RNA to make it readable. For this reason, many protein expression projects prefer to start with a cDNA clone (a cloned DNA made by enzymatic copying of the final mRNA) rather than the original gene. Second, although sequencing DNA is easier than sequencing protein, it is not always safe to extrapolate from the DNA sequence to the protein it may code, because of variations in post-transcriptional modification of the RNA and variations in the genetic code.

The flip side of protein synthesis is protein degradation, how proteins get broken down when they become chemically damaged or are not used any more. There is a complex apparatus to do this, which involves the protein ubiquitin, which is chemically linked to the protein inside the cell to tag it for destruction. This process is also important for processes such as antigen presentation in the immune system, where proteins are broken down so that their fragments can be recognized by antibodies.

Genetic disease diagnosis

A genetic disease is one which is caused by a gene, so we "inherit" the disease from our parents. (More correctly, the difference between the diseased and the healthy person is caused by the difference between the versions—alleles—of the "disease gene" that they possess.) For a true genetic disease anyone with the correct genotype (collection of genes) will display the phenotype (the physical manifestations of the genes). In practice, for quite a lot of genetic diseases the genes do not always cause the effect they are meant to, an effect called "incomplete penetrance." This makes detecting them rather complex.

Molecular genetics has made huge advances in medical genetics, particularly through making available DNA probes to detect the genes which "cause" genetic diseases even when they are not causing them—for example, when a gene is present in a carrier, or when a gene which causes a disease late in life is present in an infant. These probes have been used both to identify the gene and to diagnose carrier status in people who carry the gene but do not have the disease.

The genetic diseases for which cloned probes (i.e. probes that identify the gene itself) have been isolated include the hemophilias, thalassemias, sickle cell disease, Duchenne and Becker's muscular dystrophy, retinoblastoma, and cystic fibrosis. These are all diseases caused by single genes, also called monogenic diseases. More common diseases are usually influenced by many genes—they are called polygenic diseases.

See also entry on **Predisposition analysis, Polygenic disease.**

Genetic engineering

This is a general term for the directed manipulation of genes, and usually used synonymously with genetic manipulation or genetic modification. A wide range of technologies are involved in this, but most involve the **recombinant DNA technology** (see separate entry).

Genetic engineering falls into several different categories depending on what is being engineered:

1. *Bacteria, yeast*. This is "traditional" genetic engineering (i.e. over 20 years old). Using recombinant DNA techniques, genes are put into microorganisms to make them produce something we want, be it insulin or better beer or protein for food.
2. *Animals and plants*. These are usually called transgenic animals or plants. They pass their engineered characteristics on to their descendents. See separate entries.
3. *Humans*. Although the genetic engineering methods applicable to cows or mice are, in theory, applicable to humans, they have not been applied for obvious ethical reasons. Some experiments treating disease have been performed: these do not modify the germ cells, only the "somatic cells." This is generally called gene therapy or somatic cell therapy, rather than the more spectacular (and potentially publicly alarming) term "genetic engineering."

Genetic information

Projects such as the human genome project and the development of tests for genetic predisposition to disease have lead to much discussion about how human genetic information should or could be used. This is usually in the context of information about an individual's genes, rather than about the genome as a whole.

There is substantial concern about the use of individual genetic information, for several reasons.

Ethical. There is a strong strand of opinion that it is simply ethically wrong for someone else to know about your genetic make-up, because that is the most fundamental description of your biology. *See* **Bioethics**.

DNA fingerprinting. If you are "DNA fingerprinted," who has right to that information? DNA fingerprints are a collection of genetic markers, which could be linked to genes of medical importance—can DNA fingerprint information be used to predict how tall you are, or whether you are likely to suffer from hair loss? *See* **DNA fingerprinting**.

Insurance. The insurance industry, both for health and life insurance, is based around assessment of the risk that someone might get ill or die. Genetic information certainly can give better estimates of those risks (*see* **Predisposition analysis**). If genetic information was available, or if insurance companies could demand genetic tests, would that unfairly discriminate against people? The present consensus is not to require any sort of genetic test except for a few inherited diseases such as Huntingdon's disease if the disease is already known to run in the family, but requiring someone applying to them to tell them anything that they know about their health status. If they have had a genetic test and it turned out bad, then they must tell the company. The United States has drafted a Genetic Information and Non-Discrimination Act, which would prevent the use of genetic information for discrimination in health insurance and employment, but has been in legislative limbo since 1995.

Employment. Employing someone is a risk for a company, which has to assume that the person is suitable for the job. In the near future genetic information about the person could (in theory) provide indications of suitability for some tasks, because of medical risks from workplace hazards, like chemicals or overwork. Maybe the employer has a duty to do the test, for the benefit of their employee, and could be sued if they did not?

Property rights. Who owns information generated from my genes? Information generated from my tissues (after they have been cut out) belongs to the researcher, not to me. But my genes are still part of me.

Genetic instability

This means that the genome or a specific gene within it is more likely than average to change with the generations.

There are a number of ways this can happen, the most common being because the gene has a simple sequence repeated many times. In humans, a number of genes have regions where a three-base sequence is repeated many times, either to code for a string of the same amino acid or in the 3′ untranslated region. These are found to mutate far faster than normal through a mechanism that adds copies of the three-base sequence, making the overall stretch of gene longer. Fragile X disease, Huntingdon's chorea, myotonic dystrophy are among the diseases are caused by mutations in trinucleotide sequences, which are called trinucleotide expansion diseases (TREDs) as a result.

More generally, DNA that is made of a short sequence repeated many times is liable to gain or, more usually, loose copies of the sequence over generations. The original DNA fingerprinting markers (VNTRs—*see* **DNA fingerprinting**) were sequences that were many copies of the same short sequence joined end to end. They are useful for fingerprinting because they mutate very frequently compared to normal sequences, and so there is a good chance that any two people will have different "mutants" (versions). In this case, the sequence changes do not seem to have any medical effect. Mutation to loose copies of repeated sequences from genes is more common in bacteria than in eukaryotes, and is the reason why it is so hard to clone some genes, such as those for collagen or silk, into bacteria—they code for very repetitive proteins, and so their DNA is very repetitive and tends to have bits "lost" from the middle by mutation.

The other common genetic instability in the medical field is chromosomal instability. This is when whole chromosomes are lost or gained. Down's syndrome is the most common human condition caused by an extra copy of chromosome 21, one of the smallest human chromosomes. Extra copies of other chromosomes all cause more severe disease, and lack of one chromosome copy (out of the normal two) is invariably fatal—except the Y chromosome: having just one X chromosome causes Turner's syndrome.

Cancer cells are inherently prone to genetic instability. Mutations can be speeded up, but also the rate of chromosome loss is much higher in cancer cells, leading to aneuploid cells (cells with a different number of chromosomes that normal). Chromosomes can also be broken or fused: a standard test for a compound's ability to cause cancer is the micronucleus test, which tests for small fragments of chromosomes formed by some carcinogens. A related observation is that very small chromosomes are "lost" from cells much more easily than regular chromosomes. This particularly applies to plasmids in bacteria or yeast. This is why genetically engineered plasmids all have selectable marker genes to find the cells that have not lost them. *See* **Plasmid.**

Somaclonal variation is a change in the genetics of a clone, compared to its parent. There is not enough evidence to say definitively whether it also happens with mammals, but if it does it seems to be less common that with (say) potatoes. See separate entry.

Note well that, despite Star Trek and other science fiction films, "genetic instability" does not mean that an organism is going to mutate in front of your eyes into something strange and mucus covered. It is something that occurs over many generations, either of organisms or (in the case of cancer) of cells.

Genetic map

A genetic map is a description of how genes lie along a chromosome. It looks rather like a ladder with very erratic rungs. There are several types.

The map can show the distance between any two points in terms of how much DNA lies between them (usually measured in megabases—Mb). This is called a physical map, because it represents what the genes are physically like.

A map can also represent the recombination distance between two genes. This is the chance that two genes at those two points will be separated by recombination during meiosis. (Meiosis is the process whereby chromosomes are shuffled before being packaged into sperm or ova.) The resulting map is called a genetic map or a recombination map, with distances measured in centimorgans (cM), which is the distance apart genes have to be to have a 1% chance of being recombined during meiosis. The distances on the physical map and the genetic map are rarely the same, but the order of the genes is the same.

The landmarks on genetic maps are not always genes. They are more generally known as "markers," and can be any piece of DNA, either one that has been identified as a gene, or one that is just identified chemically by DNA probes or PCR amplification (*see* **RFLP, SNP**). The critical feature of a genetic map is its density—how many markers there are on it. The more markers there are, the more use it is in guiding our experiments, and hence the interest in the human genome project in producing a high density genetic maps with lots of markers on it.

True genetic maps can only be made by genetic experiments—breeding. Physical "genetic" maps can be made by a variety of DNA techniques, such as looking at RFLPs, making physical maps of a whole lot of cloned DNAs and then stitching them together, or *in situ* hybridization to chromosomes to see where a piece of DNA lies physically on these cellular structures. Ideally the maps made by these different techniques should match. Usually, they do not, quite.

The ultimate physical map is the DNA sequence of a chromosome. This tells where every single possible piece of DNA is. However, it may still not tell you what they do, and so is not the ultimate genetic map.

Genoceuticals

A slightly passé term for a version of gene therapy where a gene is placed into a cell and there produces a pharmaceutically active protein. This has two potential applications.

"Genetic antibiotics" are genes that have some anti-bacterial or (more usually) anti-viral activity. The genes are placed into the cells which are the potential targets of the parasite. For example, a gene for a toxin could be linked to a controller gene, which is activated by a virus: when the virus infects the cell the toxin gene is turned on, toxin is produced and the cell dies.

The other application is to insert genes which will themselves make bio-pharmaceuticals. For example, calcitonin has been suggested as a treatment for osteoporosis, a bone-wasting disease suffered most frequently by older women. However, calcitonin is a protein, and difficult to get into the body: consequently it has to be injected frequently. A genoceutical approach to osteoporosis would be to "transfect" the gene for calcitonin into some suitable cells in the individual: these would then produce the hormone steadily for weeks or months.

This is a version of gene therapy, and shares the technical problems of that field. These include delivery barriers (it is very hard to get genes into adults reliably and reproducibly), and the considerable social concern over the use of gene therapy for any application.

A version that does not share these problems is to put a gene into cells outside the body and then deliver the cells. This is version of cell therapy (*see* **Artificial tissues**), but where the cells are engineered to produce a biotherapeutic rather than something they would make naturally. Neurotech is using encapsulated cells in this way to treat macular degeneration, a condition of the eye leading to blindness and potentially treatable with long-term treatment by specific growth hormones, which the cells deliver.

Genome projects

A genome project is a project to determine the exact genetic structure of an organism's genome, that is, the DNA sequence of all its genes. There are quite a few such projects.

The **Human Genome Project** is a project to determine "the" base sequence of all the DNA in humans. See separate entry.

Bacterial genomes. Microbial genomes are typically a few million bases (megabases, the standard unit of genome size) to tens of megabases. In 1995, the first two complete bacterial genome sequences were completed, for *Hemophilus influenzae* and *Mycoplasma genitalium*, both disease-causing bacteria. *M. genitalium* has a simplified genome, because it relies on its host for much of its metabolic apparatus. Even so, it is 580 067 bases round. *H. influenzae* is a more typical bacterium, and has a genome of 1 830 121 bases and 1749 genes. Up to mid 2002, 72 bacteria and 16 archebacteria genomes have been sequenced.

Complete eukaryotic genomes as of late 2002 are:

Species	Size of genome (megabases)	Year completed
Human ("working draft")	3100	2001
Mouse	2700	2002
Drosophila melanogaster (fruit fly)	137	2000
Plasmodium falciparum (parasite that causes malaria)	23	2002
Anopheles gambiae (mosquito that carries malaria) (draft)	278	2002
Encepalitozoon (parasite)	2.5	2002
Caenorhabditis elegans	97	1998
Arabidopsis thaliana (thale cress—a weed)	115	2000
Saccharomyces cerevisiae (brewers yeast)	11.7	1996
Schizosaccharomyces pompe	12.6	2002

(This *will* be out of date by the time you read this: the NCBI is probably the best source of the latest on who has sequenced what.)

Other genome projects include rats, cows, pigs, rice, wheat, maize, apple among others.

This huge acceleration of sequencing is a result of technological improvement—it is literally true that a sequencing project that took years two decades ago, now takes minutes.

However, scientists have found the genome sequences to be more perplexing and less useful than they thought: the genes are not the "book of life," but at best the "parts list of life," and we need the plans to show how they all fit together. So much effort now goes into looking for meaning in the tens of thousands of megabases of sequence available. This comes under the general heading of **functional genomics**—seeing what genomes do (see separate entry). Much of this is done by comparing genomes to each other—*see* **Comparative genomics**.

Genomics

Different from the genome project per se, "genomics" means the application of the technology of high throughput molecular biology to many problems. Technologies include highly automated DNA cloning, high throughput DNA sequencing and gene chip technology.

Common applications include:

- DNA sequencing, especially sequencing a whole genome.

- Expression profiling. Finding out all the genes that are active in a cell or tissue at a specific time.

- Population genetics, doing very large-scale genetics studies on many organisms (often people) to discover links between genes and their effects. Companies such as Oxagen and DeCODE are pursuing this path.

See **Genome projects**.

GLP/GMP

These stand for good laboratory practice and good manufacturing practice, respectively. They are a code of practice which is designed to reduce to a minimum the chance of accidents which could affect a research project or a manufactured product. Sometimes these are referred to as cGMP (Current GMP) or cGLP (clinical good laboratory practice).

The GLP and GMP prescriptions are quite voluminous, but boil down to a few key points. The essential point of both GLP and GMP is that everything is recorded, and only established procedures are used by people who have been trained to use them. This may seem obvious, but extends to everything: in a true GLP laboratory, for example, only staff who have been trained to use a balance may use it, every weighing has to be checked by another person (who has also been trained for that specific balance) who has to sign to say that they have checked the weight of material is correct, weighing has to be done according to a written standard operating procedure (SOP) for using that balance, the protocol used has to be noted in the record of the experiment, and so on. All records are kept, and have to be archived on microfiche, magnetic tape, or (more recently) write-once CD-ROM. Samples of every batch of material that is used in the experiment or the manufacturing process also have to be archived, so they can be referred to in future should this become necessary. This is called traceability—you can trace where everything came from, and exactly what happened to it.

Using these types of procedures, it is possible to trace exactly who did what at every stage of an experiment or a manufacturing process. Thus, if there is a problem afterwards, the GLP or GMP user can either point to a specific material or standard operating procedure which caused the problem, or can point to exhaustive documentation which shows that the problem is not their fault. This can be very important in pharmaceutical development and manufacture (GLP was set up after some severe side-effects of a drug were overlooked during pre-clinical research because the protocols for an experiment were faulty). Actually working to GLP or GMP means getting certification from a suitable authority (of which there are a huge range of industry- and country-specific organizations) which is very time-consuming and costly in effort. So many biotechnology companies claim to work "to the spirit of GLP or GMP." In practice, this means little.

It is very difficult to do innovative research to GLP, where you have to define a set of standard operating procedures, train the staff formally etc. just to do one experiment which might take half a day. GLP is more relevant to pharmaceutical development (where a large number of very similar experiments are performed). GMP is a standard requirement for pharmaceutical production, and for a number of other industries.

A related term is good clinical practice (GCP). This applies the same disciplines to clinical trials (and other clinical practices, of course). Again, "working to the spirit of GCP" means "not GCP." That is not to say that it is bad or sloppy work, of course, just that the documentation and traceability is not in place.

GMP also stands for good microbiological practice, a code of laboratory practice for performing basic microbiology safely. In this sense, GMP is simply a way of reducing the change of contamination problems (contamination of the sample or of the laboratory) during a microbiological experiment to a minimum.

The International Standards Organization also has a "good practice" standard, ISO 9000. There are a number of sub-standards within ISO 9000, dealing with "good practice" in manufacturing, product development, provision of services, etc. Organizations can be certified to ISO 9000 as a mark that they provide a consistent quality of result to their customers. Certification requires that the company be regularly inspected by external, qualified inspectors—it is not a once-and-for-all event. ISO 9000 certification and GMP status are not formally linked, but have many of the same requirements for documentation and traceability. Neither GLP/GMP nor ISO 9000 specifies that the product need be any *good*, only that it is consistent and well documented: as Richard Buetow of Motorola observed, ISO 9000 allows you to make concrete lifejackets provided they are made according to documented procedures.

cGMP can also stand for the chemical cyclic guanosine monophosphate— *see* **Messengers**.

Glucose isomerase and invertase

Glucose isomerase is probably produced in larger amounts for industrial use than any other single enzyme (although the largest category of enzymes by far is the general class of alkaline proteases, used in detergents). It catalyzes the interconversion of the two sugars glucose and fructose. As fructose is slightly more chemically stable than glucose, a mixture of glucose and fructose with the enzyme will end up almost entirely as fructose. This is valuable for the food industry, as fructose is substantially sweeter than glucose, and so you get more sweetness per gram than by using glucose.

The usual use for glucose isomerase is to take glucose made by hydrolysis of corn starch and turn it into a mixture of mostly fructose with some glucose. The corn starch is broken down using amylases. The result is called "high fructose corn syrup" (HFCS).

Invertase takes sucrose ("sugar") and turns it into glucose and fructose, a mixture called invert sugar or invert syrup. Invertase was the first enzyme to be used in industrial production, in the late 1940s. Thus, in conjunction with glucose isomerase it can convert sucrose into HFCS. Invertase can also be used in its own right to convert the easily crystallized sucrose into the less easily crystallized glucose–fructose mixture. "After Eight Mints," for example, have invertase in their centers—it turns the hard sucrose core (which the chocolate coat was poured onto) into the soft center that we finally eat.

Glue

Biological glue is one of the many areas where biotechnology and medicine can overlap. Doctors are always interested in new medical techniques for repairing wounds. An obvious way is glue: however, the glue must have unusual properties. It must be able to set (cure) in a wet environment, not be broken down by watery liquids, not irritate or poison the body, not induce a immune or allergic response, and the body must be able to break it down after a time if its function is only temporary, like stitches.

The most widely used and discussed is the protein fibrin. The body itself produces fibrin, a component of the clotting proteins in blood: however, it is not a very strong glue, and, unless derived from human blood (with its concomitant risk of viral contamination), causes a strong immune response. However, it is a natural human product, and is used in several commercial medical glue applications.

Several marine organisms produce glues which could fit these criteria. Mussels and barnacles such as the Blue Mussel (*Mytilus edulis*) produce protein-based glues which could, in principle, be produced by more convenient organisms using biotechnology. Genex has produced a yeast which makes the protein (which has a very unusual amino acid composition, making it difficult for the yeast cell to make it efficiently). The protein also needs extensive and rather specific post-translational modification, which the yeast cannot perform. Thus, these proteins are some way from commercialization yet.

Many other organisms make materials which glue them onto things or things (like eggs or nest material) onto other things. However, these have not been investigated nearly well enough to make them attractive targets for medical glues.

Biotechnology is also investigating semi-synthetic glues made from natural materials. Proteins can be cross-linked into a glue-link material which can be used in wound healing and surgery, being fairly rapidly re-adsorbed by the body.

Glycation

Glycation is the non-enzymatic reaction of sugars with proteins. Many proteins are glycosylated deliberately by enzymes in the body. However, sugars can also react with the amino groups in proteins in an uncontrolled, chemical reaction. As every part of mammalian bodies have sugar in it, this means that all proteins get glycated after a while. This is much accelerated if the sugar levels are very high or if the mixture is heated. Hence, chemical glycosylation is important to protein processing and flavor formation in food. The reaction is called the Maillard reaction and is responsible for much of the characteristic flavor of foods as diverse as roast meat (where the Maillard reaction can go faster because there is less water present) and chocolate. The products can undergo further reactions during cooking to produce a positive witches brew of chemicals, which is why burned food can have extremely strong and rather unpleasant flavors and smells.

Chemical glycation is also very important in the damage done to diabetics when their sugar levels rise above normal, and to all of us as we age. Indeed, one school of thought holds that much of the damage that causes aging is due to the effects of glycation. Particularly, glycated proteins can continue to react and form complex, stable cross-links with sugars and through them with other proteins. These complexes are called advanced glycosylation end-products—AGEs. The body seems unable to remove them specifically, and so they accumulate, cross-linking collagen into a rigid, inflexible net, damaging critical proteins in long-lived nerve cells, maybe even directly mutating DNA.

Glycobiology

Glycobiology is the study of sugars and their role in biology. Usually, this is taken to mean the study of complex sugars and what their functions are, and not the metabolism of how sugars are put together and taken apart. The application of glycobiology is sometimes called glycotechnology, to distinguish it from DNA-based biotechnology.

The twin thrusts of glycobiology are the study of glycoproteins, which are proteins with sugar residues attached, and the study of molecules that interact with sugars and affect sugar metabolism, especially the synthesis of those glycoproteins (called **glycosylation**—see separate entry). Some glycoproteins have as much sugar in them as protein by weight, and the effects of this sugar on the protein can be substantial. Also sugars can be in branched chains, rather than just linear ones, so the possible combinations of a small number of sugar units is huge. Current theory suggests that the sugars on glycoproteins help in protein–protein binding (important for the mechanism by which cells recognize each other and by which viruses bind to and gain entrance to cells).

From this, glycobiology is interested in how the complex sugars on their own interact with glycoproteins, glycolipids (lipids with sugars attached), and each other. It is also interested in lectins (proteins which bind specific sugars) as tools.

In living systems sugars, both as simple sugars and as blocks of sugar residues, are joined onto proteins at specific amino acid sites by glycosyl-transferase enzymes (a process called glycosylation). Glycolipids can also be joined onto proteins by specific enzymes, a process called glypiation, producing glycolipoproteins.

Other uses of glycobiological expertise are in manipulating glycosylation in expression systems, and in analysis of carbohydrates and glycoproteins.

Glycosidases

A group of enzymes which break up complex sugars (such as starch or sucrose) into simple ones (such as glucose or fructose). About 12 000 tonnes of glycosidases are made per annum, almost exclusively for use in the food industry.

The major glycosidase enzymes are amylases (which break down starch) and **glucose isomerase** (which is used to turn glucose into the sweeter fructose). Amylases break the long chains of starch molecules and similar polymers into shorter segments, and ultimately into glucose. Amylases are commonly extracted from barley, beans, potatoes, and from a variety of fungi.

Other enzymes produced from bacteria and fungi for polysaccharide breakdown are isoamylases and pullulanases. *See* **Polysaccharide processing** and **Food processing using enzymes**.

A third group of these enzymes are the cellulases, which break down cellulose. As cellulose is probably the most common biological material in the world, using it as a raw material makes economic sense. However, it is very difficult to break down into its glucose monomer units. A range of cellulases and hemicellulases are used to process food ingredients, for example, enhancing the maceration of fruit pulp to make fruit juices. Gist Brocades, a major provider of these enzymes, claim that enzymes can increase the amount of juice you can squeeze out of apples by 25%, through opening up the cell structures. Researchers are also looking into using cellulases for biofuel production, through turning cellulose (which few organisms can ferment) into glucose (which most organisms can use). However, this is not economic now.

Glycosylation (glycoprotein)

Glycosylation is putting sugar molecules on things, almost always other molecules and usually proteins: glycosylated proteins are called glycoproteins. Most of the proteins present on the surface of cells, viruses, and in the blood in animals are glycosylated, and this is important for their function, so some biopharmaceuticals also have to be glycosylated to have the same function as their natural counterparts. Bacteria do not glycosylate their proteins in the same way as humans, so biopharmaceuticals are often produced in yeast or cloned mammalian cells to allow a more "natural" pattern of glycosylation.

Sugars can be linked onto the proteins through the amide groups of asparagine in the short peptide sequence Asn-X-Ser/Thr, or, more rarely, through the hydroxyl of serine and threonine. This means that how much a protein may be glycosylated can be predicted to an extent from its amino acid sequence, and hence from the sequence of its gene.

Such glycosylation is a form of post-translational modification, that is, modification of the protein's chemistry after the protein has been "translated" from RNA. Other protein glycosylation is chemical, and occurs whenever a protein sits in sugar solutions for a long time (*see* **Glycation**).

Other molecules can be glycosylated, especially cell surface lipids, which are therefore called glycolipids. The resulting glycolipids are important as tags to allow the body to recognize its cells, especially cells in the blood and nerves. Thus, they may be important functional components of liposomes, enabling the maker of liposomes to fool the body into thinking that they are cells. Proteins can also have lipids linked on (forming lipoproteins) or even glycolipids. The results cause very different responses from the immune system than the unmodified protein: however making such complex derivatives is much more difficult than making relatively simple glycoproteins.

Although proteins have well-defined places where sugars can be coupled onto them, whether sugars are coupled on, and what sugars are coupled, depends on many things. Among them are the cells the proteins are made in, and the metabolic state of the cells. Thus, proteins come in variants with different sugars linked onto the same polypeptide chain—these variants are called glycoforms. Cancer cells often produce different glycoforms from normal cells, usually glycosylating their cell surface proteins less. Many tumour markers are in fact glycoprotein differences which are specific to the cancer cells, and hence which are potential ways of diagnosing the cancer or targeting drugs to it.

Gold and uranium extraction

Gold and uranium are mined in commercial quantities using microbial leaching methods, because of the high value of the metals and some specific features of the elements.

Gold is usually found as metallic gold mixed with other materials. Crushing the minerals releases the gold metal, which can sometimes be separated physically, often by washing ("panning"). However, substantial sources of gold are ores in which the gold is extremely finely divided and so cannot be released by conventional crushing or milling: these are called refractory ores. Many different ore types with widely differing chemistry can contain gold, but it is often associated with sulphides, especially pyrites and arsenopyrites. Bioleaching methods digest the refractory gold ore in a tank fermentation system with a bacterium, usually *Thiobacillus ferrooxidans*, which oxidizes the sulfide to sulfate. This is usually soluble, so the gold particles are released for mechanical collection. Gold extraction using biological processing is gaining support because the alternatives—oxidation of the sulfur to sulfur dioxide or dissolving the gold out of the mineral using cyanide—are considered environmentally unacceptable.

Uranium mining follows more conventional bioleaching lines, with ores that are low in available uranium being incubated with an oxidizing bacterium to release the metal. The tetravalent insoluble uranium is oxidized by ferric ions (generated by the bacteria) or directly by the bacteria themselves to soluble uranium(VI) ions. These can then be recovered from the nutrient mix running off the ore heap. Uranium mining has taken a substantial downturn in the last few years as nuclear power has slipped from favor.

GM

Standing for "genetically modified," as in "GM crops," GM means any organism that has been modified using recombinant DNA technology. GM is used as a political rallying cry against recombinant DNA techniques, particularly in agriculture, particularly in the United Kingdom and Germany by groups such as English Nature and Greenpeace. In the United States, GM crops are usually called "GE"—genetically engineered. US farmers have traditionally been more open to GM crops, but in late 1999, six farmers started the reversal of this trend by suing Monsanto for trying to monopolize the market with GM crops that had not been adequately tested.

All plants are genetically modified, of course—the wheat that fed the Roman Empire and the maize that fed the Incas were completely un-natural hybrids that probably could never have survived in the wild. However, this is usually overlooked, on the not unreasonable basis that after 10 000 years of agriculture we have some idea what these crops will do, whereas the new ones are relatively untested.

Much of the opposition to widespread use of GM technologies is based on the "precautionary principle." This states that if we do not know what the result of an experiment or a product release would be, but have some idea that it could be bad, we should not proceed as a precaution against the possibility that the worst result comes to pass. The problem with this sensible-sounding principle is that there is no criteria for whether belief in a possible hazard is realistic, and of course it is impossible to prove that your belief could *never* come to pass—you can only prove that it has not done so yet. So anything can be banned under the precautionary principle, and therefore decisions are made on political clout, in which the introduction of new crops has to be seen in the context of a general decline in support for agriculture in Europe as a whole: farming fell from 1.6% of GDP in 1990 to 0.8% in 2000, and employed 15% fewer people.

Another apparent compromise is the idea of labeling GM food. Together with unverified scare stories that claim GM food is dangerous because it cannot be proved to be absolutely safe, this means that it is effectively banned from sale. The campaign against GM crops has been greatly helped by evocative label of "Frankenfood," and has resulted in four out of five of the largest UK supermarket chains saying that they will exclude all GM crops from their products, even including non-GM maize and soya in case they have been contaminated. This leads major suppliers such as Frito-Lay, Seagram, and H.J. Heinz to make the same public announcements. Labeling food does not guarantee informed choice, of course. In a survey in the United Kingdom in 2000, consumers wanted GM tomatoes labeled in case they had genes in. Only 40% realized that non-GM tomatoes had genes in as well.

Many studies have examined the safety of GM food. Nearly all have found GM food to be broadly as safe as other food. One notorious exception was a paper from Arpad Pustzai, a researcher in Scotland, who published that GM potatoes caused gut disease in rats, something that created huge media interest in the United Kingdom, especially as he was fired shortly afterwards. He was actually fired for reporting other people's research data without their permission (a serious crime in scientific circles), and many subsequent studies showed that it was force-feeding rats a diet of raw potato that gave them the gut disease, and whether it was GM or not made no difference.

It is not clear if the political pressure from the anti-GM lobby is generated by career political lobbyists or is driven by real consumer demand. In the late 1980s, before the lobby groups got started, consumers appeared to welcome GM products, and demand for the Flavr Savr tomato outstripped supply in the mid-1990s, before corporate and patent problems shut the product down. The product offered a clear advantage to consumers that they were willing to pay a premium price for.

Plant crops are not the only GM organisms to be deliberately released: transgenic salmon have been developed and if used widely would be a major release of organisms (*see* **Fish**), and transgenic mosquitoes have been suggested as a way of controlling malaria.

GPCR

G-protein coupled receptors are by far the largest group of receptor proteins found in cell membranes. The family of GPCR proteins includes receptors for amine neurotransmitters (targets for antidepressants and antipsychotics), opioids, histamine (targets for anti-allergics), chemokines, cytokines, many hormones.

The GPCRs are also called "7-transmembrane receptors" because their core structure has seven alpha helices which span the membrane. Part of the protein binds to smaller proteins inside cells called G-proteins, a three-peptide complex that binds GDP (guanosine diphosphate). When the receptor is activated, the G protein releases its GDP, binds GTP, activates other enzymes in the cell at the start of different signaling cascades, and then hydrolyses its GTP back to GDP, binding back to the receptor. The signaling cascades are often kinase signaling pathways, which go on to have the ultimate effect of the hormone (*see* entry on **Kinases**).

GRAS

Stands for "generally regarded as safe," and is an important category for acceptance of biotechnological products in Western countries, and especially the United States. It means that the product has a long history of use for a specific purpose.

For microbial or genetically engineered products, regulatory approval for general release is much easier if the product is made in an organism which is GRAS, as the only unknown is then the new product, not the organism as well. For isolated materials, to be accepted as GRAS in one application (e.g. as a foodstuff) can greatly help getting approval for another application (e.g. cosmetics). Pharmaceutical products do not usually get a GRAS rating at any time.

GRAS status varies from country to country. Quite a few of the biotechnological products used traditionally as foodstuffs in Japan, such as sea-weed derived materials and natural colourings like astaxanthin, were not seen as GRAS in the West because there was no history of their use here.

Growth factors

Growth factors are materials (apparently invariably proteins in mammals) that stimulate growth. They are of great interest as potential drugs (biopharmaceuticals) because they could be used to assist wound healing or encourage tissue regeneration. The growth factors not only stimulate cell division, but also influence differentiation of the cells and in some cases select which cells in a mixed population divide or differentiate.

Most growth hormones come as small gene families—there are at least half a dozen types of VEGF, for example. However, it is convenient to list them as single molecules. The ones most often discussed are:

- *Growth hormone* (usually called HGH—human growth hormone—or somatotropin). The original protein growth hormone, commercialised from recombinant DNA technology as Protropin by Genentech in 1985. It is produced by mammals in growth and early adulthood: production falls in humans after the age of 30: injections after this age cause muscle to build up and fat to decrease. Recombinant growth hormone has been approved for a variety of growth deficiencies, some where growth hormone is missing, some (like Turner's syndrome) where other things are causing excessively short stature. GH has the dubious privilege of being one of the few biotech drugs that is abused to any significant level (*see* Sports biotechnology). Serono launched its own GH in 1989, called Saizen. Recombinant growth hormone is now a biogeneric.

    BST—bovine somatotropin—is the cow equivalent, and is used as a growth enhancer in cattle. This was enormously controversial in the early 1990s for ethical as much as health grounds.

    It has been suggested that HGH may reduce, or even reverse, the reduction in muscle mass which occurs with aging, and also improve skin elasticity and muscle tone. Controlled trials suggest that any a small effect is offset by side-effects, so this use of HGH is not approved: however a flourishing industry sells "growth hormone equivalents" to consumers. None of them actually are growth hormone equivalents.

- *Bone morphogenic proteins (BMPs)*. These have bone growth stimulating effects, with potential therapeutic uses in encouraging bone to regrow and bone cells to colonize bone replacements. Recent work has shown that they also have roles in early embryonic development.

- *Epidermal growth factor (EGF)*. This stimulates a variety of cells in the upper skin to divide and differentiate. It could have a use in helping wounds to heal. EGF-related mutations are common in some cancers, breast, gastric, and colon cancers. Either the EGF receptor is mutated so that it is always "on," or the EGF gene is mutated to that the cell always makes it, stimulating its own growth (a sort of feedback called "autocrine" stimulation). The anti-cancer drug Iressa (*see* **Kinase**) works by inhibiting the kinase associated with the EGF receptor.

- *Erythropoietin (EPO)*. A factor which stimulates the bone marrow cells which give rise to red blood cells, used to boost the number of red blood cells in the blood, which is useful for leukemia or kidney dialysis patients. The gene for erythropoietin was cloned in 1985, and the protein product launched in 1988 as Epogen, a record speed from gene to product. As well as Epogen and its follow-up product with altered post-translational modification Aranesp (Amgen), erythropoietin is marketed as Procrit (Ortho Biotech) and as a generic by a number of other manufacturers.

- *Fibroblast growth factor (FGF)*. Stimulates growth of the cells common in connective tissue and the "basement membrane" which many cells are attached to. Suggested as a stimulant to the healing of burns, ulcers and bones. There is a variant called basic FGF (bFGF) which is involved in a range of wound healing processes, including nerve growth.

- *Insulin-like growth factor (IGF)*. This has a role in many growth and metabolic systems, and is believed to be a significant player in aging. Cephalon Inc. tried and failed to commercialize it in the 1990s, but Tercica Medica are looking at it again to see whether improved scientific understanding can find a therapeutic use for the protein.

- *Keratinocyte growth factor*. This stimulates the keratinocytes (skin cells) to grow. Applications are in wound healing, and in building artificial skin and treating skin thinning caused by aging and drug treatments.

- *Neurotropins (also nerve growth factors, neurotrophic factors)*. A variety of molecules that stimulate nerve cells to grow, survive in cell culture and differentiate. They include, nerve growth factor (NGF), neurotropin-3 (NT-3), which is generating particular interest because it may have potential as a therapeutic for degenerative neural diseases such as multiple sclerosis or Alzheimer's disease, ciliary neurotrophic factor (CTNF) and brain-derived neurotrophic factor (BDNF) are similar to NGF but targets brain cells.

- *Platelet-derived growth factor (PDGF)*. This stimulates connective tissue to grow, and is associated with wound healing.

- *Stem cell factor*. Any protein which stimulates "stem cells" to grow and remain undifferentiated. See separate entry.

- *Vascular endothelial growth factor (VEGF)*. This is a protein that stimulates new blood vessels to grow. It is of medical importance for two reasons. It can be used (in theory) to stimulate new blood vessels in diseased tissue, especially the heart after a heart attack. However, it is also made by some cancers to stimulate angiogenesis, the building of new blood vessels to feed the tumor. Therefore, as an anti-cancer therapy, VEGF antagonists are being developed.

All growth factors have different effects depending on the site of application, and most need specific, local delivery if are to be effective (e.g. BMP, NGF), of must be delivered at particular times of day (e.g. VEGF, HGH).

Hairy root culture

This type of plant culture consists of highly branched roots of a plant. A piece of plant tissue (an explant, usually a leaf or leaf section) is infected by the bacterium *Agrobacterium rhizogenes*. Like its cousin *Agrobacterium tumefaciens, A. rhizogenes* transfers part of its own plasmid DNA to the cells of an infected plant. This causes alterations in hormone levels, which causes the explant to grow highly branched roots from the sites of infection. The roots branch much more frequently than the usual root system of that plant, and also are covered with a mass of tiny root hairs, hence the name of the culture system.

The hairy root cultures do not require hormones or vitamins to grow, unlike explant cultures or cell cultures of plant cells, so they can grow on simple media of salts and sugars. They are also genetically stable, again unlike explant or cell cultures, so they can be cultured in bulk without the culture changing. Their most significant feature, however, is that they produce secondary metabolites at levels similar to those made in the original plant, for production of compounds such as food flavors or fragrances.

Hairy root cultures can be grown in unstirred tank reactors because they grow and metabolize more slowly than bacteria and do not have nearly such high oxygen demands, stirring is not usually necessary to obtain a successful culture. They can also be grown in a stirred tank reactor, but they are rather sensitive to being broken up by the stirring machinery.

Harvesting

Harvesting in biotechnological terms usually means collecting cells or organisms from a growth system. If the organisms are very large (like, say, GM salmon), this is not difficult. However, most biotechnology uses single-celled organisms like bacteria or yeast, which have to be actively collected. Among the ways of doing this are:

Centrifugation. Expensive, but guaranteed to collect any particle. It can be used in small volumes to purify viruses, and anything as large as a bacterium can be processed quite easily. There are two variants. The mixture can be centrifuged in a "pot," where the biomass sinks to the bottom and forms a pellet. Alternatively, the mixture can be spun in a pot with holes in it, and the filtrate is spun out while the solid remains behind. This has the advantage that more liquid can be added, so the centrifuge can be run continuously. This is called basket centrifugation it is in fact a combination of centrifugation and filtration. See seperate entry.

Filtration. There are a range of filtration systems. This is cheap and again effective, but usually has limited capacity. The reason for this is that the filter has to be full of holes that are just a bit smaller than the cells you want to collect. So after a while the cells fill up all the holes, the filter is fouled, and filtration stops. **Cross-flow filtration** (see separate entry) can help to solve this.

Flocculation. By adding an additional reagent to the reaction mix, or by altering conditions, you get the cells to stick together and settle out like snow. (Any loose association of particles into an irregular clump is called a "floc.") This is often the only practical way of removing cells from really big fermentations, such as beer brewing.

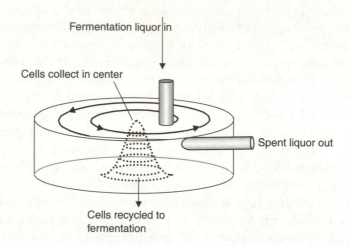

Hydrocyclone. This is also called a whirlpool separator. Here, the liquid is flowed relatively slowly round a bowl and the cells collect in the middle, like grinds in a stirred coffee cup. Cells are collected from the centre, usually to be recycled to the fermentation. Spent liquor, largely free of cells, is collected from the periphery. This is a highly energy efficient method of separation.

Herbicides and resistance

One of the earliest targets of genetic engineering used on plants was to make them resistant to common herbicides. If a broad spectrum herbicide was sprayed onto a field planted with such resistant crops, then all the plants except the crop would be killed, thus providing an effective method of weed control without having to develop herbicides specific to each weed type.

The tolerance mechanism has to be designed to fit the herbicide— consequently, different companies have been working on engineering resistance to their particular herbicide. There are two general approaches: alter the enzyme which the herbicide normally attacks so that it is no longer a target for that chemical, or add a system for detoxifying the herbicide in the plant.

The groups of pesticides which biotechnologists have examined so far are:

Glyphosate. Marketed by Monsanto as Roundup, this is a very commonly used herbicide which stops amino acid synthesis. Plants resistant to glyphosate have been created by giving them new, resistant enzymes and by selecting resistant cells and cloning them into whole plants. Monsanto has commercialised "Roundup Ready" brand cotton and soy.

Phosphinothricin (PPT). Produced by Hoechst, this herbicide blocks amino acid synthesis. Resistant alfalfa has been created by isolating alfalfa cells resistant to the herbicide and cloning whole plants from them. Plant genetic systems have also engineered tobacco and potato to resist phosphinothricin.

Sulphonylureas. These block amino acid synthesis. Mutant genes from *E. coli* have been put into plants to confer resistance.

2,4 D. A compound which mimics plant hormones, and so disrupts their growth. Bacterial genes which break it down have been put into plant cells.

Triazines (atrazine, bromoxynil). These disrupt photosynthesis, and resistant target genes are known. Getting this altered gene product into the chloroplast is, however, a major problem. Ciba Geigy (now Novartis) was working on an alternative route, putting enzymes which detoxify atrazine into several crop plants: because the detoxification enzymes work in the cytoplasm, this may be a simpler route for the genetic engineer.

High throughput screening (HTS)

This is an approach to drug discovery. It searches for a chemical that will act on a defined target, such as a receptor or an enzyme. Scientific research will have identified that target as a part of the mechanism of a disease: for example, it may be a cell surface molecule through which cells of the immune system become activated to start an inflammatory reaction, and so be a target for anti-inflammatory drugs.

There are two approaches to finding a chemical that will block that target. The first is a rational drug design approach, designing a molecule that you believe will interact with the target (*see* **Rational drug design**). Alternatively, you can search a huge range of essentially random molecules to find one that works fairly well as a blocking agent, and then optimise it. This is the HTS approach.

HTS requires a large **chemical library** to screen (see separate entry), and a highly automated method of performing biological assays on each of the compounds. The assay can be anything a biologist would do in the laboratory, but must be amenable to being done on a robot, and hence should be relatively simple, and capable of being done in small volumes of liquid. (Performing HTS on half a million mice is not practical.) Automation nearly always means using assays in microplates or just "plates." This used to be in 96-well plates, but now the 384- or 1536-well plates are standard (*see* **Standard laboratory equipment**).

Hollow fiber

Hollow fibers are tubes of a material that is porous. The tubes are very small, typically having an internal diameter of a fraction of a millimeter, and so their ratio of surface area to internal volume is very large. Hollow fiber bioreactors are versions of dialysis bioreactors or dialysis fermentors. This is any system where the cells in a fermentation are retained behind dialysis membrane (a membrane that lets small molecules pass through it, but will not let large molecules or organisms pass). The reason for using fibers as opposed to a single sheet across the bioreactor is that fibers have a very much larger surface area for fermentation substrates to diffuse across.

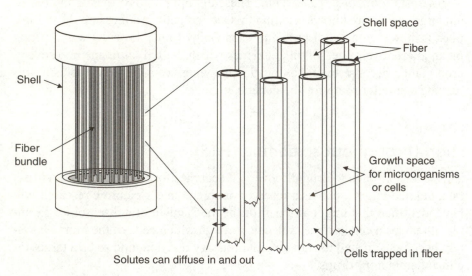

End connectors—make seal between large connector pipes and hollow fibers

Shell space

Fiber

Shell

Fiber bundle

Growth space for microorganisms or cells

Solutes can diffuse in and out

Cells trapped in fiber

This has had two types of applications.

First, hollow fibers can be used as filters. Because they have a huge surface area, they take much longer to clog up than normal filters. The filters used in artificial kidney machines are often hollow fiber bundles.

The second use is the hollow fiber bioreactor. This is a widely used type of bioreactor in which cells are kept inside the hollow, porous fibers, and the culture medium is circulated outside the reactor. The fibers have pores large enough to let the nutrients in and the product out, but not to let the cells out. The fibers are kept inside a shell: the space outside the fibers but inside the shell is called the shell-space.

Hollow fiber bioreactors are very effective for using mammalian cells to make something. They have a very large surface area for the cells to adhere to without needing a large reactor to hold them, the nutrient reaching the cells can be kept fresh, and the product can filter out into the shell space easily.

Hollow fiber reactors have been particularly useful for making large amounts of monoclonal antibodies.

Hollow fiber reactors are less use when the cells themselves have to grow, because it is hard to get at the inside of the fiber to remove surplus cells, and hard to monitor exactly how many cells you have inside the fibers. This has meant that hollow fiber reactors have limited use for bacterial cultures.

Homologous recombination

Homologous recombination is a biological process by which a living cell joins together two pieces of DNA that are very similar ("homologous"). It is a version of the general genetic process of recombination, by which any two pieces of DNA are spliced together in a living cell. Recombination occurs in all living things: recombinant DNA technology was called that because of the similarity of its gene-splicing technology to natural recombination processes.

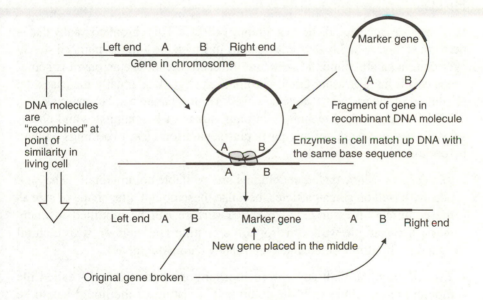

This is used as a mechanism for ensuring that a cloned gene which the experimenter wishes to put into the chromosomes of a cell is inserted into those chromosomes at a specific point (that is, at the point where the cell's DNA is the same as the cloned DNA). Because of this use, homologous recombination is sometimes called gene targeting.

Homologous recombination is used in biotechnology in three areas.

In generating new mutants of many organisms, but particularly yeast and mice, homologous recombination is a method for targeting a specific bit of DNA. The targeted gene has a piece of "junk" DNA spliced into it, preventing it from working, and so producing a mutant (*see* **Knockout**).

The second role is in manipulating large plasmids such as the Ti plasmid of *Agrobacterium tumefaciens*, which are too big to alter using recombinant DNA techniques. Genes can be spliced into them using homologous recombination between part of the Ti plasmid and a second plasmid bearing the "new" gene that the experimenter has introduced into the same cell.

The third application is in making transgenic animals (and potentially in gene therapy). Here we try to avoid disrupting any genes, and make sure that the foreign gene arrives in a suitable chromosomal environment for it to be expressed (*see* **Gene silencing**). This also offers the possibility of useful gene therapy in man, as one of the main problems with gene therapy ideas at the moment is the threat that the "therapeutic" gene introduced into a patient's cells will do as much damage as the original disease. However, homologous recombination happens much less efficiently in human cells.

HPLC

High-performance liquid chromatography (HPLC). This chromatography that is performed in very highly defined columns, in which the liquid is pumped slowly over quite a small column at very high pressure (hence the acronym is sometimes called high-pressure liquid chromatography). This greatly increases the resolution of the method, that is, how well it can separate similar materials.

HPLC is a standard technique in many areas of biochemistry and chemistry. To make it work for any particular separation task, two things have to be optimized:

- *The column*. Many different columns are available commercially. They are usually based on silica particles, because these do not alter shape or size as the column is pressurized. The silica surface can be modified in many ways, for example with a hydrophobic coating (see below), with a chiral ligand or an affinity molecule (*see* **Affinity chromatography**).

- *The solution*. As with any chromatography method, a liquid—called the mobile phase—flows over the column. The nature of the liquid has to be chosen carefully so that it dissolves what you want to analyze well, but in a way that allows some of the material to bind to the column as well. Thus, the mobile phase has to be selected for the problem and for the column you are using.

The two broad categories of HPLC are normal phase HPLC (where silica is used as the support and the mobile phase is a hydrophobic solvent like chloroform, methylene chloride or hexane), and reverse phase HPLC (where the silica is coated in hydrophobic molecules, such as long chain alkane groups, and the mobile phase is more hydrophilic, such as mixtures containing water, methanol, acetonitrile).

HPLC can be used in large scale to make pure compound at a laboratory scale—this is called preparative HPLC. However, using it as an analytical tool is much more common, for which you need a detector to tell when any molecule is coming out of the column ("off the end" of the column). Common ones are UV absorption (many organic molecules absorb UV light strongly), flame ionization (the liquid is vaporized in a flame and ionization measured by measuring current), and others.

More advanced HPLC systems use a mass spectrometer as a detector. This is called HPLC–MS (or sometimes LC–MS). The HPLC separates compounds, and then the MS tells you what they are. Companies such as Waters and Applied Biosystems provide integrated HPLC–MS systems, but they are quite expensive and complex pieces of kit.

Human Genome Project

The Human Genome Project is the descriptive name for the program to map and then sequence all the DNA in the human genome. The first phases were performed by two groups—an international academic collaboration funded by national and trans national research councils and charities, and the biotech company Celera funded by stock offerings. There was intense and sometime acrimonious competition between the two. In mid 2000 they came together and agreed on a collaborative agenda, and released a "working draft" of the sequence (i.e. one that was not very accurate). There was a "first release" of an almost complete and accurate sequence in 2001. The sequence will not be "complete" until 2003.

The academic consortium can be said to have won the race, as some of its leaders are now getting Nobel Prizes while Celera's stock price fell from around $240 in February 2000 to around $10 $2\frac{1}{2}$ years later. Only the academic consortium is now working to complete the genome.

In theory, the whole operation is under the international umbrella of the Human Genome Organisation (HUGO), but HUGO has been more of a meeting coordinator than a group with any clout.

The project is supported strongly by the biotechnology and pharmaceutical industries because it was believed that

- the genome would provide a database of information from which companies can obtain understanding of all possible drug targets in the body, and use this to design new drugs

- before that, every new gene that they discovered could be patented, thus "owning" that drug target for 20 years.

Neither of these turned out to be realistic (*see* **Functional genomics** and **Patents**). In particular, the rewards in terms of drug discovery have been disappointing, and most companies working in the area are turning to "functional genomics" as the next step in the program. However, the genome sequence is a terrific resource for basic and applied research alike.

The genome sequence is 3164.7 million bases (megabases) long. 99.9% of this is identical in every human (except that men have an extra small chromosome—the Y chromosome—that women do not possess). In 1999, it was believed that the genome contained between 100 000 and 150 000 genes. However, the final sequence showed only between 30 000 and 35 000 (only twice as many as *Drosophila*). Part of the uncertainty is identifying what part of the sequence is actually a gene. Less than 2% of the genome actually codes for protein, and of those we can assign even tentative functions to only about 50%, and at least 25% we cannot even make a good guess at what they do. Only a tiny fraction of genes have their protein function *proven* by experiments. So the sequence is far from the end of genomics.

The next stage in the program is a systematic survey of variability between people, which means sequencing (or resequencing—*see* **Gene chip**) the genomes of many more people. The genome sequenced by the public consortium was a pooled collection of many people's DNA. The Celera sequence is mostly of Craig Venter, Celera's founder.

See **Genome project**.

Hybridization

Hybridization has several meanings in biotechnology and molecular biology.

DNA hybridization. This is the formation of a double helix of DNA from two single DNA strands. The two separate strands of DNA will come together to make a double helix if their bases are complementary, so that wherever there is an A in one strand there is a T in the other, a G in one strand and a C in the other.

DNA usually comes in a double helix. To make a hybrid molecule, the original DNA helix must be separated into two strands, a process called "melting". The temperature at which a particular helix "melts" depends on

- how long it is (for helices shorter than about 50 base pairs, the shorter they are the lower the melting temperature)

- how may G and C bases it has (more G + C means a higher melting temperature)

- whether the helix is perfectly matched, or whether one or more of the bases in it is mismatched, that is, is not opposite its complementary base in the helix

- a range of chemical conditions, like how much salt there is in the solution.

The same "melting temperature" governs the temperature at which a new hybrid helix will form. Thus it is often used as a measure of how stable a new helix would be.

DNA hybridization is used as a method for using one bit of DNA (the probe) to find out if a complementary bit of DNA is present in some mixture of DNA species. It is used in "blot" techniques, PCR, gene library screening, DNA fingerprinting and a range of other techniques.

Molecular hybridization. This is forming a new molecule that has functional parts from two different molecules. It could therefore have a combination of properties from its two "parents." Examples of this approach are the new antibiotics that can be made by combining the enzymes that make two old antibiotics in one cell, and making fusion proteins by joining two functional domains of other proteins together.

Cellular hybridization. This is essentially another term for cell fusion.

Species hybridization. This is forming a hybrid between two species. It can be done by mating a male and a female from two closely related species, as a mule can be created by horses and donkeys. However, nearly all animal species cannot be "crossed" in this way (this is the classic definition of a species—that they can breed with each other but not with other species). Alternative methods of making animal hybrids include making Chimeras (see separate entry). Plants also can be hybridized by fusing the cells of two species with all the genes of both (this approach does not work for animals).

Hydrophobicity

A hydrophobic molecule is a molecule that dissolves very poorly if at all in water, but dissolves quite well in solvent such as chloroform or toluene that does not itself mix with water. They are non-polar molecules, that is, they are essentially electrically neutral all over. The opposite is the hydrophilic molecule, which dissolves well in water, or solvents that mix well with water such as DMSO (dimethyl sulphoxide) or methanol, but poorly if at all in toluene or long-chain alcohols. These molecules usually have partly charged groups on their surface, and often form ions when dissolved in water. Most biological molecules are to some degree hydrophilic, a major exception being the triglycerides ("fats"), which are hydrophobic. Hydrophobic molecules are therefore sometimes called "lipophilic."

When given a choice of environments—for example, a mixture of water and octanol to dissolve in, hydrophobic molecules will choose a hydrophobic environment (in this case octanol), hydrophilic molecules a hydrophilic environment (water). The "partition coefficient" between water and octanol (what fraction of the compound ends up in the octanol) is a standard measure of hydrophobicity—its logarithm is called Log P.

Differences in hydrophobicity can be used to separate the molecules. Hydrophobic chromatography uses this phenomenon: a mix of molecules is passed over a solid material that is mostly hydrophobic in character. The molecules that are more hydrophobic will stick to the material more strongly, and so will not be washed across the solid support as fast as hydrophilic ones.

Many biological molecules are sufficiently large to have distinct hydrophobic and hydrophilic bits. These molecules are called amphipathic. If the two regions of the molecule are at opposite ends, then the result is a *surface active* material: it will tend to congregate at the junction between a hydrophobic and a hydrophilic solvent. Phospholipids are of this type. Phospholipid membranes are arranged so that the "tails" of the phospholipids form a layer of hydrophobic "liquid" that dissolves quite different chemicals to the watery phase around it. Proteins also almost invariably have a mixture of "hydrophobic" and "hydrophilic" amino acids. The protein folds so that most hydrophilic molecules are exposed to the waterery solution it is dissolved in, and most hydrophobic ones tucked away inside the protein or exposed to lipids in a cell membrane if it is a membrane protein. Thus, the distribution of hydrophobic and hydrophilic molecules along a protein (sometimes called a "hydrophobicity plot") can be a clue as to how the protein will fold up. In particular, proteins with a large region of hydrophobic amino acids in the middle of its sequence are often associated with membranes, with the hydrophobic amino acids embedded in the hydrophobic layer in the middle of the lipid layer.

Hype

Biotechnology has been the subject of more unfounded hype than any other industry, except maybe the "dot.com boom." This is because it attempts to use very powerful and exciting technologies to solve very pressing but very hard problems, on a timescale of decades.

The typical cycle is:

1. *Invention*: New bit of science is discovered, and the potential applications realized. There is a six month burst of enthusiasm for it, and a dozen companies are set up to exploit it.
2. *First* in vitro *results*: Feverish work, and the first laboratory results a couple of years later attract the general press. Claims that it will "cure cancer" usually start around now.
3. *First clinical trials*. The first products enter clinic. Primetime TV news hears of this. The phrase that this "may be a cure for cancer, heart disease, possibly even ageing itself" is the standard formula at this stage.
4. *First clinical failures*. First early trials fail. The initial laboratory experiments are found to be mostly artefacts, or even fraud.

5. *Second generation programs started*. With little fuss, the lessons of the first wave of experiments are put into practice and more realistic, more workable products and experimental programs initiated.
6. *Second clinical failures*. Nearly all the other early clinical trials fail. Technology is decried as useless. Most of the start-up companies go bust or are bought up.
7. The only successful first generation product is launched. Brief revival of general interest in the technology.
8. Three to five years later, second generation technology products start to be launched. Technology takes its place among the armamentarium of the day.

For technology product companies such as the gene chip companies, the milestones are:

1. Invention;
2. First prototype;
3. First product launch—primetime TV talks about "a fundamental new technology for understanding disease;"
4. First products fail to sell;
5. Second generation programs started;
6. Nearly all the early technology fails to sell. Technology is decried as useless. Most of the start-up companies go bust or are bought up;
7. The only successful first generation company becomes profitable;
8. One to two years later, second generation companies go into profit. Technology takes its place among the armamentarium of the day.

The timescales are almost the same as for drugs.

This 15-year cycle is usually also synchronized with waves of stock market enthusiasm for biotechnology as a whole. This is because venture capitalists invest in expensive, speculative new technologies when the market is buoyant, and the apparent failure of those same technologies then triggers the next downturn.

Examples of such technology waves have been rational drug design, recombinant DNA drugs, antibody therapeutics, antisense, gene chips, and gene therapy.

Imaging agents

A range of proteins are being developed as imaging agents or contrast agents. This means that they are for use with the various types of body-scanners. The proteins (usually antibodies) are linked to a chemical group which allows a scanner to see them very easily. The proteins bind to specific tissue types, usually tumors, and so allow the scanner to distinguish those tissues from the surrounding tissue.

Imaging agents can be made for any of the main imaging systems:

- *CAT scanning* (*computer-aided tomography*). This techniques uses X-rays, and consequently the label tagged onto the antibody is an X-ray opaque substance, usually something made from a "heavy" metal like gold.

- *PET scanning* (*positron emission tomography*). This technique injects a very small amount of a radioisotope into the body, and then tracks where it goes by following the path of the radioactive particles. The favored isotope to tag onto your antibody for this is technetium.

- *MRI* (*magnetic resonance imaging, an application of NMR—nuclear magnetic resonance*). This uses the way that the body absorbs microwaves when it is in a strong magnetic field. Chemical groups absorb microwaves differently depending on what sort of field they are in and what the group is. A wide range of materials can be used as "contrast agents" for NMR scanning.

Immobilized cell bioreactors

Many of the plant and animal cells grown by biotechnologists are most effectively handled when the cells are immobilized onto a solid support. This helps to shield them against the stirring forces necessary to mix a bioreactor's contents, and makes them easier to move around and separate from the substrate.

There are a wide range of immobilized bioreactors. They fall into three classes:

Membrane bioreactors. These grow the cells on or behind a permeable membrane, which lets the nutrients for the cell through but does not let the cells themselves out. Variations on this theme are the **hollow fiber reactor** (see separate entry).

Filter or mesh bioreactors. Here, the cells are grown on a open mesh of an inert material which allows the culture medium to flow past it, but retains the cells. This is similar in idea to membrane and hollow fiber reactors, but can be much easier to set up, being similar to conventional tower bioreactors with the meshwork replacing the central reactor space.

Carrier particle systems. This is the most common form of immobilized cell reactor: the phrase "Immobilized cells" is often taken to mean that the cells are immobilized onto small solid particles. The particles are not much larger than the cells, often small nylon or gelatin beads. The reactor can then handle the beads in the same way as granular catalysts are handled in chemical reactions. If the particles are of neutral density, so that they do not sink or float, they can be pumped round with the liquid. If the particles settle out fast, the bioreactor can be a fluidised bed or a solid bed bioreactor.

An alternative to growing cells on microcarriers is to grow the cells as aggregates. Cell aggregates have some of the mechanical robustness of cells on microcarriers, and have much higher cell content for a given amount of solid matter. However, getting cells to grow in aggregates can be much harder than getting them to grow on suitably treated polymer surfaces.

Immobilized cell biosensor

These are biosensors, detector devices which use a biological component to allow them to detect only one thing at once, where the biological components are living cells. They are often called microbial biosensors, as usually the cells are bacterial ones. The earliest such sensor, however, was the canary in a cage that miners used as early warning for methane in coal mines.

As with any biosensor, there are two parts to immobilized cell sensors: the immobilized cell (which does the sensing and produces a very weak signal of some sort) and a detection and amplification system which takes that very weak signal and amplifies it to something detectable.

The cell used depends on what you want to detect. Analytes ("things to be analyzed") include amino acids, glucose, toxic chemicals (using any bacterium which is sensitive to the chemicals to be detected), carcinogens (using bacteria which are defective in DNA repair genes), BOD, herbicides (using plant cells or blue-green algae).

Only a few of these have been transformed into realistic sensor systems.

The readout methods can be equally diverse:

- *Gas generation/depletion*. This is a favorite one, especially measuring the amount of oxygen consumed (using an oxygen electrode) carbon dioxide produced (using a pH sensor).

- *Light production*. This uses luminescent bacteria, either ones which are naturally luminescent or ones which have been given the relevant genes. Light production is either a measure of general bacterial well-being (for toxicity sensors) or is coupled to the presence of a specific chemical.

- *Direct electrochemical coupling*. Some groups are working on hooking the electrode directly into the bacterium's own electron transport system. This is a more sophisticated version of measuring oxygen uptake. A related approach measures the current flowing through mammalian cells. This is less useful for environmental applications (the cells are too fragile), but very useful for probing what affects those currents, and particularly how the various ion channels in the cell turn on and off under different conditions. This is a variation of the "patch clamp" technology.

Bacterial biosensors are usually less specific than other biosensors, as bacteria are very diverse and complex things. However, they are very active, and so make a "signal" which is easier to detect than that produced by anti-bodies or DNA probes. The bacteria can be used as "live" cells or as spores, which are stronger and do not grow, but which can be harder to engineer to give a useful signal.

Of the few commercial "biosensor" systems, several are bacterial biosensors: two luminescence-based bacterial biosensors (for toxicity and BOD measurement) are in use in the water industry, for example.

Immortalization

Immortalization of a cell type is its genetic change into a cell line which can proliferate indefinitely.

Cells taken from a mammal, called primary cells, will divide in culture for 20–60 divisions, but then stop dividing. This is not because they have run out of room or nutrients to grow, but rather because they have become incapable of dividing any more. They usually show characteristic changes in their structure, and reduce the amount of biotechnologically useful product they produce, be it metabolite or protein. This is called cell senescence, and severely limits the manipulations that a primary cells can be put through.

To avoid this, cells are "immortalized": put through a treatment which allows them to overcome senescence and divide indefinitely, keeping whatever differentiated features they had to start with.

Transfecting the cells with one of several oncogenes, when transfected into a cell, will immortalize the cell, as will telomerase. Some genes from oncogene (tumor-causing) viruses can also immortalize cells, notably the T-antigen gene from SV40 virus.

Alternatively, the scientist can look for a spontaneous mutant in the cells you are wanting to immortalize. This is done by growing a large number of the primary cells in culture and simply looking for any that keep on growing when the others have become senescent. The rate at which this happens varies between different organisms—mice seem to generate immortal cell lines much more readily than humans, for example.

Last, you can fuse your target cell with an already immortalized cell. The result is usually an immortalized cell. This is how the technology of making monoclonal antibodies immortalizes the one lymphocyte which is to make the antibody characteristic of a hybridoma.

A related concept is transformation of cells. This is their immortalization and conversion into a more aggressively growing form of the cell. Typically, such cells will grow to form clumps of cells (colonies) in soft agar media, whereas normal, untransformed cells will not—they need a hard, charged plastic surface to grow on. You can measure the extent to which a chemical or a virus can transform cells by measuring this "clonogenic potential," by exposing cells to the material and then seeing how many can form colonies in soft agar. This is important because it is believed to represent what happens to cells when they become cancerous.

Immune system

The immune system is the system of cells and molecules which gives us protection against invading organisms. The system cannot be pre-programmed to recognise pathogens, because the pathogen would simply alter their appearance at a molecular level to evade recognition, so the immune system has to "learn" what each invader looks like, and then mount an appropriate response to it. Doing this takes a lot of different cell types and molecular interactions. These are important to biotechnology in two ways.

- *As a source of materials.* Many materials made by the immune system are valuable biochemicals, and biotechnology has invested a lot of effort into making them efficiently. Leading the list are antibodies and cytokines.

- *As a target for therapy.* The immune system can go wrong in many ways, and biotechnology has expended much ingenuity in trying to cure those diseases without destroying the beneficial functions of the system.

A discussion of the science of the immune system (immunology) is beyond this book. Some key aspects of relevance to biotechnology are:

- *How antibodies are made.* These are made by white blood cells called B cells, which themselves are stimulated by T cells (under the microscope, B and T cells look virtually identical). Each B cell makes a different antibody, from genes that are rearranged when the B cell is formed. To start the process off the antigen—the molecule that the antibody it to react to—must be "presented" to the B and T cells properly. This is done by yet another specialist cell type, the antigen presenting cell (APC), which is usually a type called a dendritic cell. Getting all this to work efficiently is important for vaccine design and for antibody production.

- *Cytokines.* These are the signaling molecules of the immune system. See separate entry.

- *Immune disease.* Many diseases result from the immune system going wrong, usually from attacking the body as well as invading bugs. These are called autoimmune diseases, and result from the immune system failing to learn what its own body looks like (immunologists talk about the system discriminating "self" from "non-self"). A related aspect is Th1/Th2 switching. The immune system can do two things to attack an invader—attack directly with T cells, or make antibodies against it. The former tends to be against viruses and parasites, the latter against bacteria. Autoimmune diseases such as asthma result from over-active Th1 responses, ones like rheumatoid arthritis from Th2 over-activity. A lot of effort is going into trying to "switch" an immune system from a Th1 to a Th2 response or visa versa, so you can turn off the disease without turning off the whole immune system.

The practical aspect of getting an immune system to make antibodies for us, either to protect the animal (vaccination) or to make antibodies that we later collect, is called immunization. Usually the animal is injected with the antigen several times, and initial dose and then several "boosters." Often the antigen is injected with an adjuvant, a chemical or mixture that stimulates the immune cells to greater activity than the antigen alone. The amount of antibody present is often called the titer of the antibody: this is a measure of how far you can dilute the original antibody and still get measurable binding. 1 in 1000 would be pretty hopeless, 1 in 100 000 would be pretty good. This titer should go up each time you "boost" your subject.

Immunoconjugate

A compound which is a combination of an antibody molecule (or part of one) and another molecule. There are several types.

Immunotoxins. These are conjugates of an antibody and a protein toxin, used to target that toxin to a tissue we want to destroy (see separate entry).

Antibody contrast agents and tracers. These are antibodies conjugated to a chemical that shows up extremely well in an imaging system. *See* Imaging agents.

Antibody-enzyme conjugates. These are complexes where the antibody has been chemically linked to an enzyme. They are widely used in immunoassays, where the enzyme acts as a "flag" to mark the presence of the antibody. A few nanograms of antibody can easily be detected if a suitable enzyme is attached. Common ones are horse radish peroxidase (HRP) and alkaline phosphatase (AP).

Immunodiagnostics/immunoassays

One of the success stories of biotechnology, these are medical diagnostic methods which use antibodies. The antibody only binds its target antigen, and does so at very low concentrations, and so can be a very sensitive test. This combination has meant that monoclonal antibodies are used in about

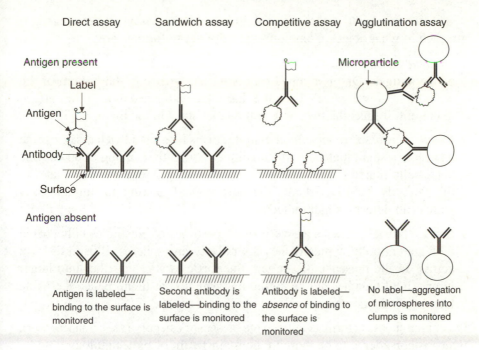

| Direct assay | Sandwich assay | Competitive assay | Agglutination assay |

Antigen present

Label

Antigen

Antibody

Surface

Microparticle

Antigen absent

Antigen is labeled—binding to the surface is monitored

Second antibody is labeled—binding to the surface is monitored

Antibody is labeled—*absence* of binding to the surface is monitored

No label—aggregation of microspheres into clumps is monitored

40% of all medical diagnostic tests. Exactly the same test technology can be used in other, non-medical applications, which are called immunoassays.

However, the antibody does not do anything obvious when it binds to an antigen, so we have to arrange the assay so that some other process detects that binding has occurred. There are various facets to this.

The Label. Antibodies can be labeled in various ways. As well as labels used for **imaging agents** (see separate entry), immunodiagnostics can use a variety of labels in *in vitro* assays. These usually have different names:

- *ELISA*—Enzyme-linked immunosorbant assay. Uses an enzyme label on the antibody.

- *RIA*—Radio-immuno assay. Uses a radioactive label on antibody or antigen.

- *FIA*—Fluorescent immuno-assay. Uses a fluorescent tag on antibody or antigen.

- *CLIA*—Chemiluminescent immunoassay. There are a variety of luminescent (light-generating) labels that can be used. *See* **Luminescence**.

- *Immunogold*. Labeling the antibody with colloidal gold particles. This can be used as a label in a macroscopic assay (where it acts as a very intense dye) or in immunohistology (staining tissue sections with antibodies), where the gold particles show up very clearly in the electron microscope.

A new approach is quantum dot technology, which is a tiny particle but which gives a strong fluorescent readout. *See* **Nanoparticles**.

The second facet is the physical format of the assay—what reagent is attached to what object. Common aspects of assay formats are:

- *Sandwich assay*. Here two antibodies are used which bind to different parts of the antigen. One is trapped on a solid surface (e.g. the bottom of the wells on a 96-well plate). The other has a label attached to it. If the antigen is present, it links the two, and so the label stays in the plate.

- *Competitive assay* (*competition assay*). This is like a sandwich assay, but the analyte is a small molecule which competes with the binding of an enzyme chemically linked to the analyte hapten (the hapten–enzyme conjugate) to the antibody. This is virtually the only way of making an immunoassay which can detect a small molecule.

- *Latex*. "Latex" particles are very small particles of plastic, usually coated with the antibody: typically they are polystyrene spheres, 100 nm to 1 μm across. In the presence of antigen, the particles stick together into larger lumps, held together by the antibodies which coat them, hence the name latex agglutination assay.

Other immunoassay formats fall into the biosensor category (*see* Immunosensor), which is considered to be more in the mainstream of biotechnology.

Immunosensors

These are sensors that use antibodies to identify and bind their analyte. This binding then has to be used to trigger something the scientist can see. This is usually a physical mass-detection system or an optical device. The optical devices are discussed under optical biosensors.

Mass detection is done using extremely small and sensitive mass detectors, usually manufactured on a silicon "chip" (and hence sometimes called "microchip biosensors"), to detect the tiny changes in mass found that occurs when an antibody binds to an antigen.

The simplest type is based on the tuning fork principle. The note a tuning fork sounds depends on the mass of the tines. If the mass increases, the note goes down. The sensors have the equivalent of a microscopic tuning fork with the antibody coated on the tines. The silicon surface from which the tines are made detects the frequency with which they vibrate. When something binds to the antibody, the note falls and the circuit picks this up.

Surface acoustic wave (SAW) devices are a variation on this theme, only the waves are lateral waves in the surface rather then displacement waves. Because the tuning fork is made of piezo-electric material, these are sometimes called piezo-electric sensors.

The problem with all classes of mass-based sensor is that anything landing on such a sensor gives a signal. Thus, despite having a very specific antibody as a biological element, they are very prone to interference.

Immunotherapy

Immunotherapy means giving the patient antibodies or immune system cells to support their own immune system.

Providing general human antibody (gamma globulins, a blood product) is a standard therapy for some infectious diseases. It is non-specific, but relatively cheap.

Making specific antibodies and using them for antibody immunotherapy has been fairly unsuccessful. The molecular target must be highly specific to the disease you want to cure, and must be accessible to the antibody—most antibodies do not enter solid tumors, for example, so an anti-tumor antibody is not likely to have any effect. The antibodies can be used "naked," using the body's immune system to attack antibody coated cells, or can also be conjugated to a "warhead," such as a radioactive atom, a toxin (*see* **Immunotoxin**) or an enzyme (*see* **ADEPT**). Bispecific antibodies can also be used, which bind the target cell with one arm and lock onto another cell with the other, bringing a killer cell to bear on your target.

Immunotherapy can also mean using whole cells from the immune system as a therapy. This latter has been tried under the title of adoptive immunotherapy, taking NK cells ("natural killer" lymphocytes, a type of white blood cell able to destroy other cells) or tumor infiltrating lymphocytes (TILs) from cancer patients, stimulating the cells using cytokines to become more energetic (or transfecting them with cytokine genes), and then injected them back into the patient. The therapy had some effect, but severe side-effects.

Immunotoxins

Immunotoxins consist of an antibody joined onto a toxin molecule. The idea is to target very potent protein toxins to the few cells that you want to kill (usually cancer cells), where a tiny dose will kill the cells.

The toxins used—toxins from bacteria *Diphtheria, Pseudomonas*, or *Shigella* or the castor bean toxin ricin—are extremely poisonous. The toxin is joined to an antibody molecule which can bind specifically to one type of target cell. The resulting "conjugate" is injected into the blood at extremely low concentration, so the systemic toxic effect is small.

Refinements use parts of the toxin molecule, not all of it. Most toxins consist of a part which enables the toxin protein to enter the cell (the A chain) and a part which kills the cell (the B chain). The A chain is not toxic and the B chain on its own cannot get inside the cell to work, and so is much less toxic. Conjugating the B chain to an antibody makes a much less dangerous material: however, it can still kill cells if the antibody binds to them, as the local concentration of B chains around that cell is so high that a few B chains get inside anyway.

Immunotoxins can be made by linking toxin and antibody molecule chemically, or by making a fusion of the genes for the toxin and the antibody: the resulting fusion protein is more stable, and can be smaller and less prone to binding to other tissues than a chemical conjugate. The antibody can also be "humanized," reducing other complications.

Induction

In biotechnological terms this means getting an organism to make a protein by exposing it to some stimulus, usually a chemical trigger (an inducer). In nature the protein is usually an enzyme, and the inducer is the substrate for that enzyme. A classic example is the induction of beta galactosidase (the lac gene) by lactose (the sugar that it breaks down) in *E. coli*—Jacob, Lwoff, and Monod won the Nobel Prize for working out how this genetic system worked. Induction involves the control of gene expression, but it is not a strictly genetic phenomenon, as no new genes or gene rearrangements are involved. It is only the expression of genes already there.

In general, an inducible gene, that is one that is capably of induction, can be induced by one or a few inducer compounds, which act via the promoter part of the gene (*see* **Gene control**). Thus, in order to be inducible, a gene needs to have the right promoter region. Some expression vectors (*see* **Expression systems**) have inducible promoters in them. They must also carry the genes for any proteins involved, of course—the inducer does not bind to naked DNA on its own.

A related term in a repression, which is the opposite effect—a chemical turns a gene off, due to its ability to make a repressor protein bind to the promoter region, or to stop an inducer protein binding there. This can be very important in biotechnology, as many genes for useful enzymes such as those that make antibiotics and other secondary metabolites are repressed by common substances such as glucose: the bugs used must either be engineered to remove this, or grown in conditions where it does not happen, if they are to produce decent yields of the antibiotic.

Induction also means a form of logic that reasons from specific examples of something to general rules about it. This is something biochemists do a lot (despite that fact that it is logically indefensible, it seems to work), but that is rarely what they mean by induction.

Industrial waste (liquid)

Industrial liquid waste is a major economic and environmental issue in the industrialized world. Most industrial processes use water as a cleaning liquid, sometimes in huge quantities. If it is pumped into the general environment with contaminants still suspended or dissolved in it, it can be a major source of pollution. However, cleaning it up is often complex and expensive.

The problem of cleaning up industrial liquid waste is a more specific one than cleaning up sewage, as usually it has a much smaller range of chemicals, but at higher concentration than in sewage. There are two general approaches

- *Integral waste management*. This aims to minimize waste production at all stages of the industrial process. This is the best approach, but is not often very practical. It may or may not have a biotechnological aspect to it.

- *"End of pipe" solutions*. This seeks to maximize the productivity of the industrial process, and then deal with the waste effectively when it leaves the plant.

Generally industrial waste falls into two categories:

- *High carbon wastes*. Many just have very high loadings of carbon compounds, that is an extremely high BOD value. A major example is waste water from food processing, from both major factories and from smaller concerns like chip shops, which both has very high BOD and is very hard to get rid of. Another is lubricating and cutting oils from engineering factories, a major environmental problem because of the scale of the engineering industry. Oxygen gas can be injected into such liquid to help bacteria oxidize it, but this is expensive. A growing trend is to use a combination of biofilter bed-based systems. These can be implemented on many scales: Viridian Bioprocessing has developed a small one for restaurants and pubs.

- *Specific pollutants*. There are a huge range of more specific chemical problems that particular industries have to solve. For example, lignin production from wood and paper processing (*see* **Wood**), heavy metals from many industries. These can be attacked with more specific solutions—*see* **Phytoremediation** and **Biosorption**.

Inhibitor

An inhibitor is anything that stops something happening, but in a biochemical context is usually a chemical that stops a protein from functioning. A few terms are important in biotechnology.

Inhibitors can be specific or general. A general inhibitor blocks all kinds of proteins. Soap or sodium docecyl sulphate (SDS), also called sodium lauryl sulphate, a very powerful detergent, is a general enzyme inhibitor—it will stop almost anything. General inhibitors usually work by denaturing the protein—they just destroy its normal structure so that it cannot function.

Specific inhibitors affect only one or a small number of proteins, leaving the others unaffected. The ideal drug is a very specific inhibitor. Specific inhibitors can work in one of two ways. They can mimic the molecule the protein is meant to bind, so that they compete with the normal ligand for the protein. These are called competitive inhibitors. Or they can bind somewhere else on the protein and make it alter its conformation so that it no longer binds to its target molecule. These are called non-competitive inhibitors. (Strictly they can be un-competitive inhibitors as well—anyone who uses this word has done a biochemistry degree course in enzymology.) Non-competitive inhibitors are also sometimes called allosteric inhibitors: strangely, you can have allosteric activators as well, which make the protein *more* active.

Competitive inhibitors can be designed—a molecule that looks like the normal ligand for a protein, or one that fits into the proteins active site very well, is likely to be a competitive inhibitor. Design of competitive inhibitors of this sort is the basis of structure-based drug design. *See* **Rational drug design**.

A related term is antagonist. This is used in the context of pharmacology, and refers to a chemical which blocks some pharmacological action. Usually, this is an inhibitor of the relevant receptor, and "antagonist" and "inhibitor" are usually used fairly interchangeably when talking about chemicals that bind to receptors. The opposite of "antagonist" is "agonist", which is the compound that makes a receptor do what it is meant to do. The hormone or neurotransmitter that makes the receptor "fire" (function) is its "endogenous agonist" (i.e. the natural chemical that makes it work), but synthetic agonists are quite possible to make, although usually harder to discover than antagonists. Finding what the endogenous agonist of some receptors is a major research project in its own right.

Inoculation

Inoculation (other than as a synonym for vaccination) is introducing a small culture of a microorganism into a new environment with the intention that it should grow there. Thus, fermentors are inoculated at the start of a run with a batch of organisms that are ready to grow rapidly in the conditions provided by the fermentor. It takes some skill to make this work, as the conditions under which the inoculant were grown are probably not the same as those inside the final fermentor, and so the organisms could be adapted to a rather different culture condition.

The small dose of organisms (typically between 1% and 10% of the number of organisms expected in the final fermentation) is called the inoculant.

This is inoculating in the laboratory or production plant. Bacteria can also be inoculated into soil (to help bioremediation or to colonize the roots of plants), or onto plant roots or seeds directly. Again, the aim is to get them to grow in their new environment.

Many biochemists think that inoculum should be spelt with more than one "n," or "c." Apparently, they are wrong.

In vivo, etc.

There are a number of Latinisms in biotechnology.

In vitro vs. *in vivo*. These Latinisms are widely used when scientists are talking about doing something "simple" in the laboratory and then taking the result and applying it to a more complicated, living system.

In vivo. Literally means "in the living," and means in a living system, such as a complete animal. It is contrasted with *In vitro*, literally meaning "in glass:" which is translated by every English language newspaper to "in the test-tube," although no-one in biology has used test tubes for 20 years. It means "in the laboratory," and is taken to mean the opposite of "*in vivo*."

Doing experiments in cultured cells is usually taken to be *in vitro*.

In situ. This means doing something in place, rather than taking out of its usual context. Sterilization *in situ* is sterilizing some piece of equipment in place in the overall machine, rather than taking it out, cleaning it, and then putting it back.

There are a number of *in situ* techniques in cytology, the study of cells and their structure. They mean that you do an analytical technique on a cell that has been sliced and stuck to a microscope slide, rather than on chemical components extracted from a cell. This produces an image of the distribution of whatever you are testing for in the cell. A commonly used example is *in situ* hybridization (ISH). This uses a DNA probe to hybridize to DNA in the chromosomes in a cell as it sits on a slide. This tells us where in the chromosome that piece of DNA lies. (Of course, the cell has to be dead to do this.) A common *in situ* hybridization technique is fluorescent *in situ* hybridization—FISH—which produces very attractive color photographs of chromosomes with brightly glowing spots on them where the target genes are.

Ex vivo. This is used in reference to medical therapy strategies, and means that you are going to treat something outside the body rather than inside it, where it normally resides. Usually this term is applied to blood or bone marrow cells, which are taken out of the body, treated *ex vivo*, and then put back in. The advantage is that only the target cells are treated, not the whole patient.

In silico. These mean "done in a computer, not in real life." Most biology cannot be done in a computer—it has to be done using the much more expensive, much messier and much slower techniques of laboratory experiment. *In silico* methods include computer analysis of large databases of information to supplement or replace experiment (such as Digital Northerns—*see* **Expression profile**), and modelling many aspects of biology that are slow or expensive to measure (*see* **ADMET, Systems biology**). The same ideas are sometimes also called "virtual biology." A now out of date equivalent phrase is *in machina*.

Ion channels

Ion channels are proteins that allow ions to flow into and out of cells. As the cell membrane is hydrophobic lipid, ions cannot usually pass across it, and a substantial concentration gradient and voltage difference can exist across an intact cell membrane. Ion channels selectively either allow ions to pass across this barrier, or actively pump them across.

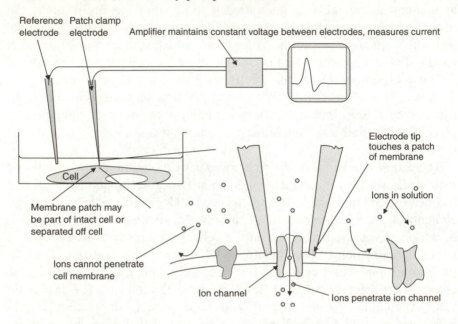

Reference electrode · Patch clamp electrode · Amplifier maintains constant voltage between electrodes, measures current

Electrode tip touches a patch of membrane

Ions in solution

Cell

Membrane patch may be part of intact cell or separated off cell

Ions cannot penetrate cell membrane

Ion channel · Ions penetrate ion channel

Ion channels are important for the control and maintenance of muscle and nerve function, and for the control of many other physiological properties. So, they are major drug targets. Many are 7-transmembrane receptors, which can switch between open or closed states, starting or stopping the ion flow. All sorts of things can "gate" the ion channel—voltage across the cell membrane, pH, simple ions, neurotransmitters, hormones, and of course drugs.

The flow of ions through an ion channel can be measured by looking at its effects on a cell. However, it is also possible to measure it directly, using a technique called **patch clamp** electrophysiology. See separate entry.

Ion channels can be fashioned into sensors. This can be in a patch clamp configuration, where the ion channel is integrated into a bilayer membrane and the flow of ions through it measured electrically (*see* **Langmuir–Blodgett film**). Or ion channels or the cells that contain them can be immobilized onto an ISFET. This has been investigated mainly as a method for screening cells for inhibitors of the ion channel function in single nerve or muscle cells.

ISFET

This stands for ion-sensitive field effect transistor. A field effect transistor (FET) is a semiconducting device in which the electric field over an n–p–n or p–n–p junction is used to modulate the current flowing through that junction. (The electric field attracts electrons or holes into the junction region, so providing charge carriers to carry current through the depletion layer.) It is a standard component of integrated circuits. Closely related in terms of its electronic effect is the metal oxide semiconductor FET (MOFSET).

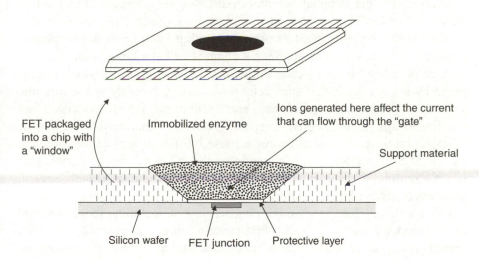

FET packaged into a chip with a "window"

Immobilized enzyme

Ions generated here affect the current that can flow through the "gate"

Support material

Silicon wafer   FET junction   Protective layer

A FET is inherently a sensor for ions. If ions build up near the junction, they will create an electric field which will switch the FET "on," and a current will flow. Thus, a device that allows ions near to the FET junction is an ion-sensitive FET, and will allow a current to flow that is dependent on the amount of ions present. If the junction is overlaid with a material that binds one ion specifically, then it becomes a specific sensor for that ion.

These devices have been suggested as method for monitoring ion concentration in a range of biotechnological processes. They have also been turned into biosensors by replacing the ion-selective layer with an enzyme which generates ions by its action. This sort of device is also called an Enzyme FET (EnzFET or ENFET). Other configurations are possible, where other proteins or even whole cells deliver ions to the FET gate.

Although in theory such a sensor could benefit from the mass production technology of the semiconductor industry, in practice the biological part has proven too hard to make reliable so far. A few exceptions use a FET as a detector for urease, the enzyme being used as a "tag" to track the presence of some other molecule such as DNA or an antibody.

Kinase

A kinase is an enzyme that takes a phosphate group from adenosine triphosphate (ATP) and adds it to another molecule. They are now a hot topic in drug discovery because many of the cell's processes are controlled by protein kinases, together with balancing protein phosphatase enzymes (which take the phosphates off again).

Protein kinases add the phosphate to a serine or threonine residue, or to a tyrosine. The tyrosine kinases are the most common drug discovery targets.

Many G-protein coupled receptors (**GPCRs**—see separate entry) act to trigger a kinase. This in turn adds a phosphate group ("phosphorylates") another kinase (often as well as phosphorylating itself), which phosphorylates another kinase and so on. Such a chain is called a kinase cascade.

A well-known kinase cascade is the MAP kinase cascade, which is triggered by many signals that start cells proliferating (notably in the immune system). In the immune system, then end result of the MAPK cascade is that it activates a protein called NF-kappa-B (NF-Kb), which turns on a host of other genes. Several other pathways activate NF-Kb as well. Thus, knocking out the MAPK pathway will not guarantee that NF-Kb is not activated. This is an example of biological systems showing redundancy, with several different pathways having overlapping effects.

Other kinases affect many other protein classes, including those involved in intermediary metabolism (the first protein kinases discovered) and ones that phosphorylate histones, the proteins that bind to DNA in the eukaryotic nucleus.

Kinases received a boost as potential drug targets with the launch of anti-kinase cancer drugs Iressa and Gleevec, which showed that anti-kinase drugs could be successful.

Kinetics

Kinetics is the study of the rate of change of something. In biotechnology, it can mean a variety of things.

Cell growth kinetics. This is the rate at which cells grow, discussed in more detail in the entry on **cell growth.**

Reaction kinetics. The speed with which chemical reactions can be made to happen is essential to many aspects of biotechnology. Among the commonly mentioned types of kinetics are

- *Binding kinetics.* How fast a molecule such as an antibody binds to its target. Usually the molecule doing the binding is called the ligand, and so these are ligand-binding kinetics. Related to this is ligand binding thermodynamics, measuring how much energy is gained or lost when a ligand binds to its target, and hence how tightly it will stay bound (*see* **Affinity**).

- *Enzyme kinetics*. This is a rigorously defined area of science, and studies how fast enzyme-catalyzed reactions can go. Key terms are the Michaelis Menten equation, and its constants K_m (the Michaelis constant), and V_{max}. The latter is a measure of how fast each molecule of the enzyme works, the former of how dilute a solution of its substrates its works on. Another kinetic measure is turnover number, which is the number of substrate molecules each enzyme molecules processes each second.

- Also part of the study of enzyme kinetics is the study of inhibitors—(see separate entry).

End product inhibition is a result of enzyme kinetics. Here the end product of a chain of enzyme reactions—a metabolic pathway—acts as an inhibitor of the first enzyme in the pathway. So when the pathway operates and the end product builds up, the first is automatically "turned off" and the pathway stops again until whatever it is making is used up. This is a form of feedback control, also called feedback inhibition. End product inhibition relies on one of the end products of a pathway being an allosteric inhibitor of one of the enzymes.

Knockouts and mutants

Knockout organisms are ones where a specific gene has been inactivated ("knocked out") by recombinant DNA techniques. They are widely used as tools to find out what a gene is: knock it out, and see what effect it has. Knockouts are made by using homologous recombination between the target gene and a piece of cloned DNA to insert a piece of "junk" DNA into the gene you need to disrupt. If the organism is haploid (i.e. only has one chromosome set) then this technique will result in that organism's only copy of the gene being knocked out. If it is diploid, then only one of the two alleles will be knocked out, and you will have to do some conventional breeding to make an organism that has two copies of the gene knocked out.

Knockout organisms, especially knockout mice and yeast, have been used extensively to try to find out what the genes discovered by genome projects are actually for. Many of the genes discovered in even such a simple organism as yeast have no obvious function, so this is the only way to know what a new gene is doing.

Knockout mice can also be used as models of human disease (*see* Transgenic disease models).

This type of knock-out knocks out every copy of a gene in an organism and its descendents. This is a problem if the gene is for something vital, so different versions seek to knock genes out in specific tissues at specific times. You transfect the organism with a version of the gene that is flanked by a gene for an enzyme that cuts DNA, and the specific places where it cuts it.

KNOCKOUTS AND MUTANTS

Examples are the Cre/Lox and Flp/FRT systems. The enzyme (the first part of each name pair) cuts DNA at a specific site (the second part of the name), so when you induce the protein in some cells those cells lose those particular DNA sequences and whatever lies between them, including the original gene. The same technology could be used to remove transgenes from transgenic organisms.

You can also "knock-out" proteins, by getting an antibody into cells. Xerion does this by tagging the antibody with a light-activated chemical, so light can be used to permanently knock out whatever protein the antibody binds to. This is a cell-by-cell method, however, and cannot be inherited by daughter cells as a genetic knock-out could.

A related technique is the "knock-in." This is similar to making a transgenic animal, but without adding a whole gene, only a gene activating element. One knock-in technique is random activation of gene expression (RAGE) (Athersys), where promoter elements are put into the genome at random to make large amounts of protein—allowing you to guess at what the protein is and what effect it has.

Langmuir–Blodgett films

These are thin films of molecules formed on the surface of water. The original Langmuir–Blodgett film was a lipid layer on top of the water, but the term is often used to describe lipid films in which both sides are in water, or these films when they are transferred to a solid surface. The films form because the hydrophobic "tails" of lipids stick together and the hydrophilic "heads" stick into the water, like a flat liposome—*see* **Liposome**. The same structure, a lipid bilayer, forms the membrane surrounding all cells, and much of the internal structure of eukaryotic cells.

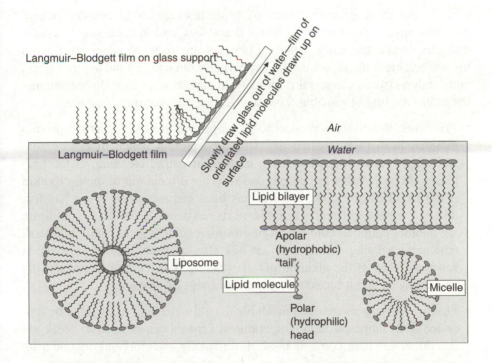

Lipid bilayer films or membranes are only one example of "liquid membranes," in which a thin layer of liquid is stabilized so that it can last for a long time in water. They all have to be stabilized by some chemical means, as otherwise they collapse into little globules of liquid, or dissolve in the water.

Biosensor applications rely on the high electrical resistance of a Langmuir–Blodgett film, or on its optical properties.

Electrical sensors are based on the ability of some proteins to carry ions across a lipid membrane. Some antibiotics, proteins from nerve cell membranes, a variety of "transport" proteins which allow cells to get materials from outside the cell into the cell without making large holes in the membrane, all these proteins can be inserted into the membrane, and their effect

233

on allowing molecules to pass through can then be measured. *See* Patch clamp.

The problem with this is that the membranes are mechanically and chemically unstable, as are most of the proteins we want to put in them. Thus, while a sensor system may be lashed up in the laboratory, none have worked "in the field."

See also Molecular electronics.

Laundering

One of the most pervasive uses of biotechnology is in laundry, where enzymes have been in wide use since the 1960s, and in domestic "washing powders" since the early 1980s. Enzymes are used to digest dirt from in-between cloth fibers, where their specificity (attacking food and not fiber) and catalysis (so they digest dirt under mild conditions, which do not damage the cloth) are highly valuable. The main classes of enzymes used are:

- *Proteases*. Relatively non-specific proteases are used to digest the protein material in dirt—it is often the denatured protein which makes organic stains hard to wash out. Most are used in industrial cleaning, as the proteases are powerful enzymes and can strip the protein out of the user's skin if not handled carefully. Subtilisin has been engineered by Novo Nordisk and by Procter and Gamble to improve its resistance to oxidation (oxidizers are another common component of commercial detergents) and "builders" (calcium chelating compounds). It has also been engineered to make it function better at the alkaline pH of washing powder, and to broaden its specificity so that it breaks down a wider range of proteins.

- *Lipases*. Most lipids are quite insoluble in cold water, and so can only be separated from fabrics at high temperatures. Lipases can be used to break the fats down into their component alcohol (usually glycerol) and organic acid, both of which are more soluble, and do so at relatively low temperatures.

- *Amylases*. These digest starchy residues, making food remains much more soluble.

Leaching

Microbial leaching, or bioleaching, is the use of microorganisms, usually bacteria, to isolate metals from mineral ores by solubilizing them and allowing them to be washed ("leached") out of the ore. Thus, it is a method of mining, and is a major component of microbial mining (biohydrometalurgy) technology.

Many ores cannot be processed economically because the concentration of metal in them is too low. Some of these ores are low-grade ores which are

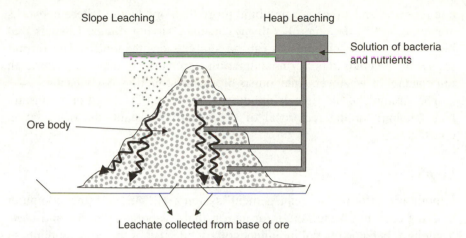

Slope Leaching   Heap Leaching

Solution of bacteria and nutrients

Ore body

Leachate collected from base of ore

discarded as waste during mining operations aimed at higher-grade ores. (The "grade" of an ore depends mainly on how much metal there is in it, but also on how accessible that metal is. Clay has a very high content of aluminum, but it is extremely expensive to extract aluminum from clay.) However, if the metal can be released as a soluble salt, then it can be washed out and collected without the ore having to be mined, crushed and smelted, as in a normal mining operation.

Leaching is done on rocks that have been broken up so liquid can filter in and between them. This can be in a heap, a spoil tip down the side of a mountain, or in its original site. Sulfuric acid and nutrients are sprayed on the top to encourage bacteria, and the eluant with the metal is collected at the bottom. In some cases (such as copper mining) the main role of the bacterium is to oxidize sulfide to sulfate, converting insoluble copper sulfide ore to soluble acid copper sulfate. In others the bacteria act directly on the metal ion, for example oxidizing uranium IV (insoluble) to uranium VI (soluble).

Bioleaching has been demonstrated to work on a number of valuable metal ores. However, only copper, nickel, uranium and gold are produced commercially using bioleaching.

Ligand

A ligand is anything that binds to another molecule. Usually, it means something that binds to a receptor or enzyme molecule, such as a hormone or protein growth factor. Thus, the antigen molecule that binds to an antibody is an example of a ligand. Enzyme substrates can be considered ligands, although they are not usually called this because, soon after they bind, they are converted to something else.

Discovering the natural ligands for receptors is a substantial part of the research into what receptors are doing. Often a ligand will bind to more than one receptor, and a receptor will bind more than one ligand. These networks are important in the control pathways in many living things. Ligands that bind receptors on cell surfaces can be agonists or antagonists. The former trigger the receptor to perform its function—signaling to the cell, opening an ion channel or whatever. Antagonists block that function. *See* Inhibitor.

The strength of the ligands' binding is measured by its affinity constant (*see* Binding), or the reciprocal of the affinity constant, the dissociation constant.

LIMS

Laboratory information management systems (LIMS) are the computer systems used to manage the information that a laboratory needs and uses. They have become topical in biotechnology because of the large amounts of experimental results produced by high-throughput, automated laboratories in genome projects and high-throughput screening.

LIMS systems need to solve several related problems.

- How to get lots of different machines to communicate with one system. This sounds trivial, and is not.

- Capturing data directly off instruments, so a scientist does not have to read it off a screen or print-out and then type it in to the system.

- Tracking samples consistently, so that each sample, analysis or run is identified, its results linked to it, and the protocol used in that run is known, and this can be accessed through one system.

- Appropriate data storage and presentation. Data should be summarized to a suitable level. A chemist may want to see detailed NMR spectra, but a molecular biologist will not, so the details must be summarized in the presentation software.

In theory, a good LIMS system can result in a completely paperless laboratory, but, like the "paperless office," this remains a dream.

A related topic is "data mining"—searching through large databases for patterns or meaning. *See* Bioinformatics.

Lipases

Lipases are enzymes which break down lipids into their component fatty acid and "head group" moieties. The lipases used in biotechnology are almost invariably digestive lipases, meant to break down the fats in food. However, they can be turned to a number of different uses.

Two more valuable uses are lipid synthesis and transesterification. Lipid synthesis aims to make the lipase put a lipid back together from a fatty acid and an alcohol. This replaces conventional acid-catalyzed chemistry, which as well as using high temperature and pressure produces a product contaminated with side-products, which must then be purified. Unilever produces isopropyl myristate (an ingredient of cosmetics such as moisturising creams) using a lipase-based synthesis. The reaction is carried out in organic solvents so that there is no water around to split the lipid into its components again. (*See* **Organic phase catalysis**.)

Transesterification is a process where a lipase is used to swap the fatty acid chains between lipids without ever releasing significant amounts of fatty acid. This allows new lipids with properties that are intermediate between the starting points to be made. A specific version of this is called interesterification.

Lipases can also be used in laundry detergents: *see* entry on **Laundering**.

Lipid

A group of molecules containing esters of fatty acids. Typically, lipids contain long chain hydrophobic "tails" linked to water soluble "heads." Lipids can be broken down into their water-soluble heads and sparingly soluble tails by alkali (a process called saponification) or by lipase enzymes. Common tails the fatty acids, of which the more common saturated ones (fatty acids with no carbon–carbon double bonds) are

Lauric acid	12 carbons
Myristic acid	14 carbons
Palmitic acid	16 carbons
Stearic acid	18 carbons,

and the more common unsaturated ones are

Oleic acid	$18:1_9$
Linoleic acid	$18:2_{9,12}$
Linolenic	$18:3_{9,12,15}$
Arachidonic acid	$20:4_{5,8,11,14}$.

(The numbers after the acids are the number of carbon atoms in them, followed by the number of double bonds, and then the positions in the chain of the double bonds as subscripts).

There are a lot of head-groups, containing combinations of sugars, choline, phosphate, glycerol and other chemical groups.

Lipids are major components of living things, and as such are raw materials for food production. They have a wide range of physical and chemical properties exploited in as wide a range of products as biosensors, drug delivery systems and cosmetics.

Liposome

Originally coined to mean small droplets of fat in a cell, this word has been taken over by biotechnology to mean a closed shell of lipid molecules. In solution, lipids tend to form droplets with the polar "heads" point outwards into the watery solution and the apolar "tails" stick together into the middle. If this is a droplet of lipid, it is called a micelle: many lipids form micelles if their concentration in solution rises above the critical micelle concentration (CMC). If the droplet is filled with water, then a two-layer structure called a lipid bilayer is formed (*see* **Langmuir–Blodgett films**). Such a lipid-walled ball is a liposome.

Liposomes have been suggested as the basis of several methods of drug delivery, especially for delivering peptides or highly toxic drugs such as the anti-cancer drug daunomycin. Liposomes tend to accumulate at sites of inflammation, and in some tumors, and so they are potential delivery vehicles for anti-inflammatory and anti-tumor drugs. The lipid bilayer keeps the drug away from non-target tissues, ideally releasing it at the place you want it. Antibodies can be linked onto the outside of the liposome would bind it to the specific cells which the drug is meant to affect (a combination sometimes called an immunoliposome).

Liposomes are also being as adjuvants in immunization—they can be used to carry the proteins of a synthetic protein to an antigen presenting cell in the immune system and stimulate it to produce an effective immune response.

Trapping things inside liposomes is a form of encapsulation, and as such can be used in many other areas. Here, however, liposomes are less favored because they can be rather less stable than other, polymer-based encapsulation methods.

A physically related structure is the colloidal gas aphron. This is (probably) a suspension of liposomes with micrometer-sized bubbles inside them. The bubbles are also lined with the surfactant that forms the liposome, and so are stabilized from vanishing as the gas dissolves out of them. CGAs are formed by very vigorous stirring of some detergents in water, are stable, and can be pumped around like water. However, CGAs are very much lighter than water, and so when mixed with water quickly separate again into CGA and "clean" water. They have therefore been used to separate cells and proteins from fermentation broths, by getting the protein (or the impurities) to stick to the very large surface area that such a large collection of small droplets possesses.

Live vaccines

Live vaccines are vaccines containing living organisms or intact viruses, rather than inactivated (killed) organisms or extracts of them. They can cause better immunity in patients, but have the potential drawback that, unless they are thoroughly "crippled" in some way, they may cause disease. The viruses used in the MMR vaccine (Measles/Mumps/Rubella) have caused contro-versy because some researchers have raised the possibility that the measles virus used is not sufficiently crippled. They can also be harder to transport and store, because you have to avoid killing the organism.

Biotechnology has generated ideas and research products for live vaccine development in a number of areas. Viral vaccines are dealt with elsewhere. Bacterial live vaccines can be developed in a number of ways.

Attenuation. When pathogenic bacteria are grown outside their host, they tend to lose their virulence genes (genes that specifically cause disease) by mutation, as they are not needed for growth *in vitro*. The result is an attenu-ated bacterium which will cause an immune response similar to the original but which is harmless. Usually, several mutations are needed to make sure that a strain is really attenuated. If the nature of the virulence genes is known, then conventional and "molecular" genetics can be used to select for muta-tions in or loss of those genes.

Gene cloning. An alternative is to place some key genes from the patho-genic bacterium into another, harmless organism. Vaccination with the safe bacterium causes the body to make protective antibodies against the patho-gen. These may be the genes for surface parts of the pathogenic bacterium such as pili or transport proteins which are seen by the immune system. The critical part of this is defining what antigen, or particular part (epitope) of an antigen, is "immunogenic," that is gives rise to a strong immune response, and "protective," so that the immune response protects against infection. Bioresearch Ltd. has produced a bacterial vaccine for furunculosis in salmon by deleting some genes that are critical to causing the disease from the bac-terium. Microscience Ltd. is using relatively harmless versions of *Salmonella* species to deliver antigens from more dangerous bacteria to the patient.

Loop bioreactors

Also loop fermentors, these are bioreactors in which the fermenting mater-ial is cycled between a bulk tank and a smaller tank or loop of pipes. The circulation helps to mix the materials and to ensure that gas injected into the fermentor (usually oxygen or air) is well distributed among the liquid. The reactors are also very useful for photosynthetic fermentations, where they allow for the photosynthesizing organism to be passed along a large number of small pipes, where the light can get to them easily, rather than inside a single volume, where only the organisms near the edges get much light.

There are lots of types of loop bioreactors, but they break down into those which have an internal loop (e.g. a stirred tank bioreactor with an internal draft tube) and those which have an external loop. Some airlift fermentors are of the first type, as are pressure cycle reactors: reactors where air or oxygen is injected into the riser half of the reactor and this drives the liquid in that half up, so pushing the flow round the vessel. A variant that is found in all of them is the jet loop reactor, in which the recycled liquid is injected with considerable force back into the main tank.

See also Airlift bioreactor.

Luminescence

Luminescence, the production of light by chemicals, is gaining increasing use as a readout system for assays. If they are carried out in a rigorously light-proof box, luminescence tests can be extremely sensitive: a photomultiplier tube can detect when only a handful of photons have been given out by a reaction, so offering the potential of detecting only a handful of DNA or antibody molecules.

There are two ways of generating light using chemicals:

- *Chemiluminescence*. This uses specific chemical groups that, when reacted, give out light. They can be attached to many other chemicals (e.g. proteins, DNA). The chemiluminescent groups are "blocked"—they do not themselves react to give out light. However, when the blocking group is chopped off or other reagents added, they react rapidly to generate light. A common chemistry allows unblocking by the enzyme alkaline phosphatase (AP). Adding chemiluminescence to such as assay enhances its sensitivity greatly.

- *Bioluminescence*. This is the use of biological systems (in this context enzymes) to generate light. (See separate entry.)

Luminescent labels are widely replacing radioactive ones in some applications. Thus the scientist will chemically link a luminescent group onto an antibody or a DNA probe where before they would have used a radioactive one: luminescent labels can be as sensitive as radioactive ones. The light can be detected by a sufficiently sensitive camera, or by exposing the gel or blot directly to photographic film, as is done in an autoradiograph. (Confusingly, this is sometimes called an autoradiograph as well, although it has nothing to do with radioactivity.)

An alternative is bioluminescence—see separate entry.

MACs, BACs etc.

MACs are mammalian artificial chromosomes. BACs are bacterial artificial chromosomes. Both are complex pieces of DNA that can be used as cloning vectors, to allow a foreign piece of DNA to replicate in a cell or organism. Together with YACs (see separate entry), they can be used to clone very large pieces of foreign DNA, in theory tens of millions of bases long.

Most cloning vectors are based on plasmids or viruses—they allow our cloned DNA to grow rapidly in their host cells. However, these are not genetically stable—the cells will lose the cloned DNA easily if we do not continuously select for it to be there. Chromosomes, by contrast, fit in with the normal gene-building machinery of the cell which ensures that every new cell gets an exact copy of the genetic material of the old cell. BACs and MACs are synthetic versions of chromosomes which use this machinery.

MACs and BACs provide an origin of replication (to make sure the DNA is copied), telomeres (in eukaryotic cells) to protect their ends, and other genetic elements which make sure that each cell contains one copy of the chromosome, and that when they are duplicated during cell division that exactly one copy goes to each new cell. Some BACs use the F1 plasmid to provide origins and other necessary sequence—others (sometimes called PACs) use the bacteriophage P1.

YACs, MACs etc are the only convenient way of handling very long pieces of DNA, that is over 100 000 bases long. MACs are being explored as ways of putting very large gene clusters into transgenic animals, and as vectors for gene therapy. BACs are used in cloning large stretches of bacterial DNA into other bacteria. Both are also used to study the control and function of genes, and of telomeres, centromeres and origins of replication.

Marine biotechnology

Seventy percent of the world is covered by sea, but only about 1% of biotechnology uses this resource. Among the potential applications of marine resources to biotechnology are:

Natural products. Marine organisms live in a very diverse range of chemical habitats, and have to prevent a huge range of different animals from eating them. They produce a range of chemicals which can be used as candidates for drug development. Bacteria, coral, sponges, seaweed and cephalopods have all been examined for "natural product" drug leads as well as fish. Halichondrin B, from sponges, is undergoing trials as an anti-cancer agent.

Enzymes. As well as making novel enzymes, marine microorganisms are adapted to a wide range of temperature regimes, from $-2\,°C$ in the lake discovered in 1996 beneath the Antarctic ice to $+115\,°C$ in black smoker vents in the mid-Atlantic ridge. Their enzymes are correspondingly adapted to work at temperatures which are of great potential value to industry. *See* **Extremophiles.**

Algae. Algae produce a range of products directly valuable in their own right. See separate entry.

The sea is also a potential source of minerals, which biotechnology has been aiming to extract for some years, but without success. *See* **seawater**.

Markers

One of the most flexible words in biotechnology, this can mean many things in different contexts.

Assay markers. The system which allows you to see the result of a chemical assay or test. Also called labels, examples include enzymes, radioactive atoms and luminescent groups. *See* entry on **luminescence**.

Genetic markers. Landmarks in a genetic map. *See* **Genetic map**.

Biological markers. These are chemical signposts that something is happening in a cell or tissue. Often they can be produced by **reporter genes**, genes we introduce into a cell to show when something has happened (see separate entry). Markers can also be chemicals produced by the cell, even things such as cell shape. "A marker for" an event is usually shorthand for "something that is convenient to measure to show that that event has happened". Marker genes are genes that do something when an event occurs: an example is the gene for GFP, which produces the readily detectable fluorescent protein when it is active.

Disease markers. Usually, this means a chemical "signpost" that someone has a disease. Thus, presence of some heart-specific enzymes in the blood (such as CK-MB) is a marker for heart attack—it shows that the patient has recently had a heart attack. It can also mean a gene that is a marker for a disease in this sense, or a gene that is a genetic marker on a genetic map. Biotechnology contributes to these by finding them, and by developing tests to measure them.

Mass spectrometry

A mass spectrometer measures the exact molecular mass of a molecule by measuring its flight path through a set of magnetic and electric fields. More exactly, it measures the mass/charge ratio. To allow them to "fly" uninterrupted, the operating chamber of the mass spectrometer contains a vacuum.

To make the molecules charged, so that electric and magnetic fields affect them, mass spectrometers include a method of ionising them. Ionisation methods include bombarding the molecule with X-rays or high-speed electrons (which breaks them down a lot), atom beams (called fast atom bombardments, or FABMS, which only breaks them a little) or spraying them out of a charged nozzle (called electrospray, which hardly breaks them at all).

A widely used technique for biological molecules is matrix assisted laser desorption ionisation (MALDI) where the sample is embedded in a volatile matrix and a laser vaporizes it. The nature of the mass of material your sample is embedded in is crucial for this. A common combination is MALDI–TOF (TOF is time-of-flight spectrometry, a fairly simple way of identifying the molecular mass of ions.) Two mass spectrometers can also be linked together in tandem MS (also called MS/MS), so that the first MS machine finds molecules with one particular size, and the second smashes them up and so gives us information about what their chemical structure is.

Mass spectrometry is being used to identify proteins directly from mixes, and to sequence proteins. A protein's identity can be proven if we use enzymes to break it into peptides, find the mass of the peptides, and then look up in the genome database what protein it could have been to give that pattern of peptide masses—this is a form of peptide fingerprinting. MS can also be used to track post-translation modifications in proteins, which genetic methods cannot do.

MS is also being used as a general analytical method. It is commonly used in conjunction with HPLC or CE: the HPLC separates chemicals, and the MS then identifies what they are. Waters and Applied Biosystems provide systems which integrate MS and HPLC for biological uses.

The ultimately sensitive MS techniques are probably accelerator mass spectrometry methods, which are used in carbon-14 dating and can detect literally a few dozen atoms in a sample—however, they can only distinguish atoms, not whole molecules, so their biological use is limited.

Messengers

Living things use many molecules as messengers to signal within or between cells. There are proteins (such as the **cytokines**—see separate entry), hormones, and other molecules usually just called messenger molecules. Notable examples include:

Nitric oxide (NO). This is generated by the enzyme Nitric oxide synthase (NOS). NO breaks down quickly in the body, so it is used for short-range signaling, especially in the circulatory and immune systems. The anti-impotence drug Viagra works by blocking NOS and so making muscles letting blood into the penis relax.

Ethylene. A plant signaling molecule responsible for controling development, especially ripening, in some species.

Quorum sensing molecules (also called "autoinducers"). Many microorganisms use them as a mechanism for sensing how many of that type of organism there are around. When there are few organisms around, they need to keep a low profile, as their target's defences could kill them easily. When a critical number is reached (a "quorum"), then the concentration of quorum

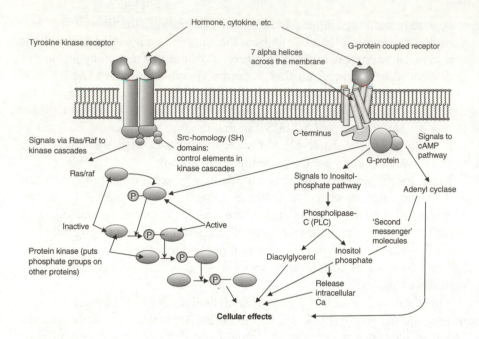

sensing molecules increases and they change their biochemical behavior for more aggressive, disease-causing chemistry. Quorum sensing mechanisms are therefore interesting targets for new antibiotics.

Intracellular messengers are often called secondary messengers, because they carry the message that a hormone has arrived at a cell to the components of a cell. Best known of these are the cyclic nucleotides—cyclic adenosine monophosphate (cAMP) and cyclic guanosine monophosphate (cGMP). Both are very important in animal cells, and are involved in many cellular processes. This process is called signal transduction (*see also* **kinases**). In general, messengers of this sort are used by different cells for different messages. In contrast, hormones have a much more specific effect, either affecting a particular cell in one way or not at all.

A related concept is messenger RNA (mRNA)—*see* **Genetic code**.

Metabolic engineering

The manipulation of cells to make them produce more of a product is sometimes called metabolic engineering. This focuses on altering the enzymes present in a cell, rather that just altering their growth conditions. Typically, a cell will be genetically engineered to produce more of a specific enzyme, or to replace one enzyme by another with more desirable properties.

It used to be thought that the amount of chemical being processed by a specific pathway (the "metabolic flux") was determined by flow through one key "rate determining" or "rate limiting" step in the pathway. If you doubled the amount of the enzyme that catalyzed this step, then you would double the amount of stuff made by that pathway. Many experiments have shown that this is not so—all the enzymes of a pathway contribute to controlling how fast that pathway works to a greater or lesser degree, and if you double the amount of one enzyme it has little effect overall.

Thus, metabolic engineering is about predicting changes in the whole of metabolism, and altering a number of enzymes and physiological conditions to bring about those changes. This treatment of the cell as an integrated system is part of the approach of **systems biology** (see separate entry).

Applications are generally in fermentation, where metabolic engineering seeks to improve the yield of primary or secondary metabolites. However, it can also be applied to plants which produce valuable chemicals, and even to human cells where the same analysis can be used to discover what goes wrong with metabolism in diseases such as diabetes and obesity. These non-microbial applications are still at the research stage.

The degree to which a single enzyme affects the flux through a pathway is called its control coefficient, a number between 0 (has no effect at all) and 1 (is the only control on that pathway).

Metabolic pathway

The chemistry that turns an organism's food into more organism is collectively known as "metabolism." The whole of metabolism is a very complex network, made up of chains of reactions in which chemicals ("metabolites") are converted one into another to get from a starting material to an end material in a series of relatively small steps. Such series of reactions are called metabolic pathways.

Metabolism is usually divided into anabolic pathways—pathways that build up molecules for use by the organism (e.g. those pathways that make amino acids), and catabolic pathways—ones that break down molecules, either for energy or simply to get rid of undesired materials. Some pathways, especially those at the center of metabolism (e.g. those that break down glucose) perform both functions, and are called amphibolic. Pathways can also be divided into "primary" and "secondary." Primary metabolism is making the basics of life that are common to nearly all living things. Secondary metabolism is making specialist chemicals such as hormones or toxins. Mutant organisms that lack enzymes of the anabolic pathways of primary metabolism are called auxotrophs: they require that the end product of that pathway be provided as food, as they cannot make it themselves. Mutants of the pathways that make secondary metabolites are called idiotrophs.

Popular press articles about "your metabolism" almost always mean energy metabolism—the metabolic reactions specifically concerned with handling molecules whose main function is providing energy, and "burning" them to generate that energy.

A growing field of bioinformatics is predicting what metabolic pathways an organism will have from its genome. This requires that you can predict the function of the proteins, but as the enzymes of central metabolism are very similar in most organisms, this is not as hard as it might be. Given the enzymes, paths are predicted from connectivity (do all the enzymes exist), and stoichiometry (the ratio of molecules involved in a reaction), which must match up for different steps. This is part of the agenda of **systems biology** (see separate entry).

Metabolome

This is the complete collection of metabolic pathways, and/or metabolites, in an organism. The rather inelegant term was coined by analogy with genomics and proteomics. The study of the metabolome is metabolomics (or metabonomics).

Metabolomics means several overlapping things. All rely on the analysis of many metabolites at the same time. The role of different genes in the metabolic network can be found by knocking them out one at a time. Studying gene expression levels in such a knock-out can find out which genes are co-ordinately expressed (go up and down together), and hence might be involved in related processes. A similar idea is to "knock them out" with a chemical (a potential drug), and see what effect that has on metabolites and gene expression, an approach called chemical genomics.

Nuclear Magnetic Resonance (NMR) is being used increasingly in metabolomics, as it can find the characteristic pattern of a chemical in a mixture without purification or processing (MS usually requires some pre-processing, which is complex and time-consuming). The drawback of NMR is that the signal of complex mixes is too complicated to work out what the components are. However, it can be used as a characteristic pattern of a metabolic state, and the NMR spectrum monitored to see if that state changes.

Microarray

A microarray is an array of chemicals in a very small area used as a test or analytical tool. Nearly all microarrays are DNA arrays, where hundreds to millions of DNA probes are bound to a surface and used to test a sample—if the sample contains DNA complementary to any of the probes, that DNA binds to the array in the specific place where that probe is located. These systems are also sometimes called DNA chips or **gene chips**—see separate entry.

Other types of microarray include

Protein arrays (protein chip). These are dense arrays of proteins on a "chip." They can be antibody arrays (designed to capture and analyze proteins) or other proteins (designed to study protein: protein interactions, find antibodies, or other research). The issues with these are how you get the proteins bound to the chip, how you get them to bind to their target, and how you detect the result: for all of these, the chemistry is different for different proteins (unlike DNA chemistry, which is very uniform), so protein chips are largely a development project at the moment.

Chemical microarrays. These are used to probe how chemicals interact with proteins—the protein is added to the chip, and it binds to potential ligands. Affymax pioneered this field with peptide arrays. Graffinity uses this approach to build combinatorial libraries on a rather large "chip" as part of a combinatorial chemistry drug design approach.

A major issue in the gene chip/microarray field is the analysis of the data. A single gene chip experiment (involving maybe a dozen samples) will generate millions of pieces of data. How this is analyzed by other than the most simplistic methods is a hot research topic.

Microbial mining

This is the use of microorganisms to remove minerals, and particularly metals, from rocks. It is a specific application of biohydrometallurgy. Microbial mining is related to the use of microbes in desulfurization, and for bioremediation: see separate entries.

Microbial mining falls into two areas:

- *Leaching*. This is using bacteria to solubilize metals in ores where they are combined as insoluble salts.

- *Purification*. Using microorganisms or microorganism components to separate and concentrate metals from very dilute solutions. The most common approach is biosorption.

These topics are covered in separate entries.

Biohydrometalurgy is used commercially to recover copper and uranium from low grade ores, especially chalcopyrite ($CuFeS_2$), covellite (CuS), chalcocite (Cu_2S), and uraninite (UO_2). A number of other metals (antimony, arsenic, molybdenum, zinc, cadmium, cobalt, nickel) can be extracted using bacteria, but this is not done on a commercial scale.

Plants can also be used to extract metals, although this is not used as an extraction technology commercially to any significant extent. Phytoextraction is the concentration of metals from the soil into the plant, usually the leaves. Some soils can produce quite toxic crops because plants concentrate elements such as selenium and cadmium in their leaves. The roots of water-growing plants such as the water hyacinth can also concentrate metals from their water supply, a process called rhizofiltration and used in reed bed biofilter technology. The only commercial use of such plants is to use metal-tolerant plants to stabilize waste tips (and to make then less ugly).

Microorganisms

There are a very wide range of microorganisms used in biotechnology. A few are mentioned elsewhere in this book under specific headings. Some are used for more than one thing, and so crop up in several biotechnological contexts.

Microorganisms, indeed all life, is classified into prokaryotes (organisms without a cell nucleus) and eukaryotes (organisms with a cell nucleus). Animals, plants, fungi are all eukaryotes. Prokaryotes include bacteria, and archebacteria, which look like bacteria but are genetically half way between them and eukaryotes, and are biochemically like neither.

Bacteria are classified into gram positive and gram negative (referring to Gram's stain) representing a substantial biochemical and genetic difference. However, they may look much the same under the microscope. Microorganisms may be ball-shaped (cocci), rod-shaped or made of very long "strings" called hyphae. Hyphae may be branched or unbranched: in either case they are often harder to grow in bulk because the stirring needed to get nutrient to all the hyphae can break them. Organisms that grow as long strings or filaments are called "filamentous."

Microorganisms are also classified into aerobes (grow in the presence of oxygen) and anaerobes (do not use oxygen). These can be facultative or obligate: facultative aerobes can use oxygen or not: obligate aerobes must have it. Obligate anaerobes are killed by oxygen. Biotechnological processes are often called "aerobic" or "anaerobic" if they need or must not have oxygen respectively.

Some of the more commonly mentioned organisms are:

- *Aspergillus niger*. A type of filamentous fungus that has been used for genetic engineering in a few cases, and which is also used to produce citric acid by fermentation from sugar molasses.

- *Bacillus subtilis*. This Gram positive bacterium is also widely used as a cloning host, especially for the expression of secreted proteins.

- *Candida utilis*. A yeast, this organism is used in fermentations to produce chemicals.

- *Clostridium acetobutylicum*. A bacterium used in the past to produce acetone and butanol by fermentation, and now used as a source of enzymes. (Acetone and butanol are now mostly made chemically.)

- *Corynebacterium glutamicum*. This widely used in fermentation processes producing amino acids for food supplements.

- *Escherichia coli*. Usually abbreviated to *E. coli* in print (and almost always in conversation), this gram negative bacterium is used in many biotechnological processes. Its genetics are the best known of any organism, and it is by far the most common host cell for recombinant DNA work. It is also used in fermentations to make many amino acids and other products, as it grows on many, very cheap fermentation substrates, grows fast, and can be manipulated genetically to accumulate many different chemicals.

- *Penicillium*. A group of filamentous fungi used primarily to produce penicillin antibiotics.

- *Pischia pastoris*. a yeast that can grow on methanol, and so is used increasingly in industrial fermentations. Piashia fermentations are used, among other things, for making several human proteins, including FGF.

- *Pseudomonas*. A group of soil bacteria that contain some extremely diverse chemical abilities, that biotechnology has harnessed in bioremediation

- *Saccharomyces*. A group of yeasts. *Saccharomyces cereviseae* is brewers' and bakers' yeast, and as such is probably the most widely exploited microorganism. Yeasts are eukaryotes, and hence have the same sort of genetic structure as humans, secrete proteins in a similar way and so on, but are almost as easy to ferment in bulk as bacteria.

- *Streptomycetes*. Gram positive bacteria that are used to produce a range of chemicals, especially antibiotics. They have also been used as the hosts for genetic engineering, in part to manipulate their antibiotic synthetic pathways.

Microorganism safety classification

As much of biotechnology involves the genetic manipulation, selection or physiological manipulation of microorganisms and their subsequent production in large amounts, biotechnology has to make sure that industrial-scale microbiology is safe.

Most of the guidelines to how microorganisms should be handled in biotechnology are derived from medical experience with possibly hazardous microorganisms isolated from sick people. (There is no evidence that recombinant organisms are any more dangerous than any other microorganism, however.) The principle of classifying the danger from a microorganism, and so of deciding how to contain that danger, is to classify the organism according to how likely it is to escape, how likely it is to survive if it does escape, and how much damage it could do if it does survive. Different countries have different rules about how this is done: the table summarizes a few of these.

If an organism is outside the Class 1/Group 1 area, then it can be contained by a variety of physical or biological methods. *See* **Clean room, Physical containment.**

Institution	Risk			
	Minimal	Normal microbiological risks	High risk to individual only	High risk to individual and community
ACDP, ACGM	Group 1	Group 2	Group 3	Group 4
EFB	Class 1	Class 2	Class 3	Class 4
WHO	Group I	Group II	Group III	Group IV

Abbreviations: ACDP, Advisory Committee on Dangerous Pathogens (United Kingdom); ACGM, Advisory Committee on Genetic Modification (United Kingdom); EFB, European Federation for Biotechnology, which has the same groupings as the PHS, US Public Health Service; WHO, World Health Organization.

Microparticle

Also microspheres, microbead, microcarrier. Small particles (0.1–10 μm) which are used to immobilize ("tie down") biological molecules or cells, or to "tag" them. If they need to be in suspension (e.g. for a diagnostic assay or cell culture carrier), then they are made of a polymer with a similar density to water: this is made by making an emulsion of monomer in water, and then cross-linking each droplet to make a bead of polymer. The suspension is like rubber latex, and hence these particles are often called "latex particles."

Specialist types of particle include:

- *Magnetic particles.* with a ferrite iron core in them. These are used widely for separations, as they can be concentrated out of solution by holding a bar magnet next to the tube. This is much easier than centrifugation. (It is sometimes called "biomagnetic separation.) You can also use magnets to put a known force on a particle, and so measure how strongly it is held in place by a binding reaction, or inside a cell.

- *Nanoparticles.* These are just microparticles, but with a diameter of less than 100 nm. They settle out of solution more slowly than larger particles, and so can be made of larger range of materials (such as glass or metals) which would settle out of suspension very fast as larger lumps. Formulating drugs into nanoparticles is a way to improve their formulation—the larger surface area allows them to dissolve faster.

- *Quantum particles/dots.* These are even smaller, 10 nm across or so, and so are intermediate between single molecules and latex. *See* Nanotechnology.

Microbeads, especially plastic ("latex") ones, can be labeled in many ways. Dying them colored or fluorescent is the simplest way. They can also have bars or slots cut into them, be build up of layers of different colors, of coated with a characteristic chemical. Such "coded" particles are used in combinatorial chemistry, to track a complex series of reaction products, with the beads being sorted at the end of each chemical step to send them to the appropriate next reaction—this is sometimes called an "encoded library" approach to chemistry.

Micropropagation

This is the use of biotechnological methods to grow large numbers of plants using tissue culture. A desirable plant is cut up into many very small pieces (sometimes single cells, sometimes clusters of a few to thousands of cells), and cultured. The culture conditions are tuned so that the cells grow into a callus, a mass of cells which look like a small mould. The conditions are then switched so that the callus starts to develop into a small plant "embryo" (*See* Embryogenesis). Once this embryo has grown sufficiently, it can be planted out as a small plant. In some techniques, the embryo is encapsulated in a protective sheath so that when it is sown it has a "shell" similar to the seed produced by more conventional breeding.

The advantages of micropropagation are that it can produce a large number of plants quite quickly, and that all the plants are usually genetically identical. The drawbacks are that it is skill-intensive, and hence much more expensive than conventional breeding, and that it can only be done on plants for which the right cell culture conditions have been worked out. So it is used mostly for propagating valuable, slow-growing plants such as trees.

Microscopy

Microscopy is a standard method of examining biological systems, both to determine their structure and examine their chemistry. Two specialist techniques used in biotechnological research are:

Fluorescence microscopy. The microscope detects the light given out by fluorescent molecules in a sample. Almost always, the fluorescence comes from a fluorescent label, attached to an antibody or DNA probe, so we can tell what part of the cell the fluorescent reagent has bound to. Several colors of label can be used, and the results are very pretty. Techniques using fluorescence microscopy include FISH and immunocytochemistry.

Confocal microscopy. Here the microscope effectively focuses on only a very thin slice of the sample at a time. So, a picture is built up not as if you were looking down through a thick sample, but as if you had sliced it into layers. Often confocal microscope systems are connected to computers which can reassemble the slices into three-dimensional picture of the original object. The approach is also useful for assays—the microscope focuses on bottom of the "test tube" (usually a well of a 96-well plate), and so only "sees" the material bound there and not all the material in solution above it.

A related technique is correlation spectroscopy. Here the "microscope" looks at only a tiny volume of solution, and measures how many molecules there are in that volume. If the molecules are small they move fast, and this number changes rapidly. If small molecules bind to larger ones then they slow down and the rate of change of fluorescence falls. Evotec commercialized this approach for screening for drugs.

See also **Scanning tunneling microscopy.**

Model organisms

Many processes that we want to understand cannot be studied in the "real" organisms for ethical, moral, legal, financial or practical reasons. Breeding experiments on people or trees, studies of individual cells in a whole cow or a pig, such experiments may give valuable knowledge to agriculture, healthcare or environmental studies, but we cannot sensibly do them. So instead we look for an organism where we can do the experiment—in the case of breeding trees, we breed *Arhabidopsis*, a fast-growing weed instead. In the case of looking at individual cells in an organism we use something like a nematode or Zebra fish, which is transparent. Some organisms have become standards for "model experiments," and so are called model organisms. They are models for some aspect of other organisms.

It is important to realize that a "model organism" is a model of *one aspect* of another organism. Mice are good models of human embryology, OK models of human cancer, bad models of human cardiovascular physiology, and terrible models of Oak tree growth.

Among the more popular model organisms are:

Caenorhabditis elegans. A microscopic nematode worm with only a few hundred cells and a very small genome. The fate map of the cells—exactly where every cell in the organism comes from and fits in the organism as it grows—has been completely worked out. This means that any new mutants of *C. elegans* can be related to the effects it has on specific populations of cells, and thus to specific biochemical effects. Thus, *C. elegans* has been a model organism for the study of the role of genes in development, a role enhanced the complete sequencing of its genome.

Arabidopsis thaliana, a small weed with "only" 80–100 million bases in its genome and short life cycle, which has become a staple of plant genetics research. It is a model for the developmental biology of more complicated and slower-growing plants like wheat and apple trees.

Drosophila melanogaster. The fruit fly, which has been a staple for animal genetics, and particularly the genetics of growth and development, since the 1920s. *Drosophila* geneticists are known for thinking of outrageous names for their mutants, such as "ether-a-go-go", "faint sausage"and "Hamlet".

Zebra fish. A small fish which is being used as another genetic model. It is the most man-like animal that can be cloned easily (mice can be cloned, but only in small clones of 8 individuals). The Zebra fish is so small that it is semi-transparent, so fluorescence marker genes such as GFP or bioluminescence markers can be seen at work inside a life animal, which makes following gene activity a much simpler task.

Mouse. The mouse is the most common laboratory animal in biomedical research, and is used for all sorts of aspects of the discovery and development of drugs as well as fundamental research. There are many inbred strains of mice, which have particular and very well known peculiarities.

Rats and *rabbits* are also used in biomedical research and drug development, but are less popular for fundamental research in part because they are not so extensively researched as mice, in part because there are not as many inbred lines as of mice, but mainly because mice are smaller and hence cheaper to rear and keep. Also, if a mouse bites you, it does not do as much damage. Rats can bite quite severely, and other animals (such as dogs or baboons) can take off your hand.

Biomedical research requires a good disease model, an animal disease which exactly mimics the human disease, so you can try the drugs out on animals instead of humans. This is not only because we do not want to test completely untried chemicals on people. It is also because the animals are more genetically and environmentally uniform than people, so that the effects of the drug are clearer, and you can kill them at the end of the experiment if you have to, to find out the effects of the drug.

Animal models can appear bizarre—much research on hepatitis B virus (HBV) is done on woodchucks, because woodchuck hepatitis virus is very similar to HBV. All early work on leprosy was done on armadillos, because this was the only animal other than man in which the bacterium would grow.

If no animal model is available, we might seek to create one with transgenic technology, for example by knocking out a gene in a transgenic mouse.

If a good animal model is not available, there is no point using a bad one. In that case, you have to make a leap of faith and "go straight to man." AZT, the first AIDS drugs, was tested directly on people because there was no animal model of AIDS available.

A popular animal rights fantasy is that scientists do experiment on animals out of sheer perverse pleasure in "vivisection." There are too many refutations of this argument to list exhaustively here—I will mention only that scientists have wives, children and pets too, and a lower record of animal or child abuse than the general population. This attitude reflects the "yuk factor"—diabetic activists are happy to inject insulin, but not willing to accept than Banting and Best killed many dogs to discover it.

Molecular biology

Much of biotechnology is based at least in part on molecular biology. Molecular biology, and its twin science molecular genetics, started in the late

1940s around a group of biologists, and physicists-turned-biologists, who were looking for a new way to attack the fundamental problems of life. Their approach was genetic, rather than the chemistry of traditional biochemistry and physiology.

The genetic approach paid off handsomely in three ways. First, it opened up whole new areas of genetics—genetics at a molecular level rather than the whole organism. This in turn allowed the investigators to start to work out the chemical mechanisms of genetics, protein synthesis, and later protein signaling and gene control.

The approach also caused a philosophical change, so that scientists started to think about life as a series of building blocks that bump into each other and latch onto and off each other in defined ways. Whereas in 1950, an enzyme was a squiggle in an equation, in 1990 it is a colored "blob" on a computer graphics display. Life became a discrete machine, and the instructions for that machine lay in DNA. Hence the centrality of DNA to much of biology today. This approach to living systems, as discrete blocks labeled "protein" and "gene", has been called "molecular Lego."

As well as this philosophical change that enabled us to imagine a "technology" of engineered genes and molecules, molecular biology gave us the basic tools of recombinant DNA technology. Restriction enzymes, DNA ligase, many cloning vectors all come directly from early molecular genetics.

Thus, molecular biology is not a science in the sense that it studies molecules, or biology—biochemistry, physiology, pathology, microbiology do that, too. It is more a way of doing biology, both a way of thinking about it and of getting the tools to do experiments. It is, in Thomas Kuhn's term, a Paradigm.

Molecular computing

Also called molecular electronics this speculative idea means making electronic or computing devices out of single molecules, or small groups of molecules. Talk about switches which are made of a single protein molecule lead to computers with greater-than-human powers which could fit in a matchbox. There are a range of biotech-related aspects. (Biocomputing is not the same—this is a term sometimes used for bioinformatics.)

DNA computers. This is the use of DNA as a computer. Information is coded in the DNA sequence, and processed by enzymes or hybridization of PCR reactions. Len Adelman is a noted pioneer in this field, but DNA is not a very obvious computing medium, as it is very slow compared to silicon. DNA is actually best suited to very long-term data storage, which is what biology uses it for. It is possible that catalytic RNA could be used instead of DNA, being able to store information and provide action.

Proteins as switches. Many proteins have charge-transfer and charge-switching properties, which could, with much greater understanding of the properties of proteins in general, be harnessed to provide some aspects of the information processing capability of a semiconductor device. Arrays of these have been assembled between electrodes and shown to carry out switching operations.

Lipid bilayers (Langmuir–Blodgett films) are known to be an essential part of the electrical properties of nerve cells, and can be made quite readily in the laboratory. They could be used together with some of the proteins in nerve cells to generate artificial switching devices.

A lot of work has focussed on using bacteriorhodopsin, a pigment from *Halobacterium halobium* (a purple, salt-loving bacterium) which has electrical and light-absorbing properties as well as being able to self-assemble into membranes. Membranes of bacteriorhodopsin assembled in the right way can store light images when exposed to electric fields, a first step to a holographic computer memory system.

The term molecular computing has largely been absorbed into the field of Nanotechnology, but remains a distinct technical application.

Molecular graphics

This is the display of molecule's shapes, usually on computer. It has gained a lot of publicity because of its application to **rational drug design** (see separate entry). Molecular graphics takes the description of how the atoms of a molecule are arranged in space from a database and draws a picture of

Different ways of drawing L-alanine

"Ball and stick": atoms are shown as small balls, bonds as single sticks

Wireframe: bonds are shown as single, double, triple

Spacefill: atoms shown at their "real" size (van der Waals radius)

Different ways of drawing the protein IGF

Backbone—only peptide backbone atoms shown

Wireframe

Ribbons—accent secondary structure (here alpha helix)

what the molecule would look like if it were made of solid balls (the atoms) or thin sticks (the bonds between atoms). Some examples are shown in the figure. Usually molecular graphics does not calculate the structure of the compound (*see* **Computational chemistry**).

Molecular graphics is a good way of allowing people to see the similarities in structures between molecules, and also of seeing if two molecules fit together well. This in turn is useful when, as part of a rational drug design program, a scientist wishes to find a molecule which will fit into the known structure of the active site of an enzyme, or the hormone-binding site of a receptor. However, these pictures can be misleading: molecules are not "really" like this, and all the pictures here are only approximations of one aspect of reality.

Molecular graphic pictures of molecules are often colored according to the "CPK" convention (named after Corey, Pauling and Kultun), in which carbon is light grey, oxygen is red, nitrogen is blue, sulfur is yellow and hydrogen is white.

Molecular imprinting

Here a polymer material, often polyacrylamide, is "imprinted" with gaps that exactly fit one and only one species of small molecule, like the binding site of an antibody exactly fits its antigen. This is done by forming the polymer matrix in the presence of the small molecule, so that the chains fold around those molecules. When the polymer has set solid, the small molecule is washed out using suitable solvents, leaving "holes" behind in the polymer. These can have quite a high affinity for the molecule that has been washed out. The results are called Molecularly Imprinted Polymers (MIPs).

The main use for MIPs being developed is as chromatography materials. They can separate chiral enantiomers, and are being used industrialy for chiral chromatographic separation.

A MIP can also be used as an antibody would in an affinity sensors, but with the advantage that, because it is basically a plastic and not a protein, it is tougher to heat, acids, solvents. The MIP may change size as it binds to its target molecule, and this can be a direct read-out for the sensor. Or a fluorescent molecule may be incorporated into the MIP, and its fluorescence monitored for fluorescence transfer effects (*see* **FRET**) or quenching when the analyte binds.

In principle, a catalytic MIP can be made just as a catalytic antibody can, by creating a MIP that binds a transition state analog. This is still a research topic.

Molecular modeling

This is the use of computers to create realistic models of what molecules look like. At one end of this range of techniques is molecular graphics, which is simply drawing three-dimensional drawings of what a molecule would look like if the atoms were large enough to see. At the other end it shades into computational chemistry—the calculation of what the physical and chemical properties of a molecule are.

Molecular modeling is the core of structure-based drug design (SBDD).

Monoclonal antibodies

Antibodies produced in the blood are made by a large number of different lymphocytes (B-cells). Each B-cell makes a unique antibody, so the antibodies which recognize any particular antigen are a mixture of molecules. This mixture is called "a" polyclonal antibody: an antibody preparation that reacts with one antigen, but which nevertheless derives from many different "clones" of B-cells. Polyclonal antibodies are problematic as reagents, because they have to be produced by many cells (in practice, a whole animal), and vary each time you make them. Monoclonal antibodies are a way around this. They are antibodies made from a single clone of B-cells which has been isolated and immortalized for growth *in vitro*. The invention of the methods for producing monoclonal antibodies won Cesar Milstein a Nobel Prize and, characteristically, very little money.

Monoclonal antibodies are generated as follows:

- *Immunization*—a mouse (usually) is immunized with the target antigen. This is done by injecting the antigen, sometimes with another material (an adjuvant) to stimulate the immune response. (*See* Immunisation.)

- *Splenectomy*—the spleen (a useful concentrated source of B-cells) is removed from the mouse.

- *Fusion*—the lymphocytes are fused with an immortal cell line (myeloma cell). This immortalizes them, that is means that they will grow for ever in culture.

- *Selection of clones* of fused cells on a special growth medium (usually HAT—Hyopxanthine/Aminopterin/Thymine—which the fused cells will grow on but the original ones will not).

- *Cloning by dilution* (*see* Cloning) to obtain pure lines of individual cells, each of which is called hybridoma.

- *Selection*—the clones are screened to find the one producing a good antibody against the antigen we want. Most of the cells will NOT be making the antibody we want.

A "good" antibody is one which binds tightly to the antigen (in chemical terms, has an affinity of 10^9 or better), does not bind significantly to anything else, and is the right class and sub-class (IgG, IgM, etc.—*see* Antibody).

If the target molecule is a very small one (like a drug molecule), then injecting it into a mouse will rarely produce an antibody response. In this case the molecule is chemically linked onto a larger molecule, usually a protein and often bovine serum albumin (BSA) or keyhole limpet hemocyanin (KLH), so that the immune system can "see" it. The small molecule is called a "hapten" in this case.

Several monoclonal antibodies are used as drugs, for example:

Name of drug	Type of antibody	What the antigen is	Disease	Year launched
Orthoclone	Mouse	Lymphocyte surface protein CD3	Prevention of transplant rejection	1986
ReoPro	Chimeric	GPIIb/IIIa protein on blood cells	Anti-clotting	1994
Rituxan	Chimeric	Lymphocyte surface protein CD20	Lymphoma	1997
Zenapax	Humanized	Interleukin 2	Prevention of transplant rejection	1997
Basiliximab	Chimeric	Interleukin 2	Prevention of transplant rejection	1998
Herceptin	Humanized	Her oncogene	Breast cancer	1998
Remicade	Chimeric	TNF alpha	Rheumatoid arthritis	1998
Synagis	Humanized	Respiratory syncytial virus	Virus	1998
Enbrel			Rheumatoid arthritis	1998
Myelotarg	Humanized drug: antibody conjugate	TNF alpha CD33	Leukemia	2000

Monoclonal antibody production

A major problem with monoclonal antibodies is how to make them. Although the average patient dose for a monoclonal antibody that is a drug is in the micrograms, and the amount used in a diagnostic test in nanograms per test, in 2000, 600 kg of antibody was produced for products, and there was a substantial shortfall in supply. Antibody production is complex—you need about 160 liters of cell culture facility per kg of antibody produced, and a multi-thousand-liter production plant typically costs $250 million to build. So no one builds them until they know the product is a success—however, you need the plant to make the antibody for it to become a success.

Monoclonal antibodies can be produced commercially in a number of ways, depending on the scale of production.

Tissue culture methods which are used to make the hybridoma in the first place can be used to make antibody—the tissue culture "supernatant," that is what is left of the medium once you have removed the cells, is a source of antibody. However, this is rarely effective at producing more than 10 mg of antibody.

Mice can be injected with the hybridoma call line which makes the mono-clonal antibody. The cells form a tumor (an ascitic tumor) which secretes the antibody into their ascites fluid (the fluid which surrounds the lungs) and the blood plasma. Either can be collected and the antibody purified from it. This requires animal handling facilities, and only produces around 50 mg/ mouse.

Immobilized cell reactors. Several types of immobilized cell reactors have been used to make monoclonal antibodies on a scale of a few grams. The most popular is probably the hollow fiber reactor. A few grams of antibody is enough for several million tests to be used for medical diagnosis, for example, and so suffices many commercial needs.

Suspended cell fermentors. The traditional biotechnology bioreactor tech-nology has been used to grow hybridoma cells in bulk. Celltech (now Lonza) has a 1000 liter airlift fermentor which can produce 100 g of antibody in a two-week fermentation with a hybridoma.

All these use cell lines, usually different from the original hybridoma. The antibody-producing genes may be engineered into cells such as CHO cells that are easier to grow than hybridomas, and may be selected or engineered for high production of antibody protein.

Antibodies can also be produced in animal milk (*see* **Pharming**), eggs (*see* **Eggs**), plants and in bacteria. Bacterial production is the most advanced of these. The genes for heavy and light chain must be spliced into one bac-terium, but when this is done the bug is very much easier to grow than mam-malian cells. This also makes genetic engineering of chimeric or humanized antibodies easier. Antibodies produced using dabs or phage display techno-logy are generated in bacteria to start with.

Motifs

These are short strings of bases or amino acids crop up time and again in different genes and proteins. Usually, they are significant because they denote some bit of the molecule has a particular function. Thus, "zinc finger motifs" in proteins suggest that the protein has a section that binds to DNA.

(Sangamo uses this technology for target validation, to turn genes on and off to find what they do.) In DNA, the "TATAA" motif is suggestive of a promoter sequence in eukaryotic cells.

"Motifs" are similar to signal sequences in proteins, and are usually "signals" to the cellular machinery as well. Being able to read such signal sequences is also helpful to the biotechnologist, as it gives clues as to a new protein's function.

Mutation

A mutation is a change in the genes. The original gene is usually called the "wild type" or, in human medical genetics, the "normal" gene. Strictly, the "mutation" is an identifiable change in one allele of the gene, one of the two copies in a cell. However, some scientists assume that if a gene is different from its form in the majority of individuals, or if the variant of the gene does not work, then it is a mutant.

Mutations can be point mutations (one base changed), deletions (a bit lost), insertions (a bit put in from some other DNA source, or occasionally from random DNA synthesis) or rearrangements (everything else). Rearrangements usually affect large regions of DNA rather than of a few bases. Chromosomal rearrangements shuffle whole chromosomal regions, switching them end-to-end (inversion) or swapping regions between chromosomes (transposition). Sometimes transpositions are straight swaps (balanced), sometimes some DNA gets lots in the process (unbalanced).

Mutations can also be defined by what they do to a protein: it may be neutral (has no effect), coding (alters the protein sequence and so has an effect of the protein sequence), or termination (changes a coding codon to a stop codon, and hence causes a shortened protein to be made). Mutations in promoter regions can affect the amount of protein without affecting its sequence.

Mutations in genes are detected by a wide range of **DNA probe** or **PCR** technologies (see separate entries). If you do not know what the mutation is, then it can be detected by sequencing the DNA and looking for differences. Gel electrophoresis methods can also detect DNA sequence differences, such as denaturing gradient gel electrophoresis (DGGE) and single strand conformational polymorphism analysis (SSCP). These will rapidly tell you if a difference exists, but not what the difference is. Gene "chips" are also being developed to detect mutations (*see* **Biochip**).

Nanotechnology

Nanotechnology is the structuring of matter on nanometer scales, a concept first described by Richard Feynman and popularized by K. Eric Drexler. Enthusiasts say that biotechnology can provide the methods necessary to build such small structures. This is because many biological molecules assemble themselves chemically into large complexes and the nanotechnologists hope to use the same self-assembly processes to get the nanomachines to build themselves.

Most of the "popular" ideas about nanotechnology are nonsense, but some practical applications are being worked up. This includes the development of quantum dots as labeling agents, nanostructured composites with material structures (and hence strength) similar to bone. The larger end of nano-technology merges with MEMS (*see* Biochip), which is a well established industrial technology.

Potential technologies with some relevance to biotechnology include:

1. *Patterning of structures around biological materials.* For example, silica can be formed with nanometer-sized pores by forming it around a network of liposomes or lipid bilayers.
2. *Nanostructures.* Molecular sized components (such as a silicon "wire" 20 nm across) can be used as sensor elements to immobilize antibodies. In these wires, virtually all the atoms are "at the surface" chemically, so binding chemicals to the surface drastically affects the conductance. So this is a new type of sensor, carrying our directly using electricity what SPR devices achieve indirectly with light.
3. *Quantum dots.* These are clusters of a few dozen to a few hundred atoms, which behave light giant molecules: they fluoresce at specific wavelengths, which can be "tuned" by altering the size of the cluster, thus making very flexible fluorescent labels.
4. *Biological macromolecules.* On the other hand, biological materials can be used to build nanotechnology devices—the most quoted example is using the large actomyosin complex from muscle or the bacterial flagella complex as molecular-scale motors.

Natural products

A natural product is any chemical or material derived from any organism, as opposed to being synthesized chemically. In biotechnological terms this often means drugs or therapeutics derived from living things, usually fungi or bacteria. There are a range of examples.

Microorganisms produce a wide range of chemicals which are used directly as drugs or as the starting materials for semi-synthesis of drugs, for example, penicillin and cephalosporin antibiotics.

Plants make many chemicals to deter animals from eating them. They include menthol from mint, limonene from lemons, caffeine from tea, and opium from poppies. Many of these have been used as the starting point of drug synthesis (such as steroids) or as drugs in their own right (such as quinine, cancer drugs vinblastin and taxol). Drugs from plants are sometimes called phytopharmaceuticals. **Bioprospecting** (see separate entry) seeks to find plants that have been used traditionally as medicines, and isolating the active chemicals from them. Shaman Pharmaceuticals and Phytogenetics were among the specialist biotechnology companies prospecting in this way, which, however, turned out to be less profitable and sustainable than they had thought.

Animals also make a variety of potential therapeutic products. "Natural products" from animals usually means chemicals from non-mammals, such as the antithrombin protein hirudin and the anti-cancer protein drug antistatin, both from leeches.

A wide range of materials such as wood and straw (traditional natural building materials) and PHB (the biodegradable polymer made from microorganisms) are also products of living organisms. However, neither the traditional biomaterials nor such other traditional products such as food is usually called "natural products" in this sense. They are often labeled "natural" to encourage the consumer to distinguish them from the products of industrial civilization.

NIR—near-infrared

Near-infrared (NIR) is radiation in the 800–2000 nm wavelength range, just longer than visible light. NIR spectra are valuable for characterizing what chemicals are in a mixture, because the peaks and troughs of the spectrum are very characteristic of specific chemical groups.

There are two types of application of this. Both rely on picking out the characteristic peaks of the chemical you are after from a complex mix of peaks, and so require complex computing techniques to use—often, neural nets or other advanced computing techniques are employed.

The first is in sensors. An NIR spectrometer can be used to show how much of a chemical is present in a mixture, and the radiation can penetrate plastic, water, and cells to depths of several millimeters. So this can be used as a "non-invasive" method for analysis. This has been applied to diabetic monitoring: a small light-emitting diode can shine a beam of NIR into the skin, and a sensor detect the absorption by glucose in the back-scattered light.

The second is as a process control aid. Here, the whole IR spectrum is used as a pattern to identify a "good" fermentation, and the amount the fermentation is different from that is measured. The advantage is that this gets an overview of all the chemicals in the broth without having to identify them one at a time. The disadvantage is that it is hard to tell exactly what has gone wrong if something goes wrong.

Nitrogen fixation

Nitrogen is an essential macronutrient (something we need a lot of in our diet) for all living things. Eighty percent of air is nitrogen gas: however, plants and animals cannot convert this into protein. They rely instead on other forms of nitrogen: ammonia and nitrates for plants, proteins and amino acids for animals. Only a few organisms can convert atmospheric nitrogen into these forms of nitrogen which can be assimilated readily, a process called "fixing" the nitrogen. The rate at which fixed nitrogen can be supplied to growing crop plants is one of the limiting factors in their growth and yield. Hence, 80 Mtonnes of nitrogenous fertilizer added to soils in 2000 (compare 1.3 Mtonnes in 1930).

Nitrogen fixing organisms are bacteria. Some live free in soil like Azobacter and Klebsiella. Some, the genera Rhizobium and Bradyrhizobium, live in symbiosis with plants. Symbiotic nitrogen-fixing organisms live in nodules in the roots of a few plants, and convert atmospheric nitrogen to ammonia for the plants in return for a supply of C4 acids (food), made by the plant from carbon dioxide. The genes which code for the enzymes which fix nitrogen—the *nif* genes—have been cloned and characterized in some detail. The bacteria also have *nod* genes, which induce the plant to make nodules in which the bacteria can live.

The simplest use of biotechnology is in producing legume inoculants, to increase the soil population of rhizobia around a growing plant. Nitrogen fixation can be limited by the rate of infection of the growing roots. Thus, dosing the soil with suitable bacteria can give a better rate of fixation. (It is controversial as to whether this is an effective economic measure.)

Genetic engineering can try to increase the efficiency of the bacteria in fixing nitrogen. BioTechnica tried an engineered *Rhizobium meliloti* in 1988, in which there were several copies of the gene for nitrogenase instead of the usual one copy. Nitrogenase is the enzyme which actually takes nitrogen molecules from the air and splits them open. The engineered bacterium was used to infect Alfalfa, but it did not result in any increased yield.

Only a few types of crop plants (legumes, clover, rice, lupins) can have nodules of nitrogen-fixing bacteria. Biotechnologists would like to force the Rhizobia to live in other plants, by introducing the bacteria to the plants in tissue culture or by engineering the cell-surface receptors of the plant root cells so that the bacteria absorb onto these roots in the same way as they do onto beans and clover. This route has been moderately successful on a laboratory scale.

Alternatively, you could transfect the *nif* genes into the plants themselves so that they did not need the bacteria at all. Fixing nitrogen takes a great deal of metabolic energy, so plants have not evolved to do this: getting it from bacteria is more efficient. It is not clear whether getting plants that do not normally fix nitrogen to do so would actually decrease crop yields, as they would divert a lot of energy away from producing the edible portions of the plant and into nitrogen fixation. The approach is thought to be technically infeasible in any case.

An alternative is to make plants better at getting nitrate from the soil, by making them express more of their specific nitrate transporter molecule, but no -one has been able to do this yet.

Nutraceuticals

Strictly, nutraceuticals are products isolated or purified from foods that are sold for their health benefits. However, the term is often used interchangeably with "functional foods," foods that themselves have some specific health benefit (other than preventing starvation). They are distinguished from "health food," which is generally believed to be good for you, but which does not claim to any more than that it will not make you ill. They are also differentiated from food supplements, which are meant to make up for a lack of an essential nutrient in your diet.

The first mass market functional food for which specific health claims were made was Benecol, a margarine containing plant sterols that block the uptake of cholesterol. Claims must be very specific—"diets rich in calcium help prevent osteoporosis" does not actually *say* that this chalk-based nutraceutical will do anything for you. However, even these claims must be submitted to the FDA before marketing to check that they do not mislead. Red wine has been claimed to have many good effects, but these are disputed, and in February 1998 the FDA limited the claims that can be made for red wine as a "health food."

Examples of the claimed benefits of food components are:

- carnitine (fights heart disease)

- cranberry juice (helps prevent urinary tract infection)

- pectin (lowers cholesterol)

- fibre (reduces risk of bowel cancer)

- vitamin E (fights cardiovascular disease).

Some nutraceuticals are scientifically accepted, others are still controversial because of a lack of medical-style trials of their effect. If trials are done, their results can also be controversial. For example, the first "double blind" study of the effect of vitamin E on heart disease, published in 1996, showed that large doses of the vitamin actually seemed to increase the death rate slightly.

Oil

Biotechnology has tried to produce or use oil for fuel applications. Two major applications are the use of oil as a feed-stock for bacterial growth (for Single Cell Protein production), and the use of organic material to make oil-like fuels or oil substitute (biofuel production). See separate entries.

Biotechnological products are also used in conventional oil extraction. Oilmen use a material called "drilling mud" to pump down an oil well to carry bits of rock drilled off the base of their hole to the surface. It is made from water, fine clay, and polymers. Xanthan gum is a favored component because it is viscoelastic—its viscosity is much lower at high shear forces than at low ones, so it behaves like water near the drilling head, not slowing up the drill, but like Jell-O when it is flowing back up the drill shaft, and so supporting the rock fragments.

Mud is also used to force oil to the surface in wells where the normal pressure forcing the oil to the surface has dropped. Water is tried first ("secondary recovery"), and then sometimes, when the less viscous water can no longer effectively displace the thick oil, "tertiary" recovery systems can be tried. These include pumping viscous "mud" into the well, including mud with biotechnological ingredients. Archaeus Technologies goes one further and pumps bacterial cultures in cheap substrates such as molasses down oil wells to make the polymers and surfactants underground. The bacterial suspension has low viscosity, and so is easy to pump.

Oligonucleotides

Oligonucleotides are short DNA (or, rarely, RNA) molecules, usually defined as 100 bases long or less. This is the length of DNA which an automated DNA synthesis machine (a DNA synthesizer, oligonucleotide synthesizer or "gene machine") can make in one go.

Oligonucleotides are usually named for their length. The naming follows the monomer–dimer–trimer scheme up to decamer (10 bases). Beyond that, the name of an oligonucleotide is generally its length as a number followed by "-mer." Thus, a 17-base oligonucleotide is called a "17-mer," pronounced "seventeen mer."

Automated DNA synthesizers use a series of chemical reactions to build up the DNA chain one base at a time. Each reaction consists of four steps, so building up a 50-base oligonucleotide (a "50-mer") requires 200 reaction steps. If one of those steps is slightly inefficient, then the overall efficiency will be very poor—this is why synthesizing oligonucleotides of greater than 100 bases becomes quite difficult. Most gene machines are completely automated, so all the biotechnologist has to do is type in the DNA sequence required and collect the DNA.

Oligonucleotides have become critical to biotechnology for three reasons.

- They can be linked together to form larger lengths of DNA which can function as completely synthetic genes. This is usually done by making overlapping oligonucleotides that hybridize to each other, and then joining them enzymatically with DNA ligase. PCR-based methods can also be used. This is technically demanding, as not only must the oligos hybridize properly, but the gene must have the right codons to work properly, not contain too many repeated sequences (which tend to get removed by bacteria) and so on.

- They can be used as DNA probes for a variety of genetic studies. In this they are particularly useful as they can distinguish between versions (alleles) of a gene which differ by only one base. Such oligonucleotides are called allele-specific oligonucleotides (ASOs).

- They are the primers for PCR.

Oncogene

Oncogenes are genes that are believed to be necessary for cancers to develop. There are a large number of them and, as would be expected from the variety of cancer types, they act in many different ways. Most are present in normal cells as Proto-oncogenes, that is, versions of the gene which are benign, and indeed are essential to the body's normal development. A mutation turns them into the malign oncogene. Oncogenes are of interest to biotechnology because of the importance of cancer as a cause of morbidity and mortality in Western societies.

Among the more talked-about oncogenes are:

- erb, a family of proteins, of which erbB-2 (also called "NEU") is associated with breast cancer.

- Myc: a protein in the cell nucleus, one of the earliest oncogenes identified. A mouse that is transgenic for a myc knockout was one of the first transgenic disease models, and was dubbed the oncomouse or Myc-y-mouse.

- Fos, a nuclear protein.

- Ras, a cell membrane protein which is associated with the protein kinase cascade, a complex series of enzymes, which regulates many of the cell's functions in growth and differentiation.

Many oncogenes have prefix letters. Thus, there is c-Myc (the cellular gene), v-Ras (a viral, cancer-forming version of Ras), H-ras (the human gene, to distinguish it from a number of homologs in other species.)

Tumor suppressor genes are genes whose normal function is to suppress some genetic activity, which could promote the development of a cancer. If a tumour suppressor gene is mutated, then it "releases" the activity of another gene and so hastens the development of the disease. p53 is a tumor suppressor gene. Mutations in oncogenes usually occur at only a few places in the gene, turning the protein product into a cancer-triggering molecule. By contrast, mutations in tumor suppressor genes can be in anywhere in the gene. This makes detecting mutations harder—in essence you have to sequence the whole gene from a patient to tell if they have potentially relevant mutation. Tumor suppressor genes can also be inactivated by speeding up their breakdown (p53) or directing them to the wrong part of the cell (BRCA1).

Tumor suppressor genes are being investigated as a therapy for cancer, using gene therapy to replace the mutated gene and so restore normal control to the cell.

Optical biosensors

A type of biosensor where the effect of a chemical on a biological system is detected using light rather than electrochemically. Several systems have been taken to commercial development in the last few years. They are based on the following principles.

Evanescent waves. Classical optics tells us that light can be completely trapped inside a prism or optical fibers by internal reflection—the light in effect bounces off the walls of the glass. However, the light is affected by the environment immediately outside the glass, as if some of it "leaks out," a phenomenon dubbed the "evanescent wave." If that wave finds a chemical there which can absorb it, then it is absorbed. This only occurs if the absorber is attached within 100 nm of the surface (the exact distance depends on many factors). Thus measuring the absorbance of the evanescent wave allows us to detect when something has stuck to our optical surface, as opposed to being free drifting around in solution.

If our optical fiber is coated with an antibody, then when that antibody captures its antigen that will change how the evanescent wave is absorbed, and so we can detect it. Variations on this idea have appeared as several near-commercial detection systems.

Surface plasmon resonance (SPR). This is a similar effect deriving from a different cause. When light is scattered off a conducting surface, the amount of light scattered at different angles depends on the exact nature of the surface and how it absorbs the light and conducts electricity. This can be used to detect material stuck on the surface.

BIACore sell a commercial sensor system based on SPR, for research use—they are useful pieces of laboratory equipment, but far from the "dipstick" of biosensor enthusiasts' dreams. They are particularly useful in dissecting the kinetics and thermodynamics of binding, and measuring on rates and off rates directly (*see* **Binding**) SPR can also be used to detect where molecules have bound to a surface, an use that Graffinity has applied to detecting when proteins have bound to a combinatorial chemical library immobilized on a "chip."

267

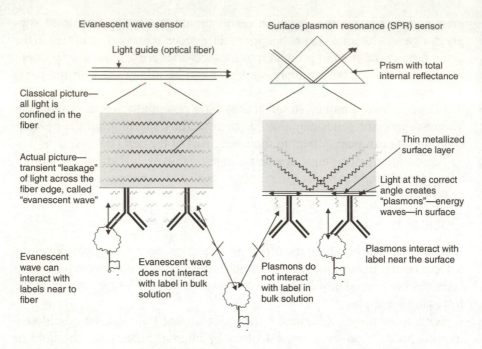

A problem with these binding-based sensors has been that they give a lot of false signals, as anything which absorbs light can stick to them and give a "positive" result. Thus, the development work necessary to make them work reliably is mainly about making them work dirty biological samples. Many optical biosensor developments have foundered on this rock.

Fibre optical chemical sensors (FOCSs) are optical sensors that measure specific chemicals: pH, oxygen, and carbon dioxide sensors are well known. They are useful for process and medical monitoring because of their strength and small size—an optical fiber can be inserted into a vein. The end of the optical fiber is coated with a plastic that is "doped" with a compound that changes color when it binds to the target analyte (such as oxygen, a specific ion, of protons for a pH sensor). If that ion is present in the solution it is absorbed into the plastic, the optical properties (absorbance or fluorescence) change, and the detector "looking" down the other end of the optical fiber can detect the change. Other ions are not absorbed and so are not taken up.

Biosensors seek to use this sensor approach by coupling enzymes to the end of a FOCS. When the enzyme causes a change in pH or consumes oxygen, the sensor detects this.

Organ culture

Organ culture is the growth of whole organs or parts of organs *in vitro*. Organs consist of several different cell types, as opposed to tissues which are made up of uniform cells.

In part, organ culture is a part of "traditional" transplant medicine. However, some scientists are also developing artificial organ systems based on cultured cells on a synthetic matrix material that mimics the extracellular matrix of the body. *see* **Artificial tissues**.

A related area is bone marrow transplant, which is half way between transplant and organ culture: here bone marrow cells are removed from one person and injected into another. However, they are often treated in between. Treatments can include

- *Stem cell selection*. Stem cells (the original type of cells from which all the others develop) are selected on the basis of their surface molecules. They are then grown up ("stem cell amplification") and given back to the patient rather than the whole bone marrow.

- *Cell ablation*. A reverse treatment is to remove cells based on their surface characteristics. Often these are types of cells that we do not want, such as leukemia cells.

- *Genetic manipulation*. Adding a gene to the cells to alter their function in some way.

Organic phase catalysis

This is the use of enzymes in solvents (liquids) other than water. Organic phase catalysis (also called solvent catalysis, hydrophobic catalysis, non-aqueous phase catalysis) is potentially useful for five reasons:

- The thermodynamics of the reaction (i.e. how the energy changes push the reaction towards one set of products or another) may be more favorable in a non-aqueous solvent, giving better yields.

- The substrate may be more soluble in organic solvents (or indeed, only soluble in them).

- The enzyme may be more stable or have an altered specificity in the new solvent.

- There will be no side-reactions involving water.

- Products may be easier to recover from a solvent (e.g. by evaporation or extraction with water).

Examples are the synthesis of peptide by proteases (in water, proteases only break up peptides). Aspartame synthesis from a "blocked" aspartic acid and phenylalanine is done in the thousand tonne scale. The transesterification of lipids by lipases is also a substantial application of organic phase enzyme catalysis (in water, lipases overwhelmingly break lipids up rather than put them together). Using lipases in organic solvents has been one of the more successful applications of this technology. Solvents such as octane, octanol, or nonane are often used.

"Normal" enzymes rarely dissolve in anything except water, and even if they do dissolve they do not work. This is, in part, because enzymes are prepared as aqueous solutions, and so a mixture of the enzyme with an organic solvent is just that—a mixture of two immiscible liquids. If the enzyme is dried thoroughly so an absolutely minimal number of water molecules stick to it, then many enzymes can be made to work in organic solvents.

The enzymes that can work in organic solvents are sometimes found to be more stable in this unnatural milieu than in water. This is probably because their 3D structure is not free to move about as much as it can in water, but is "locked" into the correct (active) shape. Their properties depend a lot on the solvent concerned: a psuedomonas lipase used for chiral resolution has an enantiomeric excess (specificity) of 2.6-fold in carbon tetrachloride, but over 30-fold in nitrobenzene.

Variations are to use supercritical fluids for the enzyme reaction, reversed phase or emulsion systems, or **bioconversion** in organic solvents (see separate entries).

An alternative approach is to genetically engineer the protein to be more stable or more active in the solvent concerned, and this is attracting some interest.

Ornamental plants

Ornamental plants, primarily flowers, are attractive commercial targets for the genetic engineer because the regulatory requirements for their release are less onerous than for something that is to be eaten, the value of a new flower is high compared to its mass, and new traits are easy to see, and hence both easy to select for and easy to "sell" to investors.

Plants are inherently easier to "engineer" through cloning or genetic engineering than animals. Crop plants can be hard to engineer because a change in them usually reduces the efficiency with which they produce their crop seeds, leaves, tubers, or other valuable parts. However, this does not apply to ornamental plants, which just have to survive and look pretty.

The world market for ornamental plants is over 30 billion dollars, making this a huge area for biotechnology. Top producers are the United States and The Netherlands—the fields of tulips are one of the latter country's major industries.

There are two major areas of applications.

Cell culture and cloning technologies. Rare variants of plants can be propagated very quickly using cloning technologies, and the same technologies sometimes generate new variants through **somaclonal variation** (see separate entry). Thus, "new" plants such as tropical species or ornamental cabbages, onions and other plants can be sold in garden centres very fast.

Genetic manipulation. Putting new genes into ornamental plants is a growing area of interest. Among the targets are:

Changing flower color. Color is partly determined by the enzymes that a flower makes to synthesize color pigments. Most of the pathways are known, and the genetic engineer can place the appropriate genes into plants to change colors. The flower pigments are usually some of a group of chemicals called anthocyanins, so putting the gene that turns a yellow pigment into an orange one into a yellow flower is likely to create an orange flower, because the other enzymes involved are already there. Antisense genes can also block color synthesis, producing white flowers or flowers that accumulate chemicals which have different colors from the normal end-products of color synthesis.

However, other aspects of the plant's physiology are also important to flower color, such as the pH of the vacuole, the small sack within the cell where the colored pigments are stored. Blue color produced by the accumulation of 5'hydroxylated anthacyanin pigments require a high vacuolar pH (alkaline conditions). One of the most sought-after blue flowers is the blue rose—however, as no rose species produces blue pigments, this will need recombinant DNA technology to create. Japanese scientists cloned a gene from Japanese morning glory that turns the purple buds into blue flowers by altering the pigment vacuole's pH—if put into roses, this might turn the currently purple roses into a real blue flower.

Pest-resistant transgenics. Plants that have inherent pest resistance are created through incorporating genes for pest-toxic proteins, such as Bt toxin (*see* **Pest resistance**). This is attractive for garden plants, some of which have very poor abilities to fight off pests.

Ethylene binding and senescence control. Cut flowers "die" largely through cell senescence—the flowers go through the same changes as they would at the end of their natural flowering season. Most of these changes are triggered by the gas ethylene, which acts as a hormone in plants. The effects can be blocked either by blocking the receptors for ethylene or blocking one of the enzymes that makes it. Antisense or RNAi can block the gene ACCO (aminocyclopropane carboxylic acid oxidase) which makes ethylene in plants. (The same effect can be achieved with chemical treatment of flowers, which many growers use.) This has been tried for the cut flower market, but could also be done to block ripening in some fruit (such as bananas) which is also triggered by ethylene. The grower would then ripen the fruit when it arrived at its destination with added ethylene.

Enhancing flower structure. The number of petals in a flower, and some aspects of their arrangement, is controlled by only a few genes. These can be transferred from plant to plant to produce unusual and complex flower shapes. (However, the majority of such new flower shapes are still random mutations that are discovered by gardeners.)

Making ornamental versions of common plants. Examples are ornamental cabbages and potatoes, which are made by manipulating genes such as the Agrobacterium rolC gene, which makes plant hormones resulting in dwarf, bushy plants that look like round clumps of leaves.

Orphan drug act

This is a legislation that gives an incentive to a company developing a drug for a comparatively rare disease. For drugs that provide novel treatments for diseases which are suffered by only a small number of people, orphan drug acts give the developer of the first drug of any one type 7 years exclusive right to market that drug. This is meant as an incentive to develop drugs for which the market would otherwise be marginal, given the intense competition in the pharmaceutical industry. It has been invoked quite widely by the biotechnology industry, as many biopharmaceuticals are so specific in their effects that they can only be used for a very narrow group of patients.

The US act was signed in 1983, and defined an orphan drug as one which would treat less than 200 000 people. Similar legislation was enacted in Europe in 2000.

Rare diseases are sometimes called "orphan diseases": this is an informal term, not embodied in legislation.

Osmotolerance in plants

Osmotolerance is a measure of a plant's ability to stand up to large amounts of dissolved material in its water supply: often this is also a measure of its ability to withstand drought. Tolerance to salt specifically is called halotolerance. Because a reliable supply of fresh water is a limiting factor to agriculture in some places, osmotolerance is an important characteristic for plant breeders to achieve.

Plants survive water stress through structural adaptation (e.g. making leaves round to reduce the surface area), physiological adaptation (e.g. developing molecular pumping mechanisms to pump water into the cells or salt out) or metabolic adaptation. Metabolic approaches to osmotolerance usually involve filling he plant cell up with something which is innocuous, which the plant can make easily, and which "attracts" water through its osmotic potential (i.e. just by being there, not because it expends any energy). A range of such compounds are known. A tentative approach to enhancing osmotolerance in crop plants to enable them to make these "osmoprotectants." This is still a research concept.

Oversight

In US regulatory contexts, this means "having regulatory responsibility over." Thus, the definition of which organisms are subject to "regulatory oversight" is a critical feature of the regulation of biotechnology—it defines which organisms must be approved by which authorities before they can be used for industrial biotechnology.

Panning

Panning is a method of screening a large population of different molecules, viruses, or cells for the few that bind to a target molecule. The target molecules are immobilized to a surface, usually plastic but sometimes glass, and the other potential binding partners are swirled over it like a gold-rush "panner" looking for gold. If any of the mobile molecules bind to the surface, then they stay behind when the others are washed off. This has been used extensively in receptor binding studies, to find which of many millions of chemicals will bind to a receptor. It is also used to select from phage display libraries of antibodies to find the antibody that binds best to an antigen. *See* **Phage display**.

Panning has been used to purify cell types. Here, the surface is coated with a ligand that binds to a receptor on the cell you are looking for, and then blood is passed over it. This is, in fact, a whole-cell version of affinity chromatography.

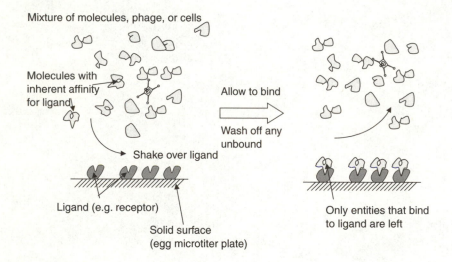

Mixture of molecules, phage, or cells

Molecules with inherent affinity for ligand

Allow to bind

Shake over ligand

Wash off any unbound

Ligand (e.g. receptor)

Solid surface (egg microtiter plate)

Only entities that bind to ligand are left

Patents

Most biotechnology is a "knowledge industry"—being able to do something is what makes a company valuable. It is, therefore, imperative for a biotechnology company to be able to protect its knowledge, known as "intellectual property," and the most common way to do this is to patent it. Deciding whether something can be patented is, therefore, very important for the industry.

In principle, patents are based on claims of novelty, utility, and enablement: a patent must describe something new, something useful, and must describe it in a way that enables someone else to duplicate it. Patent offices

274

decide whether a patent fulfills these criteria—the US Patent and Trademark Office (USPTO), the European Patent Office (EPO), and national offices in every country covered by the Patent Convention Treaty (PCT), which ensures that patent law is broadly consistent and interchangeable between countries.

These agencies can "grant" a patent. However, a granted patent is only presumed to be valid: its validity is not a definite "fact" under law, and so it is open to challenge. Much lawyer's time is expended on challenging that presumption, and so it is generally held that a patent is only as good as the last law suit it survived. Defending a patent against infringement in the courts can take years and cost millions of dollars. Most of the high-profile cases are fought in US jury trials, which means, like to O.J. Simpson trial of 1995, the jury must be educated on some extremely complex technical issues before they can even begin to understand the arguments. As in the O.J. trial, therefore, the jury often ends up putting the technical arguments aside and judging by "gut feel."

The critical part of a patent for this are the claims, which are the paragraphs at the end of the patent that specify exactly what the patent covers. Usually, Claim 1 is the broadest—"a gene making a protein that is a betahydroxylase enzyme"—and subsequent claims are more specific—"a gene according to Claim 1 where the sequence is . . . ."

Recent high-profile patent cases in biotechnology are:

- *DNA sequence*. See separate entry on **DNA (gene) patents**. Unlike the situation in the 1990s, gene patents are only considered valid now if they demonstrate substantial, credible, and specific utility. "Raw" sequence not patentable.

- *Transgenic animals*. They have been patentable in the United States for decades, but the European Union still opposes them, and tends to limit their scope. Thus the US patent on the Oncomouse was upheld in the European Union at the end of 2001, but was limited to rodent species (as opposed to any animal species, as claimed in the United States). This is more about the basic "right" to patent living things than about the technicalities of the case.

- *Festo decision*. This is a technical issue of patent law that arose in a case that had nothing to do with biotech, but affects the industry substantially. It relates to doctrine of equivalents: if you make an trivial change to a patented thing, it is deemed to still fall under the claims of the original patent, even if it is not literally claimed.

The Festo decision says that if a claim has been modified or dropped from a patent during the filing process, then the doctrine of equivalents cannot apply to whatever is claimed in that changed claim (because, if the change was really "insubstantial," why do it?) Thus, if, for example, the original version of a patent claimed "a gene for...," and this was revised to "A DNA sequence coding for," then under Festo the patent could only protect the exact DNA sequence described, not trivial alterations of it. A US Supreme Court decision in mid-2002 held that this extreme version was not valid, but nevertheless the Doctrine of Equivalents took a knock. This means that Biotech companies cannot claim all the myriad versions of a thing or a method that nature can invent, but only a generalized version of what they have found or built.

There is no formal equivalent of this process in European countries, so here, as elsewhere in patent law, different countries treat claims differently, so you can have opposite rulings on same patent in different countries.

- Related legislation is the Patent Misuse legislation in the United States, which makes it unlawful attempt to extend exclusivity conferred by patent after expiry of patent by other means. For example, a licence for the patent cannot demand royalties after the patent has expired. Several pharmaceutical companies have found themselves in court over this.

- *Percy Schmeiser case*. Schmeiser is a Canadian farmer who claimed that he developed his own roundup-resistant crop by harvesting seed from fields sprayed with Roundup. The crop had a lot of Monsanto's "Round-up Ready" seed in it, but not (apparently) bought from Monsanto. Was this patent infringement, or contamination of Schmeiser's crop's genes with pollen or seeds from nearby farms? The patent court ruled it was patent infringement.

- *Kauffman patent*. This is a very broad patent on the whole idea of molecular evolution (*see* **Darwinian cloning**). Diversa and Maxygen contest claims that it covers any use of randomized DNA to generate proteins. Others claim that such use of DNA is obvious.

- *Gene chip patents*. In the late 1980s, several groups filed patents claiming the basic idea of a gene chip—lots of DNA probes in a small space. As the patents all claimed the same thing, and the thing that many companies were attempting to commercialize, something like $40 million in lawyers fees were spent before the two principle combatants, Prof. Edwin Southern and Affymetrix Inc, settled their differences in 2001.

- Older patent battles over Factor VIII, EPO, and PCR have now largely run their course.

In addition, most countries consider that some things are not suitable to be the subject of patenting at all. Living things are not usually patentable, but biological materials obtained by non-biological means—one in which "the hand of man" has had a part, are patentable. The first patent for an organism altered by recombinant DNA technology was granted to Chakrabarty in

1980, for a hydrocarbon-degrading *Pseudomonas* strain to be used for oil cleanup. Whether something is considered suitable for patenting at all varies from country to country.

Area	Macromolecules or viruses[1]	Un-engineered microorganisms	Plant varieties	Animal varieties	Recombinant organisms
USA	Yes	Yes	Yes	Yes	Yes
Canada	Yes	Yes	No	No	Yes
EPO[2]	Yes	Yes	Yes	No	Yes
Japan	Yes	Yes	No	Yes	Yes

[1]However, there is some uncertainty about what is the difference between a recombinant protein and its (presumably) identical natural counterpart, for example.
[2]In addition to patents covering things ("composition of matter" patents), patents covering processes for making or using microbes are allowed in all the areas, although methods for breeding are not allowed by the EPO.

Even these guidelines can be overruled by TRIPS—Trade-related aspects of Intellectual Property rights agreement of WTO—which allows for patent rights to be overridden in "national emergency." In 2001, the United States interpreted this to include any patents relating to anthrax vaccines, but allows patents relating to AIDS drugs to continue in force in Africa, despite the millions of deaths caused there by the AIDS epidemic.

Major practical issues are Royalty Stacking and "submarine patents." Many biotech products need many technologies to get to work: for example, the group making the "Golden Rice" needed 70 patented technologies from 32 companies or institutions to stitch the right genes together and get them into rice. Each of those patent holders could demand a royalty payment, a fraction of the final sales price to be paid to them for the right to use the patented technology. This made the original goal of giving the rice away free to developing countries rather tough. In this case, Zeneca and Monsanto offered a free licence from the patented technologies they had contributed, and others quickly followed suit.

However, for more commercial products the stacking of royalties on many different patents makes the product too expensive to contemplate, and so the patenting of the technologies has the effect of stopping products being developed. This is especially true in the field of genomics-based drug discovery, where a drug might have to pay royalties on a gene, a protein, a HTS screening technology, and a target validation transgenic disease model before you have even discovered the chemical that will become a medicine.

"Submarine" patents are patents that someone has filed but which are not public yet. This used to be a significant problem in the United States, where a patent was not published for 4 or more years after filing. During that time someone else could be working on the technology you were working on.

When your patent is published, you pop up and said "I patented that 4 years ago, now pay me a lot of money or I will take you to court for infringing my patent" (only in legal language). In the fast-moving world of biotech, such patents are a significant worry.

Patch clamp

Patch clamp is a method of detecting the flow of ions through a single ion channel in a membrane. The membrane is stuck on the end of a micropipette, and extremely sensitive electronics amplify any current flowing: because the membrane is an insulator, the only current is that carried by the ions. The "clamp" part of the name comes because the voltage difference across the membrane is fixed (clamped) at a particular value by the experimenter, and not allowed to vary according to cell physiology.

Patch clamp is now a well-established method for studying ion channel function—what turns channels on (i.e. opens them so that they can conduct), what inhibits them, and so on. The technology for patch clamping can also be applied to some innovative sensors that can detect single molecules through their ability to move through or block channels in a membrane. The favored channel for such measurements is alpha-hemolysin, a protein that puts holes (nanopores) in cell membranes. If a voltage difference is put across the membrane, a current will flow unless the pore is blocked. Other pore-forming chemicals, such as the antibiotic gramicidin, can be used. Workers at Texas A&M University have adapted a gramicidin ion channel to detect quite large proteins (or the molecules they bind to) by their ability to alter the current through the channel.

Similar techniques have been speculatively suggested for sequencing DNA. DNA is threaded through the pore of alpha hemolysin. The different bases are detected because, having different shapes and sizes, they block the pore to different degrees and so cause its electrical resistance to change. In theory, this could "read" DNA as fast as the molecule could move through the pore—thousands of bases a second. However, the technology is still at the research stage.

See also Ion Channel

PCR (polymerase chain reaction)

Polymerase chain reaction is a method for amplifying DNA that was invented by Kary Mullis of Cetus. It takes a single copy of a DNA molecule and uses it to create millions or billions of copies of itself. Because of the specificity and accuracy of the reaction, this is an ultimately sensitive detection system.

The diagram outlines how PCR works. The key ingredients are

- Taq polymerase (a heat-stable DNA polymerase that catalyzes the synthesis of new DNA).

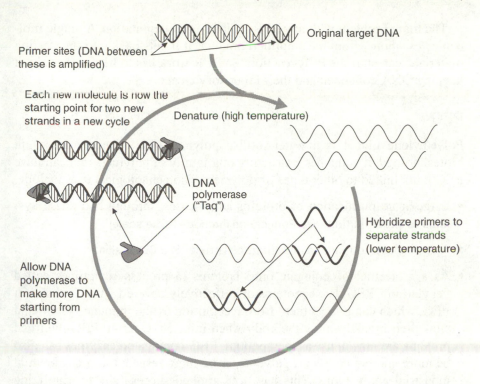

- Two primers, short DNA molecules, which are complementary to two sites either side of the piece of DNA you want to amplify, and which act as starting points for making the new DNA.

- A thermal cycler, a piece of equipment that swaps the reactions between three different temperatures in each cycle for 40–60 cycles.

PCR can be used to detect when a DNA is there, for building new DNA molecules, or for cloning DNA from sources where there are only a few molecules available, such as rare viruses, Egyptian mummies, or Dodos. Its use in genetic diagnosis is now almost universal. These are all "qualitative" applications: it is difficult to use PCR for measuring the amount of DNA (quantification) (*see* **DNA amplification**).

Variants on PCR such as single-sided PCR and inverse PCR (which rearrange the DNA before amplification so that only one primer is needed), and random PCR (which stitches synthetic DNA onto the ends of a segment to be amplified so that no new primers are needed) have been developed. A common variant is RT-PCR, which stands for reverse transcriptase-PCR and starts with an RNA target. Reverse transcriptase makes a DNA copy from the RNA, and then the PCR amplifies that DNA copy. This is widely used to clone and characterize mRNAs. *In situ* PCR seeks to perform the PCR reaction in a sample, so you can tell not only that the DNA or RNA is there but where in the sample it is (*see* **In vivo, etc.**).

The main problem with any PCR method is contamination. A single molecule "escaping" from the amplified product, if it gets back into the starting materials, can start the PCR reaction. So PCR work has to be rigorous about keeping DNA contained and their laboratory clean.

PEG

Polyethylene glycol is a water-soluble polymer used in many guises in biotechnology. It is the basic structure of a family of polymers that can have PEG units linked to other types of polymers. Biotechnological uses include

- Large-scale purification of proteins and viruses, through two-phase systems or precipitation. *See* Purification methods (large scale).

- As a fusogen to make mammalian cells fuse. *See* Cell fusion.

- As a molecule to conjugate onto proteins (a process sometimes called Pegylation). PEGylated proteins are effectively covered with a "shell" of PEG, which can protect them from recognition by the immune system and slow their breakdown by the body when injected. Over 40 PEG-modified proteins are now in use as therapeutics. PEG-conjugated enzymes can also be more soluble in organic solvents, or be more resistant to breakdown in industrial applications. This has been applied successfully to stabilizing lipases for organic-phase catalysis.

Peptide synthesis

Peptides are short protein molecules. In general, something is a peptide if it contains 20 amino acids or less, a protein if it is 50 amino acids long or more, in between, it depends on who you talk to. They are made by different routes from proteins, for two reasons. First, peptides are usually broken down rapidly by bacterial cells, so it is difficult to make them by recombinant DNA methods. Second, because they are relatively small molecules, it is feasible to make them chemically.

There are three general routes to making peptides. The first is by genetic engineering. The peptide is usually produced as a fusion protein, the peptide itself being joined onto a much larger protein. It then has to be cut ("cleaved") off the larger entity after that has been purified when the bacterium or yeast has made it. This can be difficult to achieve effectively, as you need a chemical reagent (such as cyanogen bromide, which cuts at methionine residues) or an enzyme, which cuts the fusion protein at exactly the junction between the peptide and the larger protein, but not within the peptide itself.

The second route is *in vitro* enzymology. Many proteases that break peptide bonds are known. By altering their reaction conditions they can be made to work in reverse and synthesize peptide bonds. Conditions can include making them work in organic solvents (*see* **Organic phase catalysis**), under extremely high pressure, or modifying the amino acids so that the peptide is removed from the reaction (by precipitation or because it dissolves in a second, organic solvent phase) as soon as it is formed.

In order that the protease does not assemble a whole string of amino acids, but rather adds them one at a time, the amino acids are "protected" by adding groups onto them that prevent uncontrolled polymerization. A cycle of reactions adds an amino acid, then removes its protecting group, then adds another and so on.

The third route is chemical synthesis. This performs the same sort of reaction cycle as enzymatic synthesis, but using traditional organic chemical reactions. The reactions can be carried out on a solid material (in a reaction series called the Merrifield synthesis) so that the peptide chain "grows" while attached to a support structure, or in solution, which is usually easier for large amounts but cannot make long peptides. The efficiency of each step is high, but because it is not 100% the yield is usually low after a couple of dozen amino acids have been added. Chemical routes usually require more reaction steps than enzymatic ones, but the materials are usually cheaper. Either enzymatic or chemical synthesis can produce kilograms of a peptide, and there are fully automated "peptide synthesizers" which can perform the chemistry to synthesize grams of a peptide in a few hours.

Pest resistance in plants

Genetic engineering has sought to engineer genes conferring resistance to pests in plants by several ways, as a potential alternative to using conventional pesticides.

Some plants are naturally resistant to specific pests. The plant genes often show "gene-for-gene" matching with genes in the pathogen called "avirulence genes": the virulence genes have a role in causing the disease, and the corresponding plant gene have evolved to stop them. The trick here is to find appropriate avirulence genes to transfer that block an economically important pest.

Completely new genes can also be transferred to the plant. The approaches currently used are given below.

To include the gene for a *Bacillus thuringiensis* (Bt) toxin in the plant. Different Bt strains produce 130 such toxin genes: the most commonly used one is called Cry9C. This toxin stops gut function in some insects, so if they nibble the leaf it kills them. Several commercial crop plants produce Cry9C, including the Starlink corn, marketed by Aventis, and transgenic cotton. *See* entry on **Biopesticides**.

To include an enzyme that attacks insects in the plant. DNA Plant Technologies is working on this, using Chitinase as the enzyme: chitin is a major component of insect's skeletons, and chitinase is an enzyme that breaks it down.

To include a protein that blocks a pest's usual method of attack or digestion of the plant. This has been used with good effect: the gene for cowpea trypsin inhibitor, a protein that inhibits the protease trypsin (and related enzymes), has been engineered into tobacco. This blocked the action of digestive enzymes in insects' gut, and so killed them.

To include a gene from the pathogen in the plant. This is called pathogen-derived resistance. The idea is that making an excess of one protein from a virus in the plant cells disrupts the control of how the virus is assembled when it infects the plant, and so viruses cannot propagate in the plant. This also works through an RNAi mechanism (*see* **Gene silencing**).

Phage display

Phage display is a method of making and testing proteins, usually newly invented proteins. It has been used mainly for generating antibodies or antibody variants. Antibody genes are either cloned from human lymphocytes or are synthesized on a Gene Machine. They are then spliced into an appropriate phage DNA and put into bacteria, where they make antibody fragments (such as SCAs) on the surface of the phage. If you select the phage that bind most strongly to the antigen, not only will they have the appropriate antibody on their surface but they will also have the gene for that antibody inside. This is an alternative to monoclonal antibody technology, and one that can generate human antibodies (as monoclonal technology cannot).

Phage display can also be used for a variety of protein engineering projects for other proteins, screening a large number of variants efficiently, providing that some sort of binding technique can be used to find the phage we need. Examples are finding peptides that bind to an antibody (epitopes), the sequences that bind to other proteins as signalling molecules, peptides that will bind to non-peptide targets molecules such as DNA.

Typically, the m13 phage is used for this, the protein being inserted into the Gene III. Libraries of up to 10^{12} phage can be made with "random" variants of a protein in them, and the half a dozen variants that we want identified by panning. Cambridge Antibody Technology holds the record the largest library made commercially in this way. Other types of phage can be used as well. The same process can be applied to making proteins on the surface of yeast or mammalian cells, although this is much less common, and is only used when we want to identify a specific function on the cell rather than the protein in isolation (as otherwise it is easier to use phage).

You can also display proteins on ribosomes, the protein-and-RNA packets that actually carry out "translation" inside cells (*see* **Genetic code and protein synthesis**). The trick here is to link the ribosome to its mRNA in some way so when the new protein is complete the ribosome does not fall off the mRNA, and you can select the protein and RNA at the same time: this can be achieved with some clever genetics or by chemical methods. In principle, you should be able to make and pan even larger libraries of proteins using ribosomal selection than using phage, but the ease of use of phage means that ribosomal display methods have not taken off.

Pharmaceutical proteins

Pharmaceutical proteins, also often called biopharmaceuticals and sometimes also "biologics" (in regulatory contexts), are proteins made for use as drugs. The first biopharmaceutical made by recombinant DNA technology was insulin, approved for human use in 1982, followed by human growth hormone in 1985 (both proteins had been used as drugs before). The first truly new recombinant DNA biopharmaceutical was alpha interferon, approved for treatment of Karposi's sarcoma in 1985. Around 10% of the new drugs launched between 1990 and 2000 were biopharmaceuticals.

Usually, biopharmaceuticals, which have to be human proteins to be fully effective in humans, are made from genetically engineered bacteria, as the only other source is cadavers or live human tissue. The genetic engineering of such products is covered elsewhere. Issues peculiar to biopharmaceuticals are usually the result of the stringent regulation that any drug must pass before it is allowed to go into general use.

Demonstration of efficacy. This can be hard when the protein is meant to boost a body function, and so does not have much of an effect on its own.

Demonstration that the product is free of contaminants. This is, particularly true of bacterial proteins and cell wall material that could act as a "pyrogen," that is, a material that could cause a feverish immune response in someone injected with it.

Demonstration of purity and stability. This is harder for a protein than for a small molecule drug, especially when it is made by a biological system (such as a cell line), which itself may vary with time or with growth conditions.

This means that a biotherapeutic's properties effectively include its exact manufacturing process, including the gene it was made from, organism it was made in, the growth medium they were grown in, and the purification methods. This is an issue for manufacturers who want to improve the manufacturing process, and for companies wanting to produce "generic" versions of a biopharmaceutical that is off patent. Usually this is done by proving Bioequivalence, that is, your molecule has exactly the same biological effect, as well as demonstrably the same structure, as one approved for use. This means that generic biologicals manufacturers have to carry out short clinical trials to get approval for their new version of an established protein drug, much more expensive than the ANDA (abbreviated New Drug Application) process needed for chemical generic drugs.

Companies working in the biogenerics field often set up facilities outside "The West," both to avoid the regulatory uncertainty there and to be ready to provide biologicals to non-US/European markets when patents expire. India (with nearly a third of the world's biogenerics companies) and China are major players. Thus, Dragon Pharmaceuticals produces a generic EPO that is approved for sale in China and India even though it is not yet approved in United States (where Amgen are based). Strangely, this means that these very high-tech medicines are often manufactured using technology that is 15–20 years old, simply because it is too expensive to introduce newer methods.

Pharmaceutical proteins are expensive. Costs can range from $7000 for a year's treatment with Remicade to over $50 000 for Ceredase. (Prices to the patient are higher because of the need to cover research costs as well as manufacturing costs.)

See also the entry on **Drug development pathway**.

Pharmacogenomics

This is the use of genomics technology (high-capacity molecular genetics) to study the genetic differences in how people respond to medical treatment. Many medical treatments work better in some people than others—some do not work at all in some patients. Some of this difference is caused by genetic differences. Pharmacogenomics seeks to find the relevant genes and use that knowledge to find which people will respond to a medicine.

A related term is pharmacokinetics, which is the study of known genes that are related to drug function, traditionally the genes that are responsible for variations in drug metabolism (*see* **ADME**). This applies especially to liver enzymes such as CYP2D6 and CYP3A4, for which some individuals have an inherited low level of enzymes compared to the average.

This is a hot topic in drug discovery and development for several reasons.

Pharmacogenomics could be used to classify people in clinical trials ("stratification") into those likely to respond and those not. This means you can test your drug on fewer people—the ones that are likely to respond to it—and so the trial will be cheaper and/or quicker. This presupposes that you know what the relevant genes are: you might know this if the drug's target is known to differ between people, or if other genetic components of the disease are polymorphic. This is an application being applied today.

The technology could be applied to "rescue" drugs that have failed clinical trials because of severe side-effects in some people. Finding what people are affected could (in principle) lead to a test that would be used before they were given the drug.

Related is the use of pharmacogenomic-based tests to find out what disease you really have in the first place, and find a drug or drug combination that exactly matches your disease. This is a concept called "personalized medicine," and is much talked about but as yet hardly ever put into practice. An exception is the Herceptin system (*see* Cancer).

This is itself linked to **predisposition analysis**, and the idea that medicine in general becomes much more specific to the individual.

A related idea is Toxicogenomics, the study and use of genes characteristic of toxic effects. For example, Phase I Toxicology Inc. has cloned genes expressed differently in white blood cells of people who react to penicillin, with a view to either finding penicillins that do not trigger that response or making a test for potential to react badly to the drugs. This can also be done at the protein level, hence toxicoproteomics.

Pharming

This is the use of transgenic animals to make pharmaceutically useful proteins. The aim usually is to genetically engineer animals so that the protein we want ends up in the milk. The gene for the biopharmaceutical is spliced onto a promoter and a signal peptide, which makes them express the protein in the mammary gland. Promoters for casein and lactoglobulin are effective, as these proteins are naturally produced in large amounts and only in the milk. Protein levels of up to 35 g/l have been reported. Pharmaceutical Proteins Ltd. have used sheep, and Genzyme have used goats to make alpha-1-antitrypsin commercially, for treating emphysema. Other proteins that have been made by pharming include tissue plasminogen activator (tPA), human growth hormone, urokinase, blood clotting Factor IX, and antithrombin III, the last named is in advanced clinical trials with Genzyme. The advantages over fermentation production systems are avoiding the need for sterile culture, avoiding the need for complex nutrient mixes, and obtaining the protein relatively free of other proteins, and quite free of cell wall materials or potential endotoxins.

The original idea of using cows (traditionally associated with making milk) has lost favor. Their longer breeding cycles and small numbers of offspring makes breeding them more costly and time-consuming, although Gene Pharming in Holland use cattle because of the sophisticated technology that exists for mass-producing milk from them. Dwarf goats (mature in 4 months) and sheep are commonly used: rabbits are also used, but mostly for research. Choice of animal depends on amount likely to be needed—for human serum albumin (for bulk blood replacement) a herd of 5000 milking cows would be needed to supply the world, for Factor IX (for treatment of hemophiliacs), four pigs would be enough.

A related idea is using chickens as a protein producer, directing the protein to the eggs. *See* **eggs**.

Pheromones

Pheromones are chemicals used to signal from one organism to another. They are more specific than scents—the target organism can detect them, in minuscule amounts which to other organisms are too dilute to notice. They are usually complicated organic molecules. The main value of pheromones is in integrated pest management systems, systems which use knowledge of the biology of a pest to target it with a number of diagnostic and pesticidal strategies. In 2002, a Mosquito gene that codes a pheromone receptor was patented by Sentigen, with a view to using it as a discovery target for anti-insect treatment.

They are being used as insecticides, because many insects use pheromones to signal their position prior to mating. If you load the air around your crop with pheromone, then pest insects become completely confused: at best they simply stop eating, and at worst they cannot find their mates to continue the pest infestation. Pheromone pesticides for tomato pinworm, the pink bollworm (which attacks cotton), and the coddling moth (which attacks apples) have all been developed.

Biotechnology helps in the development of pheromones by providing research tools to understand which chemicals affect which species. It can also help manufacture them, although this can also be done by conventional chemistry. A pheromone, for example, could be very effective during the mating season of an insect, but relatively useless afterwards when another type of pesticide would be needed.

Smell affects mate choice in mammals as well, including humans, but unlike insects they retain a choice over whether they respond to smells.

Physical containment

Physical containment is keeping something inside a laboratory through putting physical barriers in the way of its escape. It is the principle way by which genetically engineered organisms are kept inside a laboratory and prevented from "escaping" to the wider world. (The other route is **biological containment**—see separate entry). There are a range of physical barriers used, many of which are similar to those used in building **clean rooms** (see separate entry); however, in a containment laboratory the idea is to keep the dirt in, not out. Thus, as well as air filtration, a physical containment laboratory might have overhead UV lights to kill off stray organisms, and an autoclave to sterilize all waste going out.

National governments define several levels of containment under which different procedures have to be carried out. Typical levels would be:

- *Level 0*. Any laboratory.

- *Level 1*. "Good microbiological practice." This is taking normal care when handling microorganisms.

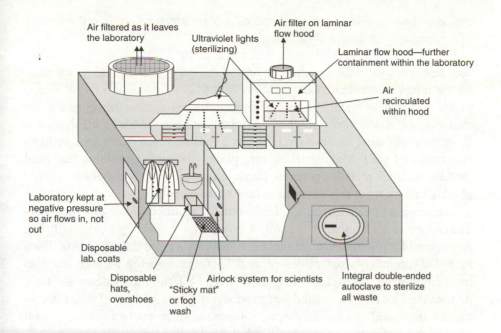

- *Level 2*. The laboratory is kept at negative pressure and air filtered. Any contaminated waste is autoclaved, so it is hard for anything other than the scientist to get out alive.

- *Level 3*. The laboratory is only entered through an "airlock" system, and all waste leaving it is autoclaved. Workers have to wear elementary protective clothing. Work on genetically engineered organisms that are expressing bioactive proteins, and on dangerous but relatively non-infectious organisms such as *Clostridia* would be done in such laboratories.

- *Level 4*. This is the ultimate containment level in most countries. Air is usually double-filtered on the way out, there is a double-airlock system for personnel with disinfectant bath to wash their shoes/boots in on the way out, and no one is allowed into the laboratory without substantial training (and no one who does not need to be there). Work on "live" AIDS viruses, genetic engineering common bacteria to express highly toxic proteins such as Ricin could be done in such facilities.

Level 4 facilities are very rare; usually even the most potentially hazardous biotechnology project is adequately contained by level 3.

Phytoremediation

This is bioremediation (clean-up of waste and environmental contamination) using plants. The use of transgenic plants has advanced rapidly here, as the concerns about the accidental escape of a plant are far less than the escape of a genetically engineered bacterium.

Work usually focuses on plants that can suck up a lot of water (so sucking contaminants out of the ground), or on the ones that accumulate heavy metals for metal contaminated sites. Heavy metals are a particular target. Some plants accumulate metals as part of their defences against being eaten, for example, the Brake Fern, which accumulates arsenic (and which Edenspace Systems is planning to commercialize as a means of removing arsenic from contaminated soil). Making transgenic plants containing proteins that bind metals, such as metallothioneins (from animals) or phytochelatins (from plants), can increase this even more (*see* Biosorption).

The same approach can be applied to collecting and disposing of chemicals that otherwise accumulate in the soil. A group from Cambridge University has engineered PETN reductase into tobacco, making a plant that can break down PETN, a commonly used explosive, and TNT-metabolizing tobacco has also been engineered. As well as being explosive, TNT (and the side-products of its manufacture and the breakdown products it forms in the soil) are toxic to human liver, and can cause cancer, so groundwater contamination around sites of explosives manufacture is a substantial cause for concern.

Although these approaches are, in principle, cheaper and more efficient than conventional clean-up, they still take years and require careful monitoring.

See also Biofilter.

Plant cell culture

Like any living organism, plants are composed of cells, which are capable of growing and dividing outside the plant given the right conditions. As with animal cell culture, it is essential to keep the cells free from any other contaminating organism like a bacterium or fungus. Although plant cells have a range of defences against infection, the bacteria or fungi can grow very much faster than the plant cells in fermentors, and so outgrow the plant cells, resulting in a large mass of contaminant and either a small mass of plant cells or the death of the plant cells. They also need a variety of hormones and growth factors: in some systems, the plant cells are grown on a callus of other plant cells, called a nurse callus, which provides a "feeder layer" to provide out cells with these.

Plant cell culture has a wide range of applications in biotechnology, in

- plant cloning, that is, the growth of plants from very small pieces of plant tissue, even single plant cells;

- plant genetic engineering;

- protein production in plants.

Plant chemicals such as scents or food flavors can also be made from plant cells in culture rather than whole plants. Plants produce a very large number of useful chemicals, but often do so only a certain times of year, and the mature plant can take years to grow and only grow well in very restricted places. If the cells from the plant could be grown in a bioreactor, then some of these inconveniences could be avoided. The main problem is that plant cells in culture produce very little of these secondary metabolites (see separate entry). This can be overcome in some cases by growing the cells with suitable elicitors, compounds, or mixes of compounds (often from plant of fungal sources) that are observed to increase the rate of production of secondary metabolites in cultured cells.

In this the plant biotechnologist is helped by the plant cell's totipotency. Most plant cells are capable of being grown back into a whole plant—they are totipotent, that is, they have all the "potency" of the original plant. This contrasts to animal cells, most of which cannot be grown into anything other than the tissue from which they came.

Plant cell immobilization

As well as the general methods used to immobilize growing cells in a bio-reactor (*see* Animal cell immobilization) there are several techniques that are relatively specific for immobilizing plant cells.

Entrapment of plant cells in gel matrices is popular: you suspend the cells in small drops of the material, which then set or harden to make little carriers. Materials such as alginates, agar, or carageenans (all of which are polysaccharides from seaweed), gelatin, or polyacrylamide have been used. Hollow fibers have been used for plant cells, but are not as popular as for animal cells. A relatively new method involves immobilizing the cells in polyurethane foam. In these "foam reactors," small lumps of foam are suspended in culture media and the cells encouraged to grow into the holes inside, where they form mini-bioreactors.

Unlike animal cells, plant cells are usually enclosed in a very tough cell wall. This means that plant cells will not spontaneously stick to a substrate as easily as animal cells will. However, you can chemically link them to one without killing them. Plant cells have been chemically linked to nylon and to polyphenylene beads using glutaraldehyde (a standard chemical for linking two biopolymers together).

Plant cloning

One area in which traditional biotechnology has been successful is in plant cloning, based on the techniques of plant cell culture and embryogenesis. The technique is an extension of the idea of taking cuttings to "duplicate" a particularly valuable plant. With cell culture techniques, the "cutting" is a single cell.

Cloning from plant cells involves several steps.

Isolating individual cells. If all you want is a number of plants, then the cells need not be rigorously separated from each other, and could be they can be small chunks of tissue (tissue explants). If you want truly clonal plants (i.e. ones derived from a single cell) then the cells must be separated carefully.

Genetic manipulation of the cells. The isolated cells can then be engineered if we need new genetic characteristics in them. *See* **Plant genetic engineering**.

Callus generation. Culturing the plant cell into a mass of cells, looking like a small piece of chewed paper. The plant cells will grow like this indefinitely, without forming any recognizable plant tissues. Callus-like growths can also be made directly from the meristem, the tip of a plant root or shoot that contains most of the growing cells. The result is a meristem culture.

Embryogenesis/organogenesis. The callus is encouraged to regenerate roots and leaves (*see* **Embryogenesis**).

Planting. Once the plant cells have generated a recognizable plant, it is safe to put it in the soil, as they will now have the mechanisms in place to fight off bacteria and fungi, and get their nutrient from simple soil chemicals.

A further step is the use of anther cultures to speed up breeding programs for obtaining homozygous plant lines. Wild plants (like "wild" animals, such as people) are heterozygous for many alleles—they will have two different copies of each gene in each cell. This makes breeding more complex, and we would like to have completely homozygous plants. In animals this is not possible, but in plants it often is. Anthers from male plants are cultured, and the haploid cells (i.e. the cells containing only one set of chromosomes, not the normal two) in the anther encouraged to grow clonally into plants. (Technically, these cells are microspores, so this is called microspore culture.) Unlike animals, haploid plants are often capable of growing in culture. As they only have one set of chromosomes, on "diploidization" (i.e. any technique which will double up their chromosomes to make a normal, diploid plant) both copies of their chromosomes will be the same, that is, they will be homozygous.

Substantial "tricks" to plant cloning are finding the right growth conditions for the callus and for embryogenesis (which are different for each plant), and keeping the culture sterile. The latter is hard to achieve for something that spends 24 h a day sitting in the soil.

The third problem of somaclonal variation arises in some species. See separate entry.

Plant genetic engineering

Plant genetic engineering is a major part of research effort in biotechnology, because of the potential it holds out for improving crop plants. A genetically engineered plant, sometimes called a transgenic plant, is the product of several technologies covered in this book. The necessary ingredients to make a transgenic plant are:

- isolating single plant cells;

- getting DNA into those cells;

- regenerating the cells into plants again; and

- in some cases making homozygous plants from heterozygous transgenics.

The first point is covered in **plant cell culture**, the third and fourth in **plant cloning**.

Getting DNA into plant cells has been difficult, because plant cells are surrounded by a robust cell wall and, unlike bacterial cells, do not have common mechanisms for acquiring DNA from their surroundings. As with all methods of making truly genetically engineered multicellular organisms, the key is not only to get DNA into the plant but to get it in a suitable amount and to have it integrated into the plant's chromosomes.

The common routes discussed are:

Using *Agrobacterium tumefaciens* (see separate entry).

Using *electroporation* on plant protoplasts (see separate entry).

By microinjection. This technique, which has worked so well in creating transgenic animals, has been applied to plants in two ways, but is not widely used.

By biolistics (particle gun) delivery (*see* **Biolistics**). This is a favoured route, and is efficient at getting DNA into plant cells.

By transformation of protoplasts. If the plant cell's wall is removed, then the resulting protoplast can sometimes be transformed simply by mixing them with DNA (in the right conditions). This has not worked with monocotyledons yet, and seems to have only limited potential.

By transformation of chloroplasts (plastids). Chloropasts are where photosynthesis takes place, and chloroplasts are only passed on by female plants, not in pollen. Transforming chloroplasts is, therefore, a route to making safer transgenics and engineering the basic energy metabolism of plants. It is hard to do, however. Biolistics is the favored technology for this.

After a gene has been gotten into a cell, the one cell among many thousands or millions which has taken up the gene must be selected for growth. This relies on a selectable gene which you have transfected into the plant cell together with the gene you want in there. This gene may be for resistance to a herbicide (which would kill the plant cell), or for an enzyme which is easy to detect using a simple assay. Some of the issues with "GM" plants have been with "marker" genes left in the plant by this process.

The methods used depend on the type of plant, and specifically whether the plant is an angiosperm or a gymnosperm, a monocotyledon, or a dicotyledon.

Angiosperms are flowering plants, that is, all plants that have real flowers, no matter how small—it includes all fruit trees, vegetables, grasses, and grains. Gymnosperms have seeds that are often carried in cones (hence the name for the common trees in this group—the conifers). The main economically important angiosperms are pine, fir, and cedar trees, which are valuable for timber and wood pulp.

Angiosperms are also divided into monocotyledons (represented here by grasses, cereals, and maize, as well as orchids) and dicotyledons (all other crops). Moncotyledons have a single "seed leaf," that first leaf that sticks up through the ground when the seed germinates. Dicotyledons have two.

The point of these distinctions is that the biology of these different classes of plant are as different as the biology of frogs and people. So a biotechnological technique that works on one dicotyledon (say, cabbage) may well work well on another (oak trees) but not on a monocotyledon (say, wheat).

Plant oils

A substantial part of commercial biotechnology is aimed at producing or modifying plant oils. The oils are stored in the plants as triacylglycerols (TAGs), that is, molecules with one fatty acid linked to each of the three hydroxyls of glycerol.

Common sources of oils include

- palm and coconut (medium-chain oils), used mostly in detergents;

- rapeseed (long-chain oils), used as lubricants, plasticizers, and for making nylon;

- castor bean and lesquerella oil (hydroxylipids), used in lubricants and coatings;

- jojoba wax used in lubricants and cosmetics;

- flax oil (trienoic), used in coating, drying agents, and to a small extent in cosmetics;

- cocoa used in chocolate and cosmetics.

Enzymatic processes involving the use of plant oils include hydrolysis (to make the fatty acid) and transesterification (to make different esters from the glycerol and fatty acids—*see* entry on **Lipases**).

Plant sterility

An important aspect of plant breeding programs is obtaining genes which confer sterility. This is, in part, so that farmers cannot breed from the seeds they are provided with (to protect the originator's patent rights), in part to assist breeding programs, but mainly so that hybridization breeding methods can work.

Here, two parent crops that do not themselves produce high-quality crops are used to produce seeds that grow into high-quality crops. This enables characteristics to be combined into one crop plant which could not be maintained by the traditional practice of keeping back a fraction of this year's crop to plant next year. However, it is essential that the grain sold to the farmer is the offspring of the mating of both parental types, and not just one. As sexing a field of wheat is tedious, this is done by ensuring that the combinations you do not want are sterile, that is, set no seed. Usually, it is the male plant which is sterilized, and so the genetic effect is often called "male sterility": a specific version of this is the "terminator gene" technology.

Making crops sterile is also useful to prevent one year's crop becoming next year's weeds (the "volunteer crop" problem).

Plant sterility technology is very politically controversial, because sterile crops cannot be used to produce seeds for the farmer to plant next year, the way that many farmers, especially in poorer countries, propagate their crop. Instead, they must buy new crop every year from seed suppliers. This could reduce biodiversity and concentrate economic power in the hands of multinationals. (It was estimated that 300 000 acres of "unregistered" GM crops were planted in Saskatchewan alone in 2000, from farmers or seed dealers growing their own seed—see the Percy Schmeiser case in patents). However, this could also be a way of stopping the spread of GM genes into other, non-GM crops (*see* Gene spread).

Biotechnology has provided a range of new ways of making plants sterile, either one sex or both sexes. It has also generated "restorer genes," genes which reverse the effect of the male sterility gene. This allows the plants carrying the male sterility gene to be cultivated on their own—without it, the line of plants would die out within one generation because of the lack of males.

Plant sterility technology is also useful for producing seedless fruit. Thus, genes that destroy cells (such as the gene for DNAse, which breaks up any DNA) can be put under the control of seed-specific promoters, so seed cells automatically self-destruct. The trick here is that seeds often make hormones that cause fruit to form, so—no seed, no fruit. So there must be a replacement source for those hormones, such as sprays.

Some plants produce seedless fruit anyway, such as pineapples and bananas, a process called Parthenocarpy.

Plant storage proteins

Plant storage proteins are proteins accumulated in large amounts in seeds not because of their enzymatic or structural properties but simply as a convenient source of amino acids and calories for use when the seed germinates. They are of interest to biotechnologists for two reasons.

Storage proteins as a source of protein. Much of the world's food comes from plant seeds or fruits, and much of the protein in those seeds is storage protein, but this is ideal for plant growth, not human nutrition. Many storage proteins are poor in some essential amino acids: a diet which relies on just one storage protein source for nearly all its protein can lead to deficiency disease, despite being quite adequate in bulk protein. Improvement of the proteins for food use would seek to engineer them to contain more of the essential amino acids, and so be better balanced, class I sources of protein.

As protein expression systems. See **Protein production in plants**.

Plasmid

A plasmid is a small piece of DNA which can exist inside a cell separate from the cell's main DNA. This means that it must be able to replicate itself inside the cell, so plasmids have the correct genetic elements in them to cause the cell's enzymes to replicate them as the cell divides.

Plasmids exist in most microorganisms. Those in bacteria are almost invariably circles of DNA. Some in yeast are linear DNAs, like very small chromosomes.

Plasmids are used extensively in genetic engineering as the basis for vector molecules. Because they are small, they are easy to manipulate: they only have only a few sites for restriction enzymes in them, and so it is relatively easy to cut them open at just one place, then to splice in a piece of "foreign" DNA and join the ends up again. They can also be manipulated to be present in many copies in the cell, rather than the one copy of normal chromosomes and plasmids. *See* the entry on **vectors** for more details.

Plasmids are a specific type of Episome, the generic name for any small DNA that can exist as an independent entity inside a cell free of the cell's main chromosomes. Some viruses can also be episomes, existing as DNA within a cell for a long time. (This does not include the retroviruses. These exist as DNA inside a cell, but their DNA is spliced into the chromosomes themselves.)

Many plasmids can transfer themselves between cells, which can be an undesirable characteristic for a gene cloning vector, which we want to stay in the cells we put it in. A first step to turning a plasmid, into a usable form is to delete genes such as "mob" (which mobilizes a plasmid) and "tra" (which allows it to transfer itself to another bacterium) from the plasmid.

Polygenic disease

Also known multigenic diseases, these are diseases where many genes play a role in whether we get the disease or not. They are an important research target in the genomic era, as genomic technology allows scientists to look at whether any one of thousands of genes are related to a particular disease. These diseases are also caused by environmental factors, which makes separating the effects of one gene out very hard.

Most common diseases are "caused" by a number of genes, in the sense that genes make us more or less likely to get them (*see* **Predisposition analysis**). These diseases are all influenced by more than one gene, and usually by a large (and unknown) number. They are multigenic (or polygenic) diseases. All of them are also influenced by environment to a substantial degree.

The main way of finding out whether a human characteristic is affected by genes is to see that it "runs in families." Family members share more genes with each other than they do with the non-family members around them, so if a characteristic is seen to be passed down from parent to child more often than chance would dictate, we can suspect that a gene is involved. A common technique is to study identical twins (which have exactly the same genes and very similar environment) and compare them to non-identical twins (fraternal twins), who have different genes and a very similar environment. Another is to look at very in-bred groups of people, where cousin marriages are common and so the chance of getting the same gene from both parents is high—genetic programs in the Amish of Pennsylvania, and the rural populations of Newfoundland, Terra del Fuego, and Saudi Arabia, have all taken this approach.

We look at the "concordance" for a disease—how often it occurs in related people as compared to unrelated ones—and from this work out the "heritability" of that trait: roughly, the extent to which the variation between people in that trait in that population is due to genes. A heritability of 100% means that differences in genes are the only factor determining the differences between individuals in the sample we have studied. From similar studies we can look at the concordance of the disease with a specific gene of genetic locus (*see* **SNP**), and hence get an idea of which gene "causes" that disease (more accurately, what differences in which gene might be part of the cause of the differences between a healthy person and a person with that disease).

Genes identified in this way could be used to make new treatments for the disease, new diagnostics, or to deepen our understanding of what the disease really is and thus indirectly improve diagnosis and therapy. So far this has had limited success, but given the complexity of the research agenda this is not surprising.

See also **Pharmacogenomics**.

Polymorphism

Many genes come in different types, slightly varying from one individual to another (*see* **SNP** and **Mutation**). They are said to be polymorphic (literally,

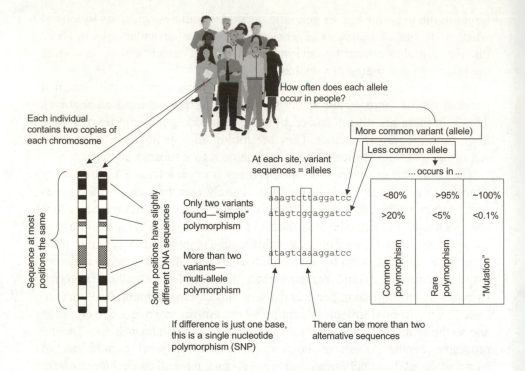

many-shaped), or alternatively to be genes containing polymorphisms (i.e. differences in their DNA sequence), in contrast to genes that are always the same in different individuals of the same species. "Polymorphism" occurs in populations; thus, Europeans are polymorphic for mutations in the Cystic Fibrosis (CF) gene, because a substantial fraction (between 1% and 2.5% in Europe, depending on the country) have at least one CF gene that has a mutation in it. On an individual level, if both your copies of a polymorphic gene are the same then you are said to be homozygous. If they are different, you are called heterozygous. The individuals concerned are called homozygotes or heterozygotes.

Polymorphisms can be variations in genes that have a drastic effect on the individual (such as the CF mutation), ones which have a noticeable but unimportant effect (such as hair color), or ones which apparently have no effect at all. There are probably millions of this last class, which can be used for genetic studies (*see* **SNP**). All but the first are sometimes called "genetic markers."

The collection of your specific genes and gene variants is called your genotype. "Genotyping" means to find out what your genotype is, usually with respect to a specific gene of interest. Thus, a scientist might say "we genotyped the patients to determine whether they carried this particular gene variant."

Polysaccharide processing

A common use of industrial enzymes in the food industry is in processing complex polysaccharides such as starch and pectins. Enzymes are involved in several processes:

Liquefaction. The dispersal of starch granules into a gelatinous suspension (similar to that when cornflour is boiled). The starch is also hydrolyzed into shorter molecules by enzymes such as pullulanase and alpha-amylase. Because liquefaction is often carried out in hot solutions, one valuable biotechnological product is heat-stable alpha-amylase and pullulanase, isolated from thermophilic bacteria, which will work at 80 or 90 °C.

Saccharification. The formation of low molecular weight sugars, often mainly glucose, from liquefied starch. Acid treatment of starch will break it down into a mixture of glucose and higher sugars, called Glucose Syrup. The amount of hydrolysis depends on the amount of acid, the time taken, the type of starch involved, and so on. The degree of breakdown is measured as Dextrose Equivalents (roughly, the fractional extent to which the starch is broken down completely to glucose).

There are also a variety of enzymes which will do this: amylases and pullulanases to break down the starch, invertase to break down sucrose, and glucose isomerase to convert glucose into the sweeter fructose.

Debranching. A chemical rather than a process term, this is the removal of the side branches from the long starch or pectin molecules, leaving long, straight molecules which are easier to break down further. Branched and unbranched polysaccharides also have different gelling properties, and so confer different mechanical properties on food. Some types of pullulanase and isoamylase can perform debranching.

Post-genomic era

Broadly, these are the things we do now that the human genome is sequenced, but by inference everything that happens in molecular biological research after 2001. In reality, of course, genomics is not "over." But the spirit of the "post-genomic" concept is that genome sequencing is now routine, most of the sequence of the human genome is now known, and the scientific cutting edge has moved on.

Post-genomic technologies include proteomics, functional genomics, and the various other "-omic" technologies such as transcriptomics and metabolomics. Technologies that are based on the assumption that the whole human genome sequence is known, such as some proteomics technologies, are called "post-genomic."

Part of this is that "genomics" companies, originally basing their business plans on the premise that the value of the genes they discover would keep them going for ever, have tried to shift to being drug discovery companies that have, as part of their platform, a strength in genomics. Gene hunting company Millennium was the first to do this. Incyte, deCODE Genetics, HGS, and Celera have all followed. This is the "post-genomic business model"—it remains unclear if it will produce any profits.

Post-translational modification

This is a blanket term to cover the alterations that happen to a protein after it has been synthesized as a primary polypeptide. They include:

* *Glycosylation (see separate entry).* This is one of the critical post-translational modifications for biopharmaceuticals.

* *Removal of the N-terminal methionine (or N-formyl methionine in bacteria).* Nearly all proteins are made with a methionine as their first amino acid, and this is usually removed. The enzyme in bacteria (peptidyl deformylase—PDF) is a target for antibiotic design, as it is essential to the bacterium but does not occur in animals.

* *Signal peptide removal.* Peptides which are to be inserted into membranes, secreted into special cellular compartments (such as the mitochondrion or into vacuoles or lysosomes) have a short string of amino acids at their front called the signal peptide. This signals to the cell where the protein is to go, and is chopped off as part of the mechanism for getting it there (*see* **Expression systems**).

* *Acetylation, formlyation.* These and a few other modifications put relatively unreactive groups onto more reactive ones. They are often put on the terminal amino group of a protein, producing a "protected N-terminus."

* *Amino acid modification.* This is the chemical modification of amino acids after they have been incorporated into the protein chain. There are relatively few examples of such modification, but they can have critical effects on the protein's function. Examples are the modification of glutamate to form gamma carboxy glutamate by a vitamin K-catalyzed reaction in mammalian liver, and the hydroxylation of proline to hydroxyproline in collagen in animals.

Predisposition analysis

Predisposition analysis can be thought of as the diagnosis of **polygenic disease** (see separate entry) before the person actually has got the disease.

The interest to biotechnology of this genetic predisposition is threefold:

* First, if there is a gene involved, we could hope to use DNA technology to find that gene and then use it as a diagnostic to detect who was predisposed to that disease. Linked to this is the idea of disease stratification—finding that (for example) what we call "depression" is actually many different

diseases with different genetic predispositions, which should be treated differently. *See* **Pharmacogenomics**.

- Second, we could hope to find out what the gene does and hence design a therapy to counteract it.

- Lastly, we could use our knowledge to identify what aspects of the environment trigger the disease, and hence improve our environment to everyone's benefit.

Analyzing whether someone has a gene that predisposes them to a disease is even harder than finding one that is related to a disease that they actually have, because many people with the gene variant will not get the disease anyway. Despite this, several genes have been suggested as "predisposing genes" for disease, including

- Apolipoprotein E4, a blood protein variant, suggested as predictor of Alzheimer's disease.

- BRCA1 and BRCA2, genes that are linked to some forms of partially inherited breast and ovarian cancer, and possibly involved in other breast cancers as well.

- HNPCC (a gene whose mutation is a cause of the rare cancer hereditary non-polyposis colon carcinoma), which similarly has been suggested as relevant for colon cancer.

There are obvious ethical and legal implications of using human genetic information in these ways. They are principally concerned that the person whose information is being used gets appropriate use of that information, is not confused or frightened by it, and that no one else gets to use the information without the patient's understanding and consent. *See* **Genetic information**.

Probiotics

Probiotics is the use of live bacteria as a health-adding part of a food or (occasionally) other treatment. The average human body contains 10 times as many bacterial cells as human cells and they exist in competition with each other, mostly in our guts and on our skin. Disease is often caused by too many of a harmful bacterium ("overgrowth" by that bacterial type), and so probiotics seeks to promote the growth of other bacteria to out-compete the "bad" ones.

Probiotic dairy foods are well established, and probably do have health-conferring value in keeping "bad" bacteria from our intestines. They can also actually support normal gut function. Gene chip analysis of germ-free mice show that colonization of gut with Bacteriodes species (normal gut-living bugs) significantly affects expression of genes involved in food absorption, blood vessel formation, and intestinal development. Germ-free mice need about one-third more food than mice with bugs. This type of work is aimed at optimizing the bacterial mix in a probiotic so that it not only displaces the "bad" bacteria but also enhances host physiology.

This idea has been extended to the growth of bacteria on any food, which can be suppressed by other bacteria. Some experiments have tried spraying "good" bacteria onto food to block the growth of "bad" bacteria. This has been marginally successful, but has not been accepted by regulators as a safe treatment method.

In principle, probiotics could be used for skin diseases as well, but this has not been exploited significantly.

Process control

Really an entire discipline in its own right, process control is how you measure what it happening in any complex industrial process, and use that information to control the process. It is relevant to biotechnology because of the complex and tight control that most large-scale biotechnology processes need if they are to work at all. This is especially true of fermentation systems, where organisms have very specific growth requirements, which must be met if the biotechnologist is to obtain useful products at the end, and which often vary during different phases of the fermentation (*see* **Cell growth**).

Aspects of process control include

- Monitoring—simple physical parameters (temperature, pressure, amount of mixing) and chemical ones (CO_2, O_2, pH, cell density), and more complex ones (metabolite levels, specific protein levels). The latter usually require off-line monitors, that is, ones where a sample is taken from the fermentor and analyzed separately, as opposed to continuous, on-line or in-line monitors, which all measure things in the fermentor as they are happening.

- *Analysis*—now very software intensive, attempting to model what goes on in the fermentor from the clues that monitoring gives you. This also includes lots of simulation—running "what if" scenarios in the computer.

- *Control*—changings the conditions so you get back on track.

See also **Bioreactor control**.

Prosthetics

Prosthetics are mechanical objects implanted in people, usually to replace parts lost to disease. They include joint replacements (especially hips and knees) and intraocular lenses (lenses implanted in the eye). "Prosthetics" often also covers contact lenses and artificial kidneys, although these are not implanted in us.

Biotechnology has contributed to the design of new materials for all sorts of implants and prosthetics. Usually these are plastics. They must be mechanically suitable, and must also be biocompatible, which means that they must both integrate with the body and not cause a reaction from the tissues they abut. Anything placed in the body will become coated with protein, and may cause the cells around it to react. Such reactions must be controlled.

Typical materials used are

- MMA (polymethylmethacrylate, known by the trade name of Perspex)

- poly-HEMA (2-hydroxyethylmethacrylate), a hydrogel that is converted from a hard perspex-like material to a soft, flexible one in water

- polyvinylpyrolidone.

Biotechnology is also developing new types of prosthetics, often also called **artificial tissues**. See separate entry.

Proteases

Proteases are enzymes which break up proteins. There are four distinct uses for proteases in biotechnology. Their use depends partly on how cheap they are to make, and partly on how specific they are, that is, whether they chop up all proteins indiscriminately or only a few proteins at specific points.

The vast majority of proteases by mass are for industrial use: 8000 tonnes of proteases from microbial and fungal sources are made each year, and most of them are used in detergents (*see* **Laundering**) and the food industry (*see* **Food processing using enzymes**).

Proteases are also used as drugs, such as thrombolytic enzymes (break up blood clots after a heart attack) or for wound debridement (removal of the thick coat of protein material that forms on the surface of wounds and can slow healing and encourage scarring). Proteases such as ACE (angiotensin converting enzyme) and HIV protease are also targets for structure-based drug design.

Proteases can also be used either as food supplements or in the preparation of predigested foods for people in hospital. Here the enzymes have to be of pharmaceutical purity.

The last use of proteases is in biotransformation reactions. *See* especially **Organic phase catalysis**.

Protein crystallography

A key part of most ways of finding out a protein's 3D structure, and hence being able to use that structure to design drugs, is making crystals of the protein. The arrangement of atoms in a crystal can be found by shining a beam of X-rays through it and studying the pattern of diffraction. "Layers" of atoms in the crystal diffract the X-rays by an angle depending on their separation. Refracting at many angles provides information from which a 3D map of where the atoms can be drawn (electron density map). Because of the importance of knowing a protein's structure to doing structure-based drug design, several companies such as Syrryx, Structural GenomyX, and Astex are now doing this commercially at very high throughput.

To do this, a crystal is needed. Making crystals of proteins is difficult. The trick is usually to perform the crystallization using very pure protein, very slowly, and in exactly the right solutions—finding the right solutions can take a lot of expertise and time. This is studied in many solutions in parallel (analogous to the high-throughput screening technology), to explore the "crystallization space"—the combination of all possible conditions that affect crystallization. Novel approaches to protein crystallization include crystallizing under high pressure or in free fall (i.e. in orbit). Crystallizing in free fall (also called microgravity) means that the crystals do not have to touch the side of the container they are in, and so their growth is not affected by that container. Eight companies and 10 research institutes had protein crystallization experiments on the Space Shuttle Columbia's mission in January 1990.

It is sometimes possible to get X-ray structure from powders, in an approach called powder diffraction. This bounces the X-rays off tiny "microcrystals" in the powder. This avoids the need for protein crystallization, but is too hard to achieve for most proteins.

For structure-based drug design it is also valuable to find the structure of a protein with a potential drug bound to it. The scientist can then see exactly what atoms are important for drug binding. Finding the structure of these complexes of chemical and protein is much faster than finding the original protein structure, in part because the crystallization conditions are often the same, in part because knowing the structure helps to interpret the X-ray pattern.

Finding a protein's structure is also increasingly being done by nuclear magnetic resonance (*NMR spectroscopy*). This is slower and not of such high resolution as X-rays, but do not need crystals, provide conformation in solution, and can do membrane proteins analysis. It can also be used to screen small molecules as they bind to proteins, an approach being used by Triad Therapeutics.

The systematic and automated discovery of protein structure is sometimes called Structural Genomics, meaning the use of a "genomics" philosophy (of doing everything at once in parallel) to find the structure of every protein coded by the genome. The prediction of protein structure is the other strand to this structural proteomics agenda (*see* **Protein structure prediction**).

See also **Electron microscopy**.

Protein drug delivery

Delivering drugs to their site of action is an important part of turning a chemical into a medicine. For proteins it is particularly hard as proteins cannot normally be taken by mouth—they would be broken down in the gut, and any that survive would not penetrate the guy wall.

The current solution is parenteral delivery (i.e. injection): this works well, and has been the way of giving patients insulin (a protein drug) for decades. However, it is invasive, expensive, and carries a continued risk of infection or tissue damage. Thus, several biotechnology companies have set out to find a better way of getting proteins into the blood. There are several approaches:

- *Oral delivery*. Taking the drug by mouth, but with some other materials which help it survive the gut. These can include protease inhibitors (to block the digestive enzymes), or carrier materials which protect the proteins but which dissolve at the right time to make them available for absorption. Other tricks include linking the protein to something (like Vitamin B12) which is actively taken up from the gut, so the protein is absorbed along with it. This is very speculative at the moment.

- *Nasal/pulmonary delivery*. This is delivery through the nose or lung, where the wall of cells between the blood and the outside is thin and easier for drugs to cross. Several companies are trying to make inhaled insulin.

- *Transdermal delivery*. Using a method of getting the protein across the skin without knocking visible holes in it. Methods used include iontophoresis (using electric fields to push the drug across the skin) and high-pressure jets of fluid.

- *Protein remodeling*. This approach attempts to remodel the protein chemically to protect it against the difficulties of entering the body. This can be done by encapsulating it (as above), by embedding it in a variety of carrier materials such as dextrans, albumin, xanthan gums, or synthetic polymers such as polyethylene glycol (PEG), or chemically modifying it with these or other materials. This often enhances their survival once in the body, but has not had much success in getting them in the first place so far.

A related problem is getting proteins across the blood–brain barrier (BBB). There is a barrier layer of cells that stop chemicals getting from the blood to the brain and into the cerebro-spinal fluid (the clear liquid that bathes the brain and spinal chord). The BBB is virtually impenetrable to proteins, so proteins targeted to the brain must be injected directly into it, a complex and potentially dangerous procedure. Approaches to breaching the BBB include joining proteins to molecules that are known to get across (such as ferritin), or making the BBB "leaky" with other drugs.

So far, protein drug delivery systems have been much hyped, but not very effective. It is not clear whether they ever will be, or whether discovering new, small molecule drugs to replace them will be successful before the drug delivery systems ever are launched into general use.

Protein engineering

Protein engineering is the design, production, analysis, and use of altered, non-natural proteins. This can be a Herculean task if we do not use a natural protein as a starting point, so usually protein engineering involves modifying existing proteins.

Almost all protein engineering is actually the engineering of genes using recombinant DNA techniques, and then expressing the proteins from the genes. There are two basic approaches:

- *Point mutation.* This is replacing one or a few amino acids with other ones. It can be directed (also called site-directed mutagenesis—you change just the one you want for another one) or random (you change things at random and see what happens—*see* **Darwinian cloning**). Site-directed mutagenesis has become much more practical in the last decade as more and more proteins have had their 3D structure determined (*see* **Protein crystallography**). It has also been a major application of phage display technologies.

- *Domain shuffling.* Here whole regions of a protein are swapped around. Sometimes these are whole "domains"—"blocks" of structure that are fairly stable on their own and often have distinct functions. This is not so sophisticated as site-directed mutagenesis, but is easier to do.

Protein engineering has a number of aims:

- *Improving protein stability.* See entry on **Protein stability**. Protease enzymes that have been genetically engineered for greater stability are in widespread use in detergents.

- *Altering antibody properties.* There are a range of specialist technologies around antibody engineering, discussed in the entries for chimaeric/humanized antibodies and Dabs.

- *Altering the substrate specificity of an enzyme.* Most enzymes catalyze only a very narrow range of reactions, and it would be helpful to be able to alter that range so that they acted on other, more commercially useful products. A spectacular research success was the conversion of Malate Dehydrogenase to Lactate Dehydrogenase, two enzymes which catalyze similar types of reactions on different substrates. Engineering of proteases such as subtilisin for use in detergents has been a more practical example, in this case broadening the substrate specificity (i.e. making it less specific). However, most other projects of this sort have failed.

- *Altering pharmacological action.* Much protein engineering is aimed at biopharmaceuticals (*see* **Protein drug delivery**). In this field it seeks to alter the biological activity of proteins which have effects which can be harnessed as drugs, by making the effects more potent, more specific, coupling them to targeting mechanisms so that they only effect a few cells or cell types, improving their survival time in the patient or reducing side-effects. This can include altering the way that proteins aggregate. Therapeutic insulin tends to form multi-molecule complexes, which slows its rate of absorption by tissues. Replacing proline 28 in the B chain of insulin by aspartate stops it aggregating in the vial, and allows it to be absorbed much more rapidly.

More ambitious aims of protein engineering involve the production of completely new protein structures or sub-structures. So far these have been academically successful—producing short peptides which fold in predictable ways—but have not produced practical products.

Protein production in plants

Plants are attractive organisms for making proteins in bulk because

- they can grow in very varied conditions, and do not need special, sterile culture media like animal cells;

- they accumulate very large amounts of proteins in seeds;

- they are eukaryotes, and so make proteins from eukaryotic genes (with introns), and add glycosylation in more eukaryotic ways, both of which are convenient for the genetic engineer;

- growing plants (agriculture) is probably the biggest industry in the world by mass of product processed, and gathering, processing, and distributing plants is a very efficient industry ready for the biotechnology to "plug in to."

- there are no issues about contamination of the product with viruses, BSE, or bacterial toxins.

Tobacco is a leading plant for protein production, in part because the tobacco companies have developed very sophisticated technology for harvesting and processing the leaves under very reproducible conditions, and in part because the same companies are increasingly desperate to find another business to get into. Tobacco can be made to make a foreign protein by infecting it with an engineered tobacco mosaic virus (TMV), which has no effect on animals but which infects tobacco very fast, or by making a transgenic tobacco plant.

Maize and sweet potato have also been studies extensively for producing "foreign" proteins. A favorite route is to make the gene express in a "storage compartment"—part of the plant that usually stores protein, such as a seed (in maize) or a tuber (in sweet potato). These organs make specific storage proteins in large amounts relative to other proteins. Several workers are seeking to make the plants produce other proteins in as large amounts (up to 60% of the total seed protein, 15% of the total seed weight) and in as convenient a form. The favored route, being tried by Plant Genetic Systems, is to splice the gene for a desired protein into the middle of a plant storage protein gene. This construct will then produce a fusion protein in the seeds, which can be chopped up to yield the desired product afterwards. This has been achieved with the small 2S plant storage protein in *Arabidopsis thaliana* and in *Brassica napus* (oilseed rape). A more radical approach would be to use the promoters of a storage protein to drive expression of the foreign protein. This is difficult, as storage proteins have specific signal peptides (not well understood) to take them to the right part of the cell for storage—the protein we want to make instead will not have those signals.

In practice, getting this technology to work reliably has proven as expensive as conventional bacterial or cell culture expression systems. This may change if there is a demand for truly large amounts (thousands of tonnes rather than kilograms) of a protein.

Protein sequencing

This is now generally rather old technology, as it is far easier to determine the sequence of the gene from which a protein is made than to sequence the protein. However, there are still a few applications, especially in proteomics, finding genes in un-sequenced organisms, or the analysis of post-translational modification where the amino acid sequence of a protein has to be determined directly.

Classically, cycle of reactions called the Edman Degradation chop one amino acid off a protein at a time, after which the separate amino acid is analysed to see what it is. There are several machines which perform these quite complex sequences of reactions automatically, and can sequence up to 40 amino acids off the end of a protein before accumulating chemical side-reactions bring the process to a halt.

Other methods include using mass spectrometry, especially Fast Atom Bombardment mass spectrometry, which is gaining popularity as it can (in theory) cope with N-blocked proteins and unusual **post-translational modification** (see separate entry).

Protein stability

Proteins are not very stable in chemical terms: they are easily denatured (i.e. converted to inactive forms). Denaturation occurs when the protein chain of amino acids, usually folded into a specific, tight coiled conformation, unfolds: the carefully arranged 3D structure of its surface is lost, and usually whatever its function was is lost with it. This change can be brought about by heat, acids, alkali, and by some chemicals such as urea and guanidine, which are known as chaotropic agents because they are extremely effective at opening up the structure of water-soluble macromolecules into a floppy, "random coil" that has no real fixed 3D structure.

If enzyme reactions can be carried out at higher temperatures, or antibodies made more stable so that they can last longer, biotechnologists would be very pleased. So there is a lot of work in trying to improve protein stability. The lines of work are:

- Using another more stable enzyme, especially from a thermophilic bacterium (*see* **Thermophile**).

- Engineer the protein, by:
  - Increasing the number of disulfide bonds within the protein. These bonds, formed by linking the sulfur atoms in two cysteine residues that are near to each other in the protein once it has folded up into its proper shape, help to lock it into that shape.
  - Increase internal hydrophobicity. Often the amino acids that end up inside the correctly folded protein are water-hating (hydrophobic) amino acids: if the protein is unfolded, that exposes them to water which needs energy to do and so tends not to happen.
  - Add other stabilizing interactions. A wide range of other interactions of amino acids with each other help to hold a protein in its correct state. These include hydrogen bonds and ion (or salt) bridges.
  In all these three cases, the protein engineer aims to add or alter amino acids to increase the number of stabilizing interactions in the protein. This needs a detailed understanding of the 3D structure of the protein, information which can be very difficult to obtain (*see* **Protein crystallization**).

- Adding specific stabilizing agents to their preparation. Very few enzymes are sold as pure protein—most have many other materials in their "formulation" to stabilize them. Some of these can have a dramatic effect, extending lifetimes from hours to weeks. Exactly what is in each stabilizer depends on the enzyme concerned. In extreme cases, antibodies can be used to stabilize a protein—they bind to its properly folded shape, and so "lock" it into that shape, helping to prevent denaturation. (Of course, the antibodies can be denatured too, so this only works at moderate temperatures.)

Folding and stability is also important when a protein is to be made by recombinant DNA technology. Foreign proteins made in large amounts often precipitate inside the cell as an inclusion body. Thus, part of the purification procedures for many recombinant proteins involve steps which partially unfold the protein and then refold it again, this time under conditions which allow it to fold up properly. (This can also help purification, by selectively unfolding and refolding the desired product: contaminant proteins fail to unfold, or fail to fold up again, and so can be distinguished from the product.) Clearly, it must relatively easy to fold up the protein if this strategy is to work—some proteins cannot be refolded into their native structure once they have been unfolded.

Protein structure

Proteins have a very specific structure, both a sequence of amino acids in a chain and how that chain is arranged in space. Both of these contribute to their function, by making sure the right amino acid side chains are in the right place to perform their function.

Proteins' 3D structure—how the amino acid chain is arranged in space—is a consequence of their sequence: different sequences usually have different structures. However, two completely different sequences can have very similar structures, and at least one case—the prion protein—is known where the same protein sequence can adopt radically different structures. So, while a protein is conveniently described by its sequence, that is not the end of the story, and knowing its structure is more useful if we want to understand how it works.

Protein structure is divided into several different levels:

Primary structure. This is the chemical sequence of amino acids in the protein. It is unique and definitive for that protein. Implicit in it is that the

....alanine-methionine-isoleucine-methionine-aspartate-glycine-threonine....

Primary

... AMIMDGTTM ... ← Usually abbreviated to one-letter code

Secondary Beta sheet Alpha helix

Super-secondary (Only backbone atoms shown—whole structure too confusing)

Tertiary

Plus three more chains in one complex

Quaternary Often shown schematically → A C B D

primary structure is a string of the 20 "biological" amino acids joined by peptide bonds.

Secondary structure. This is how the primary structure folds into small, distinct shapes, usually consisting of 3–10 amino acids. These units are very stable, and occur in many different proteins. Common ones are alpha helices (short rods), beta sheets (flat, twisted plates), and a variety of turns which join them. Alpha helices are often shown in graphics of protein structure as solid cylinders, beta sheets as parallel arrays of flat bands or arrows. Anything from 30% to 90% of the amino acids in a protein can be part of a secondary structure feature.

Super-secondary structure. Some of the secondary structures assemble into groups of three or four elements that again are quite stable and common among many proteins. The most common example is the helix bundle, a tight bundle of four alpha helices.

Tertiary structure. This is how the whole protein—secondary structure features and other amino acids—fold together in space to make a molecule. Quite often the tertiary structure is made up of structural domains, regions of the protein which are quite distinct and have a distinct function. Thus, there are recognized domain "shapes" for binding nucleoside triphosphates, which turn up in many different metabolic enzymes that use ATP as a substrate. The function of proteins can sometimes be discovered by working out what its domains are most similar to. Most proteins have a handful of domains, although some contain a huge number of modules, such as the polyketide synthases (*see* **Antibiotics**).

Quaternary structure. Many functional proteins are assembled from several amino acid chains, that is, are actually collections of different molecules sticking together. How the different chains (polypeptides) assemble together is their quaternary structure. Depending on how many polypeptide chains there are, these are called dimers (two), trimers (three), tetramers (four), pentamers (five), and so on. If the clumps contain only one type of polypeptide (e.g. three of the same polypeptide chose to clump together), the result is a homo-dimer, homo-trimer, etc. If there are more than one type of polypeptide in the clump, then they are hetero-dimers, hetero-trimers, and so on.

Protein structure prediction

In principle, the 3D structure of a protein should be predictable from its amino acid sequence, and doing this is a major scientific effort (as the alternative is to discover the 3D structure using the costly and time-consuming methods of X-ray crystallography). This is particularly important when genome-scale DNA sequencing provides the identity of proteins far faster than crystallography can determine their 3D structure. However, this is not yet achievable.

There are two basic approaches: homology modeling and *ab initio* calculation.

Ab initio modeling seeks to model what a protein will do just from knowledge of the chemistry of the atoms that make it up. As there are tens of thousands of atoms in a protein, and hundreds of thousands in the water surrounding it (which affects it structure profoundly), this is an immense and at the moment unachievable task, although the IBM supercomputer BlueGene is planned to have a crack. A half-way house is to model each amino acid as a unit, rather than each atom, but although faster this is no more successful.

Homology modeling looks for similarities with other, known protein structures. A common approach is "threading"—conceptually threading the amino acid chain through a known structure to see where it fits. In addition, programs look for conserved "motifs," such as alpha helices and beta sheets. They can also look for whole protein domains which look like other proteins, such as beta barrels (a common type of structure in enzymes) and 7-helix transmembrane proteins, based on their sequence similarity, either of the whole protein or one of its structural "domains." It is estimated that there are between 1000 and 5000 stable distinct polypeptide "folds" (basic tertiary structure shapes) in nature, with about 800 known already from experimental work: it is assumed that, as all proteins are related in the evolution of life, the more protein folds are uncovered the more likely it is that a new protein will have a fold essentially the same as one we already know.

Homology models matched up in this way are then usually "optimized" or put through "energy minimization"—the computer juggles each of the atoms to make sure there is not a very similar but more stable configuration it can find. This is a much simpler task than *ab initio* calculations.

Proteomics

The "proteome" program, also sometimes called the "phenome," is the name given to highly parallel, high-throughput analysis of all the proteins in a cell, tissue, or organism. It is an extension of the technical philosophy of the genome—study everything at once as using high-throughput, automated analysis—to protein biochemistry.

Proteome technology has advanced enormously in the last decade, and it is now practical to identify every protein present above a limiting concentration in a cell by

- *Separating the proteins*. This is usually done by 2D electrophoresis, a method that spreads them out into spots on a square slab of gel. This is a technically tricky method, and one that is not all that reproducible between labs, which makes comparing 2D protein patterns extremely hard. However, many proteome programs just seek to compare such 2D gel pictures to compare the proteins in different samples. Thus, new approaches use capillary electrophoresis to separate the proteins.

- Isolate the proteins from the gel (usually by cutting them out—spot or band excision—and mashing the gel up in a buffer solution).

- *Identify the proteins.* This is usually done by breaking them up with enzymes into short peptides, and then using mass spectrometry to identify the peptides—each protein has a characteristic peptide "fingerprint" which can be used in conjunction with a genome database or all possible proteins to identify the original.

A problem with this is sensitivity—only the more abundant proteins in a cell can be identified, as there is just not enough of the rare ones to detect. So many proteome programs pre-purify the proteins, for example, by only looking at proteins that are in the nucleus, or in the ribosome.

A new approach is to build a "protein chip" that analyses all proteins in parallel as a gene chip analyses all DNA. The chip would have to have many antibodies on it, one specific for each protein. So far this is a research concept only.

Protoplasts

Plant, fungal, and most bacterial cells are surrounded by a tough, thick cell wall. A protoplast is such a cell from which we have removed the cell wall, leaving the cell naked and surrounded only by its plasma membrane.

There are a number of reasons for wanting to do this, but all involve the strength of the cell wall itself. Often plant breeders wish to fuse the cells of two quite different plants which cannot be cross-bred by conventional methods. However, the cell wall gets in the way. Getting DNA into plant or yeast cells for genetic engineering is extremely difficult, with the cell wall being essentially impervious to any large molecules. (Getting DNA into bacteria is an exception because bacteria have mechanisms for absorbing DNA from the medium surrounding them.) Getting DNA into protoplasts is rather simpler.

Plant and yeast protoplasts are generated by dissolving their cell walls with appropriate enzymes, which will digest the carbohydrate (plant) and chitin (yeast) in the cell wall without affecting the lipid-and-protein cell membrane.

Protoplasts are very fragile to mechanical and chemical shocks, so if the scientist wants to turn them back into normal yeast or plant cells, which can be used in a commercial process, then he/she needs to treat them with great care.

Purification methods: large scale

One of the central parts of the downstream processing of any biotechnological production process is purification of the product from whatever was used to make it. Large-scale purification methods are used to take a crude fermentation supernatant or cell homogenate and isolate the product from it in a fairly pure form. Industrial enzymes are often sold in this semi-pure form as a bulk product. If they need to be really pure, then they have to go through a second purification step (*see* **Purification—small scale**). Purification of the cells from a culture is usually called **harvesting** (see separate entry), and relies on rather different methods.

For expensive biotechnology products, the cost of purification is the majority of the manufacturing cost: growing *E. coli* in a tube of broth is very cheap, but purifying a recombinant drug from them needs time, skill, and materials, all of which cost money, especially as the products are governed by strict regulations that demand not merely purity but proof of purity (*see* **GMP** and **QC/QA testing**).

In general, then, the less purifying you need to do the better, which usually means that the more concentrated the material is to start with the cheaper it will be.

There are a range of purification methods which are cheap enough to use on large volumes of material. These include:

• *Salt precipitation*. Adding salt so that a particular group of proteins precipitates from solution. Also sometimes called "salting out." Simply adding water to the precipitate usually makes them dissolve again. Ammonium sulfate precipitation (adding concentrated ammonium sulfate to a protein solution) is often used in preparing proteins at small and medium scale. A low concentration of ammonium sulfate is added to precipitate some unwanted proteins, the precipitate removed, then a higher concentration added to precipitate a partially purified mixture containing the protein we want. Such a preparation is sometimes called an "ammonium sulfate cut."

• *Crystallization*. Reduce temperature and solubility goes down, so crystals form. However, this is usually impractical for proteins, as they do not crystallize easily.

• *Liquid–liquid separation*. Also called two-phase separation, this uses the fact that the material you want will dissolve well in one solvent while most of the impurities will not. The two solvents are intimately mixed, and then separated (by allowing to stand, by filtration systems, or by gentle centrifugation). This only works, of course, if the two liquids are immiscible. This can be performed several times, reducing the amount of contaminant in the sample phase each time. For large-scale preparations it is essential that the two phases are cheap, as it rare that they can be recycled efficiently. One is usually water (as that is the basis of the culture medium) and so the other is a material like benzene, ether, or oil.

• *Two-phase aqueous extraction*. Here the protein is shaken up with a polymer-based mixture which, on leaving to stand, separates into two distinct layers [e.g. polyethylene glycol (PEG) and salt will do this trick]. The two layers contain different amounts of salt and polymer, but both are water-based. The conditions are arranged so that the product ends up in one layer, most of the contaminants in the other.

• *Polymer precipitation.* Some polymers, particularly PEG, can bind gently to proteins and make them precipitate of their own accord. This is, in fact, a variation of the aqueous-phase partitioning—one of the phases in this case is the precipitated protein.

• *Heat denaturation.* This is simple and effective if your protein is heat stable (thermostable): you just heat up the mixture and most of the proteins denature, and so "coagulate" and settle out of solution. Yours remains soluble. This only works with some proteins. It can also be used in some conditions to separate proteins from non-protein products (e.g. metabolites).

• *Isoelectric point separations.* Most proteins are quite insoluble at a particular pH (their isoelectric point or pK_i). If you add acid or alkali until the pH (the "acidity") of the solution is at this isoelectric point, then those proteins will precipitate. Again, adding water usually redissolves the precipitate.

• *Dialysis.* Here the protein is put on one side of a semi-permeable membrane, and water or salt solution is on the other. A semi-permeable membrane is one which lets through small molecules, but not large ones. So salts, impurities, etc., in a protein solution are let out, the protein remains in. The process can be accelerated by an electric current in electrodialysis, which drives the ions in one direction. Electrodialysis is used to de-ionize (i.e. remove ions from) several industrial-scale protein materials, including whey.

• *Supercritical fluids to extract biomolecules.* Supercritical fluids are gasses which have been compressed to the density of liquids above the temperature at which they can form liquids. Their ability to dissolve chemicals depends strongly on their temperature and pressure, and so can be adjusted to dissolve what you want. Some proteins can even function as enzymes in supercritical fluids (*see* **Supercritical fluid enzymology**).

Purification methods: small scale

Many biotechnological products have to be extremely pure for use as drugs or in producing fine chemicals, so the relatively crude purification methods which isolate them from large-scale culture are not good enough. A further step of purification is needed.

Most small-scale purification methods are based on chromatography. The mixture is passed down a tube which is packed with some material to which some components in the mixture stick and others do not. It does not matter whether the product you want sticks or not, providing the contaminants do the opposite. As chromatography resins cost up to $1000 per liter, with industrial-scale purification systems requiring 50-l columns or bigger, this is only a method that can be used on a small scale or for very high-value products.

Examples of chromatography methods include:

- *Affinity chromatography*. See separate entry.

- *Gel filtration*. This is a chromatography method in which the molecules are separated by size. *See* entry on **Chromatography**.

- *Ion exchange*. This separates molecules according to their charge. As the charge of a molecule depends on the pH, a combination of varying pH and ion exchange chromatography can prove very effective at purifying proteins.

- *Hydrophobic chromatography*. This type of chromatography uses the different affinity that different molecules have for hydrophobic materials, that is, for materials which are "water-haters" like plastics (as opposed to hydrophilic—"water lover"—materials like paper).

- Popular versions of all chromatographic separation methods are FPLC and HPLC, which have been scaled up from laboratory tools to production methods in some cases. *See* entry on **HPLC**.

QC/QA

QC is quality control, testing a product so that it meets the required quality. QA is quality assurance, making sure that the process you are using to make the product is appropriate to make a product of the required quality. Both are general concerns in all industries, and cannot be covered in depth in this book. Both are very important to a wide range of biotechnological processes, especially ones producing pharmaceuticals or foods.

The most important aspect of this is having a method for measuring what you want to control. This means that analytical methods are central to QC and QA. Such methods have to be validated before they can be accepted for use in QC, especially for GLP or GMP processes—a validated method is one which has been proved (and documented) to detect what you want to detect, and also any impurities you think are important. Well validated analytical methods are critical for process control, clinical trials and drug monitoring, environmental monitoring, etc.

Whole processes can also be validated, which reduces the need for analysis. Here, the process has been so well characterized that it is know that, if it is operating within its normal parameters, then the product quality is assured and you need not test it. This approach is called parametric control or parametric release (when you release a product for use based solely on the parameters of its manufacturing process, not on QC measurements on the product itself). Parametric control only works when you can control all relevant parameters, which means you know what they are. For sterilizing by heating, for example, we know that the key variables are heat and time, so if I can be absolutely sure that something has been heated to 129 °C for 40 mins then I need not test whether there are any living bacteria inside—I know that all are dead. However, for more complicated examples you cannot be 100% sure that you understand the effect of all the parameters on the process.

Quantitative trait locus (QTL)

A quantitative trait is a trait or characteristic of an organism that varies continuously across the population, and hence which is almost always controled by a combination of several genes and environmental factors. Human height is one such trait: we clearly inherit a general tendency to be tall or short from our parents, but many genes contribute to this, as well as nutrition. A quantitative trait locus (QTL) is a gene that contributes towards a quantitative trait. A synonym is metric trait locus.

QTLs are important in agricultural breeding programs. Traits such as plant productivity, protein quality, growth rates in plants and animals, fat content in animals, etc. are all under the control of QTLs, and so breeding an animal or plant with the right collection of alleles at their QTLs is a major aim of the breeding industry.

Mapping QTLs is therefore a major concern. A critical aspect to be discovered is the variance due to the QTL. How much of the variability of the trait is due to genes as a whole, and to that QTL in particular? Is the variance additive with other genes (e.g. if you find two gene alleles that will add 6 in. to your height, will someone possessing both "tall" alleles of both genes be 12 in. taller than average?)

Many human genes could be called QTLs, but generally these are referred to as "predisposition genes" instead, predisposing us to develop a specific disease. *See* **Predisposition analysis**.

QTLs are mapped using genomics technology—*see* **SNP** and **polygenic disease**—as well as by traditional breeding programs.

Rational drug design

This approach to drug discovery seeks to model the molecular structure of the target of the drug, and then design a drug molecule which will fit it. This is therefore also called structure-based drug design (SBDD). This contrasts to the alternative, which is to screen a large number of compounds for drug activity, choose the most promising and make a whole lot of variants, choose the most promising of them and repeat until a suitable drug is found. An intermediate approach is to randomly find some molecules that are effective in a test, and then to try to find the features that are relevant based on the structure of those molecules, an approach called quantitative structure–activity relationship (QSAR) mapping.

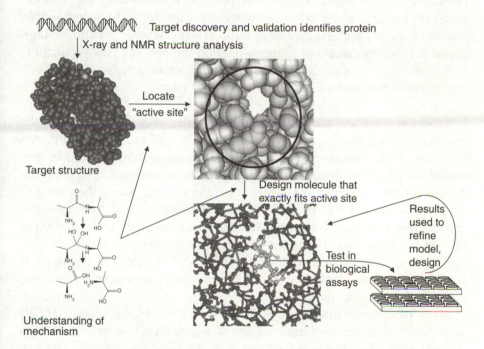

Target discovery and validation identifies protein

X-ray and NMR structure analysis

Target structure

Locate "active site"

Design molecule that exactly fits active site

Results used to refine model, design

Test in biological assays

Understanding of mechanism

Rational drug design (RDD) involves knowing the chemical structure of the drug's target, usually a protein, which is a limiting step for many programs (*see* **Protein crystallography**). Once you have that structure, there are several approaches to using it.

- *Docking approaches*. Here a computer "docks" molecules into the binding site of the protein to see which fit. De Novo pharmaceuticals specialises in this approach, which takes some sophisticated algorithms to work.

- *Fragment-based and reverse synthetic approaches*. Very small pieces of molecules are put into the binding site, and then the computer works out a method of stitching them together to make a molecule.

- *Manual design*. This is when the expert chemist takes over, which is the usual last step.

The Roche drug Saquinavir, an inhibitor of the HIV protease approved in 1996, is a successful example of a drug "designed" almost entirely by RDD methods. However, this is unusual, and most drug discovery programs use RDD methods to complement more traditional biological and chemical screening approaches.

Receptors

Cells signal to each other by chemical means—only nerve cells, as far as we know, use direct electrical signals. The proteins which receive the chemical messages are generically called receptors. They are favored targets for drug discovery, as they are usually very specific to one cell type, but also are important in fields as diverse as nitrogen-fixation biology and cow rumination. Cell receptors and other cell surface molecules are responsible for cell–cell recognition—the mechanisms by which a cell "knows" which type of cell it is next to.

Receptors generally are on the outside of cells, where they can "see" the chemical signals. (Exceptions are receptors for steroid and thyroid hormones *inter alia*, which are inside the cell nucleus because these fat-soluble hormones can penetrate the cell membrane.) They therefore need a "signal transduction" system to carry the signal from the outside where the chemical binds to the part of the cell that receives the message. Two types of signal transduction systems are common. Second messengers are small molecules that act as "hormones" inside the cell, triggering other proteins to act. One common such second messenger is cyclic AMP (cAMP). Other type is **protein kinases** (see separate entry). Often second messenger systems and kinase systems interconnect.

Many receptors are members of the G-protein coupled receptor (GPCR) **gene family**. See separate entry. Some receptors are meant not only to identify molecules outside the cell but to bring them into the cell—some of these are called cargo receptors. This process is called receptor-mediated endocytosis—*see* **Transporters**.

A related idea is an orphan receptor. This is a protein whose sequence "look like" a receptor, but for which the ligand and function are not known.

Receptor binding screening

This is one of the biotech-based methods for discovering conventional (i.e. "chemical") drugs. The method relies on the fact that many drugs act by binding to specific proteins (receptors) on or in cells. Normally, these proteins bind to hormones or to other cells, and control the cell's behavior, although they may also be enzymes or structural elements of the cell. The drug interferes with the normal role of the protein.

Finding a drug that has a particular effect on a cell or animal involves exposing the cell or animal to the drug and then looking for the often subtle effect. Receptor binding assays isolate the receptor protein, and then search for chemicals which latch onto that receptor. The ones that do may be good drugs, but the ones that do not are pretty sure not to be, so you have narrowed the field.

To do this you have to know what the relevant receptor is. You also have to clone and express the receptor gene, as these are quite rare proteins. Then you have to make an assay for the receptor, as just having a molecule bind to a protein is hard to detect, so that binding must be used to trigger some other, more detectable event. Usually, such assays are cell-based—the receptor is expressed in a mammalian cell which releases a chemical, activates a gene or otherwise does something obvious when the receptor is activated.

Recombinant DNA: bits and kits

There are a number of pieces of the technology of DNA cloning that are referred to commonly without further explanation. The more common ones are:

- *Adaptor / linker*. These are short oligonucleotides that are used to join disparate DNA molecules together. To do the actual joining, DNA ligase is needed.

- *DNA polymerase*. An enzyme that makes DNA. To do so, it must have a DNA molecule to copy (the template) and a short DNA molecule to start with (the primer). It then adds bases onto the primer, copying the template until it gets to the end.

- *DNA ligase*. Sometimes, also called T4 DNA ligase. This enzyme joins two double-helical DNA molecules together to make one longer one.

- *Restriction enzymes*. These are enzymes which cut double-stranded DNA at very specific base sequences, and nowhere else. Thus, they cut cloned DNA into a few pieces only. The place at which they cut is called a restriction site, and the map of all such sites on a clone is called a restriction map.

- *Reverse transcriptase*. An enzyme that makes DNA, but uses an RNA template to do so, not a DNA template.

- *RNA polymerase*. Enzymes that make an RNA copy of DNA. It needs a template but does not need a primer. SP4 RNA polymerase is a commonly used one.

- *TAQ polymerase*. Another DNA polymerase made from *Thermus aquaticus*, and an enzyme that is stable when heated to 95 °C.

Recombinant DNA technology

This is the blanket term for the technologies of artificially breaking, re-joining, and amplifying DNA. It is also sometimes called biomolecular engineering, especially in France (Ingenieur Biomoleculaire). Recombinant DNA techniques allow the biotechnologist to isolate and amplify a single gene out of all the genes in an organism, so that it can be studied, altered, and put into another organism. The technique is also known as gene cloning (because you produce a lot of genetically identical genes), and the result sometimes called a gene clone, or simply a clone. Often scientists talk of "cloning a gene into organism XX," when they mean making a clone of that gene in that cell or organism (as in "I cloned it into *E. coli* and screened for expression...") A related term is "sub-cloning"—taking a cloned gene, and breaking a bit out and making a clone of that on its own.

An organism manipulated using recombinant DNA techniques is called a genetically manipulated organism (GMO).

Central to recombinant DNA are the enzymes that cut and join DNA molecules, respectively, called restriction enzymes and ligases. Other entries in this book discuss other components, including **vectors, hybridization, PCR, and recombinant DNA "bits and kits."**

There are a lot of "kits" on the market, collections of reagents, enzymes, DNAs, and even organisms that have been developed as a package that works together to process the purchasers samples. Among the more common are *in vitro* transcription and translation kits (which carry out transcription and translation in a "test tube"—*see entry* on **Genetic code**), kits for site-directed mutagenesis, labeling DNA with radioactive, fluorescent or chemical labels, and so on. Having done molecular biology both with and without kits, I think that kits have a lot going for them, allowing the scientist to concentrate on doing creative experiments rather than making all the reagents he needs.

Regulation

Biotechnology sometimes complains that it is a heavily over-regulated industry, but in practice it is no more regulated than many other industries, and especially those which rely on relatively new technology. Several aspects to the regulation of biotechnology are covered in this book.

- Patents and intellectual property rights. *See* **Patents**.

- Safety of the microorganisms and genetically engineered constructs. *See* **Microorganism safety**.

- Safety of genetically engineered organisms to be released into the outside world. *See* **Deliberate release**.

- Drug development and launch (*see* **Drug development pathway**).

- There is also the area of the ethical right that individuals or humans as a species have to perform any specific part of biotechnology. This is a contentious area of debate, in which sincerely held views vary from the belief that biotechnology is one of the most inherently beneficial technologies ever invented, to the belief that the technology threatens the whole structure of life on Earth. *See* **Bioethics** and **Yuk factor**.

- A future area of growing control and regulation is likely to be technologies that could be used to make biological weapons.

An issue with any sort of regulation is compliance—how can you prove that your processes or products comply with the relevant regulations. This is particularly relevant to very complex products, such as proteins or cell lines, where regulations put stringent requirements on the manufacturing process as well as the final product. The various good practice guide (*see* **GLP/GMP**) are aimed at addressing this—they allow a process to be thoroughly documented.

Regulatory authorities (United States)

The United States has the most advanced biotechnology industry in the world, and is seen as the leader in regulating that industry. The dominant player for most biotech companies is the Food and Drug Administration (FDA). The FDA oversees and regulates all medical drugs and devices, and new food and cosmetic products, assuring that they work, and are not harmful. An autonomous agency, it is the principal regulatory agency which any company must appease to launch a new drug or medical device onto the market. In general, FDA regulations set the pace for other countries in biotechnology, because the US market dominates biotechnology products and so all companies want to make sure that their products and processes fit FDA regulations. FDA regulations cover what the efficacy of a drug is (and hence how its trials as done), how it is manufactured (*see* **GLP/GMP**) and how it is formulated. It is notable that, since 1958, the burden of proof that a drug or food additive is safe falls on the producer, and the FDA is not responsible for proving that it is not safe.

Other agencies that affect biotechnology include:

- **EPA**. The Environmental Protection Agency, which responsibility over deliberate release of organisms into the environment.

- **USPTO**. The US Patent and Trademark Office, which handles all patent applications, a highly contentious area. *See* **Patents**.

The FDA is partly funded by contributions from industry, which is a political hot potato as industry thinks it should get more lenient treatment and faster approval for its dollars, whereas in fact the FDA is the most rigorous approval agency among the major nations. European equivalents are entirely state funded.

Reporter gene

This is a gene which we introduce into a cell to signal when some cellular event is happening. Often they are used to signal the activation of a whole lot of other genes. The reporter gene makes an enzyme which is easy to detect—often luciferase, the enzyme which catalyzes luminol-based biolumines-cence, or green fluorescent protein (GFP). The gene for the "marker" protein is joined with a controller gene which responds to what we want to measure. Thus, if we want to measure the effect of steroids on a cell, we might tie the luciferase gene onto a steroid-sensitive promoter—when steroids got into the cell, the cells make luciferase and will glow. When we succeed in blocking the effect of steroids, the glow will die. Other reporter gene approaches might signal when the cell itself does something, so we can distinguish "active" cells from "inactive" ones.

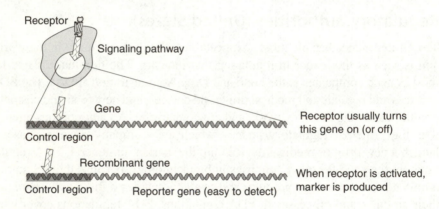

Examples of the use of reporter genes are studying how signal mechanisms affect gene control in cells and tissues (e.g. **Kinase** cascades—see separate entry), detecting viral infection of cells, showing what effects a new gene or mutant has, studying the different effects of hormones or protein factors like cytokines on different types of cells. Reporter genes are responsible for pictures of glowing plants and bacteria, which make good press.

Resistance to anti-infectives

Many infectious organisms—viruses, bacteria, and parasites like the malaria trypanosome, are becoming resistant to drugs that were once very effective at killing them. This is a major issue for healthcare, and hence a challenge for biotechnology to overcome.

Resistance to antibiotics is now well known. Bacteria mutate and vary as all living things do, and as a result, by chance a few will be more resistant to any given drug. As that drug is used, its sensitive siblings will be killed off, but the descendents of the resistant bug will survive. After a while, all the bugs of that sort will be descendents from the resistant organism, and that strain will be resistant to that antibiotic. Bacteria are also good at swapping genes, and can exchange genes that confer antibiotic resistance.

As a result, resistance to a new drug can spread very fast around the world. This has become a major problem with some bacteria. Some strains of *Staphylococcus aureus* are now resistant to almost all antibiotics (called MRSA—methicillin-resistant *Staph. aureus*). Only one antibiotic—vancomycin—works against them, and there have been a few reports of vancomycin-resistant *Staph. aureus* (VRMA) as well, against which there is no defence. Multi-drug resistance of this sort is also rising among tuberculosis (TB). In Spring 2000 Linezolid, a completely new type of antibiotic, was launched: the first case of resistance to the drug was reported less than a year later.

This problem is made much worse is the antibiotic is used indiscriminately, so every bacterium in the world has a chance to "see" the drug and evolve resistance to it. Resistant strains of bacterium usually arise in Asia, where some countries have far laxer controls on sales on antibiotics than in the West, and in economically deprived TB sufferers who cannot afford to take a full course of drugs to kill off their TB and so trickle in medicine when they can afford it, giving the bugs a chance to recover, grow and evolve between each dose. For the same reason, giving antibiotics routinely to farm animals to help them grow faster is now banned in most Western countries.

The problem can be even worse with some viruses. For HIV, for example, resistance can evolve within one patient as they are treated with a drug, so a therapy that worked at the start of their disease becomes ineffective after a few months. This is because some viruses have a very high rate of mutation and polymorphism. So in any one patients there is a high probability that there exists a virus that can resist any one drug. When the patient is given that drug, the resistant virus survives and flourishes, and soon all the virus particles in that patient are resistant. Giving the patient several drugs at once reduces this chance, although it does not eliminate it completely.

Malarial resistance to drugs is also a major problem. Not only does malaria kill more people worldwide than any other infectious disease, but most of those people live in countries that cannot afford new high-tech drugs. Partly as a result of this, the pharmaceutical industry is making most malaria research a charitable objective rather than a commercial one.

Biotechnology addresses resistance in three ways:

- *Discovering new drugs*. A bug is rarely resistant to a drug that no one has used before (although it may be if the "new" drug is actually very similar to an old one). So discovering new anti-infective agents is one route to overcoming resistance. Of course, in time, resistance to the new drug will appear as well.

- *Discovering drugs that block resistance mechanisms.* Bacteria, in particular, have specific mechanisms to "disarm" drugs, by chemically transforming them or pumping them out of the cell using ABC transporters (*see* Transporters). A drug that blocks such mechanisms could, in principle, make the bacteria sensitive again to conventional antibiotics. Of course, the bacteria could become resistant to *that* chemical as well.

- *Discovering drugs that target the mechanisms of pathogenesis.* These are whatever mechanism the bacterium uses to cause disease (*see* Antibiotics).

Retroviruses

Retroviruses are viruses whose RNA genes are copied onto DNA as part of their life cycle. The DNA is then usually inserted into the DNA of their host cell, and it can remain there for many cell divisions as a "provirus" until some signal triggers it to be transcribed onto RNA and so be translated into viral proteins, making more virus. The only thing distinguishing the provirus from any other DNA in the cell is its base sequence.

The retroviruses are of interest to biotechnology for two reasons.

Several retroviruses are of medical importance. The AIDS virus HIV is a retrovirus, as are several other immune-system targeting viruses (the HTLV family), and some viruses which can cause cancer in laboratory models (the oncogenic retroviruses). Thus, retrovirus biology is very important to the search for treatments and cures for AIDS.

The ability of retroviruses to infect a cell and then insert their DNA copies into the chromosomes of that cell has also been harnessed to make DNA cloning vectors which can get foreign DNA reliably integrated into mammalian chromosomes. These have been used to transfect mammalian cells, and to create transgenic animals by infecting EC cells with retroviral vectors (*see* Chimeric animals). The vectors have to have only part of the viral DNA in them, because otherwise they would simply produce infectious virus. Thus the ideal retrovirus-based vector has those genes that are needed to insert the DNA into the chromosomes but no others. Sometimes this requires that the engineered vector is infected into the cell along with a "helper virus" which provides some of the genetic functions necessary but which does not itself get into the cells.

Reverse genetics

Reverse genetics is the type of genetic analysis which starts with a piece of DNA and proceeds to work out what that does. By contrast, normal genetics ("forward genetics") starts with the phenotype—what the organism looks like—and proceeds to work out what the genetic structure is that makes it look that way, ultimately decoding the DNA itself.

Such feats of gene cloning as the isolation and characterization of the cystic fibrosis gene are often called reverse genetics: however, although these use an impressive panoply of recombinant DNA techniques it still starts with an observed phenotype (the disease) and works via ever more detailed genetic techniques to a genetic explanation of what is going on. Reverse genetics has been used, for example, in understanding the genetic structure of a range of viruses, including the AIDS virus. Here, the DNA structure is known in detail, but what it does is not known. So mutations are found or made in the DNA, and then their effect on the phenotype is discovered. In this way, the function of those bits of the gene is worked out.

Reversed phase biocatalysis

Some enzymes work on reactants or products which are almost completely insoluble in water. Others work using water as a substrate, and it would be useful to be able to remove the water from the reaction to make it "run backwards." In both cases, it is useful to be able to operate an enzyme reaction in a solvent other than water.

Organic phase catalysis and supercritical fluid catalysis offers ways of doing this (*see* **separate entries**), but an alternative which is not so radical is reversed phase biocatalysis (also biphasic biocatalysis), in which an enzyme dissolved in microscopic droplets of water is suspended in an organic solvent containing the reaction substrate and/or product. The enzyme's substrate diffuses out of the solvent in very small amounts, is acted on by the enzyme, and diffuses back into the solvent. Because the droplets are very small, the rate of diffusion is very fast and so the reaction proceeds at a useful rate.

A variation is to use a solid support to hold the enzyme in a totally organic solvent. The solid support has a single-molecule layer of water adsorbed onto its surface: the enzyme sticks to that, and is simultaneously immobilized (so that it is easy to remove as part of particulate solid once the reaction is done), activated by the water and stabilized by immobilization. Inorganic materials such as silica are usually used.

These systems have the advantage that you do not have to dehydrate the enzyme so thoroughly before the reaction (organic phase catalysis needs a thoroughly dehydrated enzyme to work properly), and so can be much easier to get working.

RFLP

This abbreviation stands for restriction fragment length polymorphism and usually pronounced "riflip." It means a piece of DNA which varies between two individuals: whether the DNA has a function or not, or whether the variation is important, is irrelevant. The term refers only to the way of detecting the variant, which is by the use of the very specific DNA-cutting enzymes called restriction enzymes. The essence of an RFLP is that one variant DNA is cut by a particular enzyme at one site, the other is not. This means that the fragments produced by that enzyme on those DNAs have different lengths.

		Length of fragments produced	On DNA electrophoresis gel fragments appear as...
Enzyme cuts DNA at sequence GAATTC			
DNA from individual 2	tattttgaattcgagatcgaggaattcgacccaccgaattctatctaaa	15 14	Individual 1 Individual 2
DNA from individual 1	tattttgaattcgagatcgaggaaatcgacccaccgaattctatctaaa	29	

A related term is the allele-specific oligonucleotide (ASO). This is an oligonucleotide which will hybridize to the DNA from one individual but not to that of another, because the DNAs differ by one or two bases. The variant forms of the DNA are called alleles. Both RFLPs and ASOs have found substantial use in human genetics, and in animal and plant breeding programs.

RFLPs and ASOs have found wide use as "marker" genes for genetic studies. Here, the RFLP is used to detect when a piece of DNA has been inherited by an individual from one parent (rather than the other). *See* **Genetic map**.

RFLPs are now generally used less frequently than SNPs, because SNP technology is better suited to extremely high throughput, "genomic" technology.

Ribozyme

Also called catalytic RNA, these are RNA molecules which catalyze chemical reactions, often the breakdown of other RNAs. Their discovery in the mid-1980s overturned the idea that only proteins could be biological catalysts.

Ribozymes have potential in two areas. They are widely toted as potential pharmaceutical agents, as their action against other RNAs can be extremely specific. They could, for example, attack a viral RNA without affecting the normal RNAs in a cell, thus acting as anti-viral agents. Oncogene-targeted catalytic RNA could be an anti-cancer treatment and so on. Ribozymes as therapeutics are still a research technology, however. Although very specific in the test-tube, like antisense RNA they can have unexpected effects when introduced into cells. There is also the problem of how to get them into cells.

The other area is to use ribozymes as industrial catalysts, selecting suitable catalytic activities through *Darwinian cloning* (see separate entry). RNA can be selected by cycles of binding and amplification: this has been done to make RNAs which will bind specific low molecular weight chemicals very tightly. The next step is to find one which binds a transition state analog for a reaction to make a new catalytic RNA. Noxxon is using an approach of this sort to select RNA that is made with D-sugars (i.e. is the opposite enantiomer of normal RNA)—*see* **Aptamer**.

In the last year, the enthusiasm previously reserved for antisense and catalytic RNA has largely transferred to RNAi (*see* **Gene silencing**), although this is at an even earlier stage of development

Scale-up

Scale-up is the process of taking a biotechnological production from laboratory operations to a scale of operation at which it is commercially useful. A few biotechnological processes can be run on laboratory scale systems (e.g. the production of reagents for research use, such as monoclonal antibodies). All others have to be done on much larger installations than a research laboratory can handle.

The difficulty with scale-up is that a tonne of any biological process seldom behaves in the same way as a gram of the same process, unless it is divided into a million separate tubes. Differences occur about the way materials diffuse into and out of the reaction (especially gasses), how much energy needs to be put into stirring, the forces involved in stirring and moving the reaction, how fast it can be started and stopped, and how temperature, pressure, and other physical parameters are controlled in a large mass of material.

Because of this, it is not usually possible to take the procedures which have worked well in the laboratory and apply them to an industrial process. Instead, they are gradually adapted to even larger scales of production, each step usually between a 4- and a 10-fold increase in size over the previous one. At each stage, the optimum amount of various chemical and mechanical parameters must be determined, based on the biotechnologist's experience with previous production systems and a general knowledge of scale-up procedures. There is some mathematical modeling available to help this, but even so experimentation is essential.

It helps if the initial technology is compatible with scale-up processes. Some processes, especially those involving animal tissues or very skill-intensive steps, simply cannot be economically scaled at all: they must be re-invented from scratch. This is a particular problem for consumer products, foods, or agrochemicals, where the cost of goods is a significant part of the final price.

Scanning tunneling microscopy (STM)

This is a type of microscope in which, in essence, an ultra-sharp needle point is scanned slowly over something, and either the force on the needle or the electric potential of the needle tip is monitored. When the tip encounters an atom sticking out above the general surface, the extra force/current is measured. By scanning back and forth across a surface, a picture of the hills and valleys can be built up on an atomic scale. If force is measured, this is an atomic force microscope (AFM)—if electric potential or current, a scanning tunneling microscope (STM).

There are two application areas in biotechnology, neither has advanced beyond a laboratory curiosity stage. The first is directly detecting the physical shape of complex molecules, getting round the need for pure crystals that

X-ray methods need. Arscott and Bloomfield at Minnesota University have produced pictures of the double helical structure of a synthetic DNA using STM. By hitting molecules under the STM with light (and so altering their shape), something can be deduced about the chemical nature of individual bits of a new molecule as well as their size and shape. A possible far-future application of this is "reading" the sequence of DNA directly from the molecule.

The other, even more radical idea is to use STM as a way of actually moving atoms around, creating new chemical entities. So far this has been confined to drawing letters with individual atoms on crystal surfaces, the atoms being xenon (at IBM in San Jose) or sulfur (at Hitachi in Tokyo). In principle this could lead to the direct fabrication of new biomolecules which would be enormously hard to make by any conventional method: however, this is definitely "Buck Rogers Stuff" at the moment.

SCP (single cell protein)

Coined in 1966 at Massachusetts Institute of Technology (MIT), the term single cell protein refers to protein biomass used as a food additive for animals or people. Either isolated protein or whole bacterial cells (suitably processed) may be called SCP.

The drive to develop SCP came from the realization that the "food shortages" seen in many Third World famines were primarily shortages of protein, not of food bulk per se. Similarly, the limiting factor in many animal feeding systems is how much protein is available for animal growth, not the total calorific content. The idea behind SCP technology was to use bacteria, growing on a cheap carbon substrate and with a cheap nitrogen source such as ammonia, to make protein fit for human, or at least animal, consumption.

The key to making SCP economic is to find a carbon source cheap enough. Oil and natural gas have been tried, but are only marginally economic even when the oil and gas prices are low. Methanol made from natural gas is a good potential substrate as bacteria find it easier to use (they need less oxygen to grow on methanol than on methane, and it is very soluble in water). ICI developed a large-scale biomass process based on the methanol-using bacterium *Methylophilus methylotrophus* to produce a partially purified protein product ("Pruteen"). The production scale plant had a volume of $1000\,m^3$ and a capacity of $70\,000$ tonnes of SCP per annum: despite the economies of scale, however, this was at best only marginally economic, despite ICI's use of genetic engineering to improve the effectiveness of the bacterium's metabolism in using ammonia to make the protein.

Cellulose, wood, waste starch, paper processing effluent, and other complex sources of carbon have all been suggested as potential SCP substrates, using *Cellulomonas* and *Alcaligenes* organisms: however none of these are efficient enough to be economic. Algal products such as *Chlorella* and *Spirulina* use sunlight and CO_2 as raw materials, which are cheap enough: however, it is rarely economic to isolate protein from their tough cells, and so these are used as products in their own right.

Most microorganisms have a much higher nucleic acid (DNA and RNA) content than animals or plants, which can cause health problems if eaten raw. So the nucleic acid must be removed, together with toxic materials made during fermentation, and the cells themselves may be extremely indigestible or allergenic. This has limited the use of SCP in human food, and meant that most effort has gone into using it as an animal feed supplement. This competes directly with soybean meal and fish meal, which are extremely cheap

One SCP-like food that has gained general acceptance is the fungus-derived meat-substitute Quorn (*see* **Biotechnological food**).

Seawater

As well as working with organisms that live in the sea (*see* **Marine biology**), biotechnology has occasionally flirted with the idea of extracting valuable chemicals from seawater itself. There have been many and varied plans for extracting metals from seawater, often lured on by the idea that a cubic mile of seawater contains over a thousand tonnes of gold. In fact it does not—on average it contains 40 kg—and that is spread in a very large volume of water. No process yet devised is cheap enough to get the gold—or anything else other than salt and a few other chemicals—out of it.

Biosorption and bioaccumulation is a biotechnological route to obtaining value from sea water. The idea is to use bacterial cells to accumulate a specific metal from the water: all you have to do is pass the water over the cells, and then afterwards "wring them out" into a much smaller volume, resulting in concentrated gold solution. Despite this attractive sounding proposition, it is never economic to do this when the actual costs are worked out, including for example the cost of pumping 4 billion tonnes of seawater through your extraction apparatus, and replacing it regularly as it is eroded by the salt water.

Secondary metabolites

Primary metabolites are the chemicals commonly found in most living things, and which are essential for them to live. Compounds such as glucose or glycine would fit into this category. Secondary metabolites are chemicals that are usually unique to one organism or class of organisms, and which are not essential for the basics of cell survival. They perform more specialist functions, like signaling specific stages in the organism's life cycle, degrading unusual food sources or (often) fighting off other organisms. Many of the chemicals that plants or microorganisms produce which are of biochemical interest, including antibiotics, are secondary metabolites.

Unlike primary metabolites, which most organisms contain most of the time, the production of secondary metabolites is very dependent on the environment of the organism. Thus, small changes in culture conditions of an *actinomycete* (actinomycetes are the most commonly used sources of new secondary metabolites) will dramatically alter how much of a particular chemical they produce.

Plants often produce secondary metabolites as defenses against infection or being eaten: caffeine in coffee, atropine in foxglove, and the vinca alkaloids in Madagascan Periwinkle are examples of quite poisonous compounds made to ward off attack. These secondary metabolites are usually not produced efficiently in isolated, cultured cells. However, their production can sometimes be stimulated by specific compounds or mixtures, which are called elicitors.

Secondary metabolites are used for many purposes. The two most common are:

- *As drugs*. Many drugs were discovered when a plant or fungal extract was found to have pharmacological activity. Almost invariably this activity is due to a secondary metabolite.

- Flavors and fragrance compounds, other than sweet flavors and salt, are usually secondary metabolites, often from plants. (Meat flavors arise rather differently, from chemical reactions between fats, protein breakdown products and sugars in the meat.) Several companies such as Universal Foods and Universal Flavours and Fragrances are working on using plant cell culture and cloning methods to produce flavor or fragrance chemicals by fermentation.

Signal peptides

Most proteins have to be transported around the cell to get to the site where they are meant to function. The cell achieves this by having short peptides in the protein that act as signals. There are several known.

- *Secretion signals* (sometimes just called "signal peptides"). Secretion is the active export of a material from a cell or organism. Proteins that are to be secreted from a cell have a short peptide on their front end—the signal peptide—which acts as an export label. The signal peptide is chopped off the protein as it is exported (during a step called "processing"), so the final protein does not have this extra peptide on it.

- Related is the "stop transfer" sequence that tells a protein to stop transferring out of a cell, so it hangs up half in and half out. These become the membrane proteins, proteins that are set in the outer membrane of the cell, such as receptors.

- *Nuclear location signal* (NLS). This is the signal that says that a protein must go to the nucleus of the cell (where the DNA is). The nucleus has a sophisticated barrier around it, and this can in fact select many types of molecule for transport in (such as specific protein, thyroid hormones) or out (such as RNA).

Biotechnologists can use such signals to direct products to specific parts of the cell. Proteins that are not normally secreted can sometimes be sent to the export department of a cell by stitching a signal peptide on the front. An expression vector that has a signal peptide next to the site for a foreign gene to be inserted, so that the protein is automatically made with a signal peptide on its amino terminus, is called a secretion vector. Similarly, proteins can be directed to the nucleus with an NLS. The product need not be a protein, and indeed the signal can be engineered so that it is not a peptide either. Thus a gene therapy DNA can be tagged with a steroid as an NLS, as steroids are hormones that target the nucleus.

Sewage treatment

Sewage treatment is one of the most widespread biotechnological processes in urban Western societies, which produce huge amounts of human and animal waste, and is of great economic and environmental importance.

After removing large solid objects (sticks, paper, gravel), sewage is invariably processed biologically to remove the organic material. In countries with lots of space and/or a high tolerance of smell, such as the United Kingdom, sewage is usually treated in open, shallow trickle beds where the bacteria get the oxygen they need from air by diffusion. In many European countries such as Holland this is being replaced by a version of the deep shaft system, the activated sludge process. This is more akin to a fermentor. Oxygen or air is pumped into a column of waste, and absorption is enhanced by the high pressure at the bottom. The result is a more compact and faster-acting digestor system: however, it is more complex (and hence more expensive) to build, and to run. Activated sludge digestors usually have the cells in them in free suspension, mixed with the liquid waste.

Variants include:

- The upflow sludge blanket system, where the cells form flocs which are kept in suspension by the upflow, but remain together.

- Biofilm airlift fermentor technology, where the bacteria are grown on small solid particles suspended in dilute solution. This is more useful for contaminated farm run-off, where a trickle bed system would not develop a consistent biofilm.

An important aspect of sewage processing is the reduction of the amount of organic carbon compounds in the sewage, expressed as biological oxygen demand (BOD). BOD is the amount of oxygen that is needed for microorganisms to use up all the nutrient sources in the water. BOD value are the amount of oxygen in milligrams per liter of water that is consumed by microorganisms' respiration in 5 days. Typical BOD values are:

Water source	BOD
Outfall from a sewage treatment plant into rivers	<45
Farm "dirty water"	300–1000
Farm manure slurries, raw sewage	>3000

In conventional sewage, the organic material is metabolized by the microorganisms in the treatment plant, ending up as carbon dioxide and biomass. The biomass material (sewage sludge) is usually burned, composted, or used as fertilizer. Alternative methods generate methane (biogas) from this material, but this is not a common usage. If carbon is not removed from waste material before it is dumped into the environment, the result is eutrophication, making the environment over-rich in nutrient or minerals at the expense of oxygen. This leads to rapid growth of anaerobic bacteria, depletion of oxygen in water, and death of oxygen-requiring organisms like fish. Eutrophication was a common fate of rivers in urban areas in the middle of the last century.

Sewage treatment also has to remove nitrates and phosphates. Some of the bacterial processes developed for nitrate removal from water have been marketed as kits for domestic use, to remove excess nitrogen from ponds to reduce pond scum and algae, which is a sign of eutrophication.

See also **Digestor**.

SNP

SNP (pronounced "snip") stands for single nucleotide polymorphism. A **polymorphism** means that the different copies of a gene present in a population are not all the same. A SNP is a case where the difference between the genes is just one base, and means a genetic site at which one base is different between two otherwise copies of a gene.

SNPs can be used as genetic markers (*see* **Genetic map**), and have become the markers of choice for large-scale studies of population genetics (human or other) because they can be measured easily using high throughput sequencing or gene chip technology, whereas other markers (such as RFLPs) are harder to analyze in very large numbers. There has been a lot effort identifying SNPs, and over a million are known. A public consortium of industry and academic laboratories, called the SNP Consortium, collects, validates and maps SNPs, and provides databases of them to the larger bioinformatics centers.

The only drawback with SNPs is that any one SNP can only be one of two bases. So the possibilities of distinguishing two people are pretty limited, and often they will have the same SNP by chance. This can be overcome by finding combinations of SNPs which "run together" in families, like a genetic bar code. Such SNP combinations are called haplotypes. (This term is also used in immunology, to indicate someone's combination of tissue types.) Haplotype mapping is now a major part of any large-scale gene mapping program. Haplotypes can spread over a small region of DNA or a very large one, depending on the type of study being carried out.

SNP-based gene mapping is a major part of any large-scale gene discovery program. Often the initial mapping is done on pooled samples. The DNA from (say) 100 sufferers from a disease is mixed into a "pool," and the same done for the DNA of 100 non-sufferers. If a SNP is more common in one pool than another, that suggests that it is linked to a gene that is relevant for causing that disease. You then look at the DNA of individuals to check the association. Pooling much reduces the number of genetic tests you have to do in the first stage.

Soil amelioration

This is the improvement of poor soils, usually using bacteria or fungi. (This contrasts to bioremediation, which is the cleaning up of toxins, usually in soils.) Amelioration includes breaking down organic matter, forming humus (i.e. good soil structure), making minerals such as phosphates in the soil available to plants by solubilizing them, fixing nitrogen, and sometimes an element of bioremediation as well. Symbio Ltd produces a fungal product aimed at soil amelioration on lawns and grass playing fields, with fungi that can grow in relatively anaerobic, waterlogged soils and secrete enzymes into them that release nutrient that allow other aerobic bacteria to grow, improve soil structure, and so allow roots to grow deeper and access the nutrients in the deeper soil.

Soil improvement methods were touted as being the way to "green the deserts" in the 1960s: however, they have not worked all that well, mainly because the deserts are not very promising material to start on for climatic as well as chemical reasons. Most of what was then called soil amelioration is now included in bioremediation.

Solar energy

There has been quite a lot of interest in using biotechnology to generate fuels or power from sunlight. This is what plants do all the time, of course, but getting them to do it for man has proven difficult.

The simplest method is to grow plants, and then turn them into fuel: this can be by extremely traditional routes (burning wood) or by growing organisms with high oil contents to make fuel oil (*see* Biofuel).

More speculative schemes plan to use the electrochemistry of photosynthesis directly to generate electricity. This can be done either using intact cells (analogous to a bacterial biosensor) or by isolating protein complexes from the photosynthetic apparatus and using them as chemical reagents. Protein complexes considered have included the photosystems (I or II) which transduce light energy into electrochemical potential in the chloroplast, and more specific parts of the photosynthetic apparatus such as the antenna complex which actually captures photons and passes them to a reactive center. Power outputs to date have been vastly exceeded by the effort and energy needed to make the materials needed for the experiment, and the complexity of the photosynthetic apparatus inside a cell makes this an outside chance for a workable system. Chemical systems or "conventional" solar cell technology is more practical.

Solid phases

Many biotechnological processes rely on immobilizing something—a cell, an enzyme, an antibody—on a solid phase. This means anything that is not a liquid or a gas, but in practice usually means silica (glass) or plastic.

There are many reasons for doing this, discussed elsewhere. In general, it is to allow easy separation of two molecules which would otherwise both be in a solution together. Thus, enzymes are immobilized onto solid phases in bioreactors so that they can be separated from their substrates and products, which remain in solution. The principle is the same as a making coffee from ground beans—the solid coffee grounds are easy to separate from the liquid coffee because they are large lumps, and so can be separated by settling or filtering.

Immobilizing enzymes can also sometimes stabilize them, so that they are active for longer. This is thought to be because the solid material, which is never completely uniform, provides microscopic "pockets" which have an environment that is unusually conducive to the enzyme or antibody. However, no one really knows why this works.

The rate of reaction of a molecule that is immobilized on a surface with others in solution can be quite different from that when they are both in solution. A critical factor is the boundary layer or diffusion layer. The enzyme can convert all of its substrate within range very fast, but then has to wait for some more material to move to its molecular neighborhood before it can act on that. There is in effect a layer of substrate-poor liquid near the surface, and one which is virtually unstirred in all but the most violently stirred liquids. Such a reaction is said to be diffusion limited, because no matter how good the enzyme is the rate of reaction is only as fast as the substrate can diffuse to the enzyme from bulk solution.

Somaclonal variation

This is the variation seen between the individuals in a clone. It applies particularly to plant clones, that is, plants that have been grown up from the cells of a single "parent" plant. Gardeners will be aware how easy it is to grow a whole new potato or dandelion plant from a tiny bit of root left in the ground. In theory all such plants should be identical, and for a few species this is so. However, for most (particularly for potato) there is substantial variation between the offspring, often more variation than usually arises between plants of that species. The variations are heritable, so some at least are due to mutation. This is called somaclonal variation.

This is a reflection of genetic instability, but it is not a feature of the whole plant, which may be bred using normal methods quite well, and so must be an effect of the cell culture system. Why it happens is not understood.

This can be a problem or an opportunity for the plant breeder. It is a problem if you want to use plant cloning technology to grow a lot of your prize plant: the offspring of most cloning methods will not be the same as the parents. The opportunity is that this is another way of generating new plant types which would be harder or impossible to generate by conventional plant breeding.

Sports and biotechnology

Despite the fact that recreation, and specifically sports, is as big a business as medicine and approaches the chemical and agricultural industries in size, biotechnology has consistently ignored the lighter side of life in preference to healthcare or process industry products. The only major exceptions seem to be discussions of the potential abuse of biotechnology products for sporting advantage. Two specific cases are widely discussed: these may or may not be actual rather than potential abuses, as "authoritative" rumors abound but there is almost no factual support for them.

- *Growth hormone.* Growth hormone applied at the right time during development can make some abnormally short people taller. It is believed, although not proven (and actually not very likely) that it could make anyone who was going to be "short" (i.e. below average height) taller, or make a normal person unusually tall.

Kabi Pharmacia placed advertisements in the medical literature in late 1991 suggesting that growth hormone could be a "cure" for the childhood "condition" of being short (not pathologically short, but just in the lower few percent of the normal human range for children of that age). This could be defensible on psychological grounds.

The abuse of growth hormone by adults who try to use it to increase muscle mass is fairly well established. Rumors that people have tried to acquire growth hormone to give to their children to make them grow into 7-ft basketball stars are widespread—whether this is an urban myth, along with the women who put poodles in microwaves and the people who discover rats in hamburgers, or is based on any true event is not clear.

- *EPO (erythropoietin).* This biopharmaceutical was developed to boost blood production in a number of diseases such as anemia and kidney failure, where patients lack red blood cells, and where other therapies, especially for leukemias, have depleted the bone marrow and so the patients develop an iatrogenic anemia. A few athletes have used EPO to boost their levels of red blood cells above the normal level to give their blood greater oxygen-carrying ability. This may give them greater endurance in long-distance running. It would almost certainly be dangerous, increasing the viscosity of the blood and hence the risk of heart attack, stroke, or hemorrhage. Dutch cyclist Johannes Draaijer died of a heart attack at 27 years of age in 1990 under suspicion of abusing EPO.

Other drugs such as IGF could also be abused in this way. The problem is that these drugs are very hard to detect, because they are used in small amounts and because they are similar or identical to natural proteins, so you cannot prove that someone with extraordinary hGH levels does not have them naturally.

Standard laboratory equipment

There are a few pieces of kit which all biotechnologists seem to use and refer to, called by trade names analogous to "Hoover" or "PC." Among the more common are:

1. *Multiwell plate.* Also 96-well plate or microtiter plate. A postcard-sized plastic dish with eight rows of 12 small round wells in it. Used extensively in cell culture and molecular biology for doing reactions where you need to perform the same action on up to 96 samples at once. Machines for washing and for detecting color in 96-well plates automatically are common. In the past decade, the 96-well plate has been joined by the 384-, 1536-, and 3456-well plates (respectively with 16 rows of 24 wells, 32 by 48, and 48 by 72, all in the same sized plastic card), used for high throughput screening.

337

2. *Gilson*. Any type of micropipette, a device which will measure small volumes (i.e. 1 μl–1 ml) of liquid routinely. goes with Gilson Tip, the replaceable tips that go on the end of the pipette unit and are the only part in contact with the liquid.

3. *Eppendorf*. A centrifuge the size of a "mini" hi-fi deck that sits on the bench: also the disposable 0.75 or 1.5 ml plastic tubes that fit inside the centrifuge.

4. *Universal*. A cylindrical tube with screw cap holding about 20 ml and now usually made of plastic.

5. *Kit*. A set of reagents and materials that allow a scientist to perform a research step without the need for any other materials other than his research subject. There are, for example, DNA labeling kits, PCR kits, *in vitro* transcription/translation kits, cloning kits etc., all of which contain everything you need to perform that particular protocol (including instructions) except for the DNA you are going to do it on. Older molecular biologists mumble about how good it was in the old days when they had to make their own reagents, forgetting that the process of getting a specialist to make things for you is what got us down from the trees.

Stem cells

Stem cells are cells that can divide to make more stem cells, and also develop to make other types of specialist cells, a process called differentiation. Nearly all discussion of stem cells is about mammalian stem cells.

Usually this is a two-step process. The cells become "determined"—they are biochemically switched to become a specific cell type, but do not obviously change—and then "differentiated"—they take on the biochemical characteristics of the final cell type. Differentiated cells can sometimes divide (like liver cells), sometimes not (like nerve cells), but in either case they cannot turn into other cell types. Cells which cannot change at all any more (like nerve cells) are called terminally differentiated. This does not mean that they are about to die—nerve cells can last for decades apparently. To keep stem cells growing in culture without becoming determined or differentiated requires some very specific growth conditions and stem cell growth factors, which are different for different types of cell.

There are many different classification of stem cells.

Pluripotent vs. *totipotent*. Not all stem cells can make all other cell types. Totipotent stem cells can be turned into any other type of cell at all—the cells of the fertilized zygote are stem cells of this sort. Pluripotent stem cells can turn into several other types of cell, but not into anything.

Embryonic stem cells (ES cells). Embryos contain many types of stem cell that are probably rare or absent in the adult. They have different biochemical characteristics as well. Whether any of these cells survive to adulthood is unknown. Embryonic stem cells are usually determined into one of the three main cell lineages of the body: ectoderm (skin, nerve, brain), mesoderm (muscle, bone), endoderm (liver, guts).

Tissue-specific stem cells. Many tissues in the adult are able to renew themselves if damaged through growth of new cells from stem cells within the tissue—the bone marrow and the lining of the gut are examples. Stem cells for the blood system are the best characterized. They are called hematopoietic cells (blood-making), and sometimes CD34+ cells (as the CD34 antigen is a characteristic cell surface marker by which they can be discriminated from other bone marrow cells). These stem cells are restricted—they can only produce cells for that one tissue, but that still includes many different types of cells (e.g. the bone marrow produces all red and white blood cells). It is hotly debated whether adults have nerve stem cells at all—some initial results that suggested that we do have cells that can grow new brain cells turned out to probably be due to cell fusion happening after cells were injected into the brain.

Embryonic carcinoma cells (EC cells). These are stem cells derived from teratomas (a benign cancer of the reproductive cells). These cells are totipotent, but being derived from a cancer they are not considered useful for developing therapeutics. However, if a few EC cells are mixed with the normal cells of a mouse embryo, then they can be incorporated into that embryo: the resulting mouse has cells derived from the EC cells in many tissues. They can also be encouraged to differentiate *in vitro*.

Stem cells are a hot therapeutic topic because, in principle, you could use them to replace any damaged cell in the body—for example, to grow more brain after a stroke, or replace a liver destroyed by hepatitis. So far this medical application has not been realized, although animal experiments are very encouraging. Fetal brain tissue, containing stem cells and some differentiating nerve cells, has been used to treat Parkinson's disease (which is caused by death of cells in a part of the brain called the Substantia nigra). A systematic test of this therapy was done by the US NIH ending in 2001. It showed that fetal cell transplants worked in some younger patients, but could cause severe and irreversible side-effects.

Stem cells can also be used as a source of cells for biological research or drug discovery. However, both research and therapeutic applications need a large supply of stem cells, and although they will divide for a while they will not divide for ever. So to be practical products, stem cells have to be immortalized (*see* Immortalization).

Working with embryonic stem cells is an ethically problematic area, because the use of stem cells is often confused with the debate over abortion. Responsible workers in the area make sure that there is a "firewall" between them and the hospital where abortion is performed to provide the embryos from which human stem cells must be isolated, such that there is no way that what they are doing can influence whether a woman has an abortion or not. Even so, President George W. Bush ruled in August 2001 that no stem cell work would be allowed in the United States using stem cells made after that date, so that no more embryos could be used in stem cell research. This has meant that the focus of stem cell research had moved to Europe.

Sterilization

There are a number of established ways of sterilizing equipment and materials for biological use. Many biotechnological processes require that no unwanted living organisms are present, because they would disrupt or contaminate the process, infect the patient, or just gum up the works with extracellular matrix. Thus, sterilization is an important part of many biotechnological processes.

Sterilization is not necessarily the same as cleaning. Cleaning means removing "all" the dirt from something, usually all the contaminants that you can conveniently detect. One bacterium is not usually detectable, but if a sample contains one live bacterium, then it is not sterile. Thus, a plate taken out of a conventional dishwasher will be clean, but probably certainly not sterile (not after you have touched it, anyway). A sterile material is something that has no living organisms in it. However, it can be very "dirty," as long as the dirt is sterile too. Thus, a recently run car engine block will be sterile (because the heat will have killed off any bacteria on it) but definitely not clean. Often, biotechnological "cleaning" systems aim to sterilize as well.

There are four generally used approaches to sterilizing things.

Heat. All organisms are susceptible to heat, although some are more susceptible than others. Heat can be dry heat or wet. Wet heating to 121 °C in an autoclave (essentially a large pressure cooker) is a very popular way of sterilizing equipment and reagents, as it is cheap and easy to do. Lesser levels of sterility can be obtained by boiling, or by heating to 75 °C for a short time (pasteurization). The most heat-resistant biological entity known is the BSE agent, which is reputed to be able to stand heating to 300 °C (*see* **Transmissible encephalopathies**).

Chemicals. Many chemicals are inimical to life. Extremely corrosive material such as chromic acid are used to strip all biological residues off glassware, including all living organisms. However, more benign biocides—chemicals which kill microorganisms but leave most other things intact—are more commonly used. Many are used as cleaning agents, as, unless swallowed, they are relatively harmless to humans. A variation on chemical treatment is treatment with a biocidal gas, usually ethylene oxide. This has the advantage that equipment being sterilized does not then have to be dried out. Usually biocides are not suitable for sterilizing liquids, because there is no way of getting them out again afterwards. Exceptions are ozone and hydrogen peroxide, which break down on their own to oxygen and water.

Irradiation. Gamma rays will sterilize anything, but are dangerous and relatively expensive to produce. UV light is an effective sterilizing agent, and somewhat safer. However, UV does not penetrate very far into most liquids or solids, so it is usually only useful for sterilizing surfaces. UV wavelength of 254 nm is the most effective (it is absorbed efficiently by DNA), but sterilizing lamps usually have a broad spectrum output between 220 and 300 nm. Recently, very intense white visible light has been demonstrated to be a surface sterilizing agent.

Filtration. This is suitable for liquids or gasses, but is extremely effective: usually a 0.2 μm filter (i.e. a filter with holes in it about 0.2 μm across) will remove all living things except viruses from a fluid. However, if the fluid is viscous or has a lot of particles in it, filtration is not practical.

You must select the appropriate sterilization method for the materials to be sterilized. Thus, many plastics are discolored and rendered brittle by gamma-rays, and melted by excess heat. Many fermentation and cell culture media cannot be autoclaved, as this would destroy some of the essential nutrient in them.

The effect of sterilization is measured by measuring viable cell counts or bioburden. These are both measurements of the number of living cells in the sample. Process industries have developed very sensitive methods for detecting just a few live bacteria in large samples of drugs or sterilized foodstuffs. Traditionally, this has involved culturing a sample of the food in bacterial culture medium ("broth") and seeing if anything grows. Some of this can be replaced by ATP luminescence tests (*see* **Bioluminescence**). The drawback to this method is that many other things in the food can also contain trace amounts of ATP, or can stop the enzyme working, giving false positive or false negative results, respectively.

Stirred tank bioreactor

This is the simplest type of continuous bioreactor. A reaction vessel has liquid medium pumped into it, gas pumped in (usually through a sparger at the base), and a stirring system to keep the ferment uniform. The stirred tank bioreactor is the most common type of system used for fermentations, and is used in every size from a few hundred milliliters in a laboratory system to hundreds of cubic meters in a large production plant. It is suited to many microbial growth fermentations.

Many fermentations are carried out in continuous stirred tank systems: material is added and taken out continuously (like a shower) rather than in batches (like a bath). This means that the system must be started gradually, scaling up the fermentation, unless the bacteria can grow in "clean" broth (which some cannot—they need the presence of other bacteria to flourish).

There is also a careful balancing act between putting material in and getting it out. By contrast, enzyme reactions tend to be carried out in plug-flow reactors, where the material passes through the reactor in a "plug" that is processed all at once, rather than as a continuous process.

The variations on the stirred tank bioreactor include the size and nature of the stirring systems (paddles, turbines), whether there are baffles inside the reactor to break up the fluid flow, and how the gas and liquid are introduced. More complex modifications tend to result in fundamentally different types of bioreactors.

See also **Tank bioreactor**.

Strain (cultivar)

A strain of an organism is a type that is genetically distinct from other representatives of the species to which the organism belongs, but which is not different enough to be called a new species. Members of a strain are much more genetically similar to each other than to members of other strains. The word "strain" is normally used of microorganisms to describe a particular organism that has been isolated or engineered to have some property, like growing well or making a lot of a product. Isolating and improving strains of microorganisms is a major part of the process of making them suitable for an economic biotechnological process (*see* **Strain development**").

How you define a strain of bacterium can change how you name it. Strains or types that are defined by their reaction to a characteristic antibody are called serotypes. Types that produce a characteristic chemical profile (usually of secondary metabolites) are called a chemotype.

For animals, the term "breed" or, sometimes, "race" means much the same thing—a genetically homogeneous collection of animals, usually derived from one pair of parents, which is significantly distinct from other animals of the same species. Breeds or races can interbreed with each other, where animals of different species almost never can. Thus, there are a large number of different "breeds" of dogs (huskies, labradors, poodles, and so on) which can interbreed to form a "generic" mongrel dog. For plants, the term "cultivar" and "variety" have a similar meaning. "Strain" is sometimes used for plants, rarely for animals.

Human "races" are not really equivalent to animal "races" or "breeds," as humans of different "races" are in fact genetically almost identical. The human idea of "race" is as much to do with culture and social training as with biology. However, there are a few objective differences between humans from different parts of the world (apart from cosmetic ones like color). Most people of Japanese descent, for example, have a much lower ability to metabolize ethanol than people of European descent, and do not generally have the ability to digest large amounts of lactose as adults, with the result that they get drunk more easily and are less tolerate of milk products. These differences mean that it is important to carry out clinical trials of new drugs in different places in the world, just in case reaction to the drug varies geographically as well.

Strain development

Also called strain improvement, this is the general term for improving the genetics of an organism so that is carries out a biotechnological process more effectively. Usually, this refers to bacteria, sometimes to plants.

The aims are to create a strain of the organism that makes what you want, makes it in large amounts, does not make much of anything else (so that you can purify your product easily), uses cheap and easily obtainable things to grow on, and does not require excessively careful control of culture conditions. The idea of an "improved strain" is exemplified by coniferous trees used for wood pulp production: they grow almost anywhere on soil, air, and water, and you can make most of their mass into the product simply by pulping it. (This is why wood pulp is cheaper than interferon.)

There are many routes to strain development.

- *Incremental selection*. This takes many individuals of the current strain and looks at them to see if any have acquired a mutation which makes them more productive. Sometimes you speed this up with mutagens (mutation-inducing chemicals). This is a time- and labor-intensive operation, but frequently the most useful route to improving the production of chemicals such as antibiotics or amino acids from fermentations. The key to success lies in how many potential variants can be screened rapidly and automatically, that is, the system's "screening capacity." This is a combination of handling many organisms, and of finding a clever way of identifying the one you want.

- *Hybridization*. This is taking two strains and combining them genetically. It has been used extensively in agriculture, but, because the organisms used in biotechnology are so diverse, often cannot be used here so successfully. A variant which is more applicable to bacterial systems is conjugation, the bacterial equivalent of breeding where one bacterium passes some of its DNA to another.

- *Conjugation*. Here only a few "desirable" genes are transferred between one strain and another.

- *Genetic engineering*. This seeks to alter the genetic make-up of an organism by directly introducing genes into it. These could code for a more efficient enzyme, or block the action of an enzyme which destroys the product you require. A more complex and costly path, but one which may be the only route open if "traditional genetics" has failed.

Often, the key to successful strain improvement via any route is finding a selection procedure. This is a set of conditions under which the strain you want has an advantage over all others. For finding strain which makes an enzyme that breaks down one particular compound or group of compounds this can be straightforward. For example, an oil-eating bacterium can be selected by growing a population of bacteria in a medium when the only carbon source is oil. Thus, the only bacterium to flourish will be one that can metabolize the oil, and the faster it can metabolize it the faster it can grow. Strains of *Serratia marcescens* that produce large amounts of biotin were found by mutating the organism and then growing it in the presence of acidomycin, an analog of biotin that blocks its action. The only bugs to survive are ones that produce far more biotin than normal, so overcoming the effect of the acidomycin. However, such relatively straightforward selection procedures are rarely available.

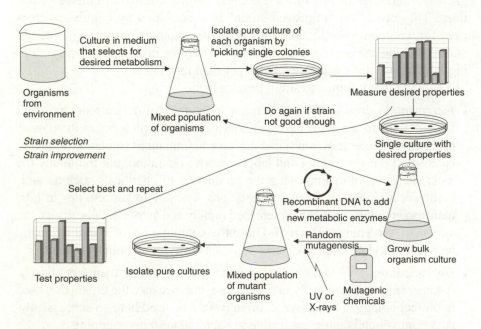

Strain isolation

This is the isolation of any bacterium, yeast, or indeed animal or plant, from the outside world. The trick is to find the organism that does what you want. In general, there are two approaches to strain isolation for microorganisms.

- *Large-scale sampling*. Nearly all biotechnologically useful microorganisms are isolated from the soil, which holds between 1000 and 1 000 000 000 microorganisms per gram. The microorganisms which exist in a particular place depend on the local soil ecology, and clearly this varies greatly. Thus, one approach to finding the ideal organism is to sample as many soils as possible. Many major chemical and pharmaceutical companies have (or had) programs under which any staff member traveling to exotic parts of the world brought home a small sample of soil to be used in the in-house screening programs. The key to success here is not the number of samples you can collect, but the variety of environments from which they can be collected. Hence, the biotechnological interest in "black smoker" vents, hot springs, arctic seawater organisms, and so on.

• *Environmental selection*. You find an environment in which organisms will have had to develop the characteristic you want in order to survive. Favored sites are the effluent paths or waste tips for chemical plants, which tend to accumulate microorganisms which can break down all the chemicals which are present in their local environment. But many other possibilities exist. The organisms which break down methane, for example, were originally isolated from the soil around a cracked gas main, and Diversa Inc. searched for new thermophilic bacteria on the hot springs in Yellowstone Park, USA. *See also* **Bioremediation**.

An important part of this is to develop a method of selection or "enrichment" of the organism you want. This involved making an enrichment medium which will select for the particular set of characteristics you are looking for. Exactly the same approach is used in Strain improvement (*see* **Strain development**).

The terms enrichment media and selection media are also used when talking about the selection of mutant organisms, or even mutant cultured cells. Thus, the HAT medium used to select for hybridomas, the cells that produce monoclonal antibodies, is a selection medium, because it allows hybridomas to grow but prevents the growth of other cells.

Despite all the efforts that biotechnologists have put into developing recombinant DNA methods for optimizing bacteria for biotechnology uses, as often as not it is the original selection method that has the greatest impact on whether the organism is going to be the basis of a commercial process or not.

Supercritical fluid enzymology

All materials have a critical temperature (T_c) above which their gasses cannot be turned into a liquid by compressing them. For carbon dioxide, for example, the critical temperature is $31\,°C$. Thus, at room temperature, if you compress carbon dioxide enough (as in a gas cylinder) it will turn into a liquid. Above $31\,°C$, no matter how hard you compress it it will not liquefy—it will just become a denser and denser gas.

This highly compressed gas behaves in part like a gas, in part like a liquid. It is called a supercritical fluid (SCF), and has some useful properties for chemical and biotechnological processes. Diffusion in supercritical fluids is usually much faster than in liquids, so diffusion-limited reactions (which covers quite a number of enzymatic reactions) can occur faster, and boundary layers are less of a problem. For gases such as CO_2, the pressures and temperatures involved are not damaging to many biopolymers.

SCFs have been used for several model enzyme reactions. In general, it helps to include a small amount of water (which also dissolves in some SCFs) to assist enzyme stabilization: it is also essential if the enzyme uses water as a substrate.

Against the advantages, of course, is the disadvantage that SCFs must be kept at high pressure. One of the much advertised advantages of enzymes is that they work at mild temperatures and pressures. Working at 100 bar pressure in SCF removes one of these advantages. This SCFs are only useful for enzymatic catalysis if some other feature of using SCFs clearly compensates for the additional complexity of working with pressurized gas.

The solubility of chemicals in SCFs depends very sensitively on pressure. Thus, reagents can be dissolved or products removed by precipitation when you alter the pressure. Some chemicals that are only sparingly soluble in water can be made extremely soluble in SCFs by choosing the right pressure and temperature. The same properties of supercritical fluids are also used to purify materials, using supercritical fluid extraction (SFE). SCFs can have very specific abilities to dissolve chemicals, which can be "tuned" by altering pressure and temperature. Thus, SFE with CO_2 has been used commercially to extract caffeine from coffee, leaving virtually all the other chemical components of the coffee behind. The CO_2 is removed by reducing the pressure, when it all turns back into gas.

Systems biology

This is a grand scheme to treat biology as an integrated system rather than a collection of parts: it is often contrasted to genomics, which it characterises as looking solely at genes as if they were independent building blocks, a view parodied as "molecular Lego." Biology clearly *is* an integrated system. Arguably, what distinguishes life from a pile of reactive chemicals is that living systems are integrated so that every component plays a part—usually several different parts at once—in the organism's overall function. However, this general argument does not take us nearer understanding what the parts do or, for the biotechnologist, how we could use them.

A core part of systems biology is that single genes, proteins or other elements are not the most important thing to look at when trying to understand biology. Rather, it is the "mode" of operation of the network. This is a pattern of operation, rather than a single function. A mobile phone can have a "make call" mode and a "receive call" mode: the same chips, usually the same components in the chips, are operating in both, it is just the way that they are operating and how they relate to each other that distinguishes which you are doing, and hence how you respond to, for example, call-back or billing requests. Finding such modes of operations and how they are controlled is the long-term aim of systems biology.

Practical implementation includes:

- *Genomics-scale analysis*. The most immediate and popular use of the term is in the analysis of genome-scale expression and proteome profiles. Rather than look for single genes or proteins that are associated with specific diseases, a systems biology approach looks for patterns of proteins that are coordinately changed in a particular cell or disease. These are characteristic of a specific mode. Some early success in using this to characterize lymphomas has encouraged people to believe that this will work. Leroy Hood's Institute for Systems Biology is at the forefront of this type of analysis.

- *Chemical genomics (also chemogenomics).*—The use of chemical probes to disturb the gene expression network, and so find which genes are regulated together. For both these ideas, *see* **Expression profiling**.

- *Metabolomics*. The whole metabolomics agenda, but particularly the use of the parallel analysis of many metabolites using NMR methods, is based on the idea that it is not the genes that are important, but rather what they are doing.

- Biochemical systems approaches. These attempt to model the individual steps in biochemical systems into a whole network, and so understand how the network functions. There are many versions:
  - Genetic circuits analysis. This is a bottom-up attempt to work out how many genes interact with each other.
  - Biochemical systems theory and metabolic control analysis. These are mathematical methods for analyzing networks of chemical reactions and finding out what happens if you alter some of them. One of the results of such analysis is that the idea that metabolism has distinct "controlling steps" is wrong—under the right conditions, any part of metabolism can control almost any other part.
  - Cell simulations. This attempts to bring all the components of the cell into a huge mathematical simulation. Physiome Inc. have done this quite successfully with a simulation of the electrophysiology of the heart, but many other attempts (such as Entelos' physiological models, Simulations Plus' GI model and the E-cell project) are very approximate attempts at best.

The problem with this whole approach is that, if you leave something out, then the whole simulation will fail, and almost always there is something you do not know, so you leave it out.

The systems biology idea is not only a philosophical one: it has been applied to a range of expression profiling applications and the use of proteomics and metabonomics to detecting toxicity and disease. But the long-term aim is to understand the nature and connectivity of biological networks, and why it is that they work so robustly when the much simpler, and apparently much more logical, circuits and networks in computers crash so easily.

Tank bioreactors

Bioreactors, also called fermentors, are the vessels in which fermentation takes place. Tank bioreactors are vessels in which the microorganism is grown in a large volume of liquid. This contrasts to fiber/membrane **bioreactors** and **immobilized cell bioreactors** (discussed in separate entries). The large majority of bioreactors used in biotechnology are tank bioreactors, and most tank bioreactors are stirred tank bioreactors, because stirring helps to distribute gas and nutrient to the growing organisms effectively.

The bioreactor must provide a mechanism for introducing reagents and the microorganism into the reactor vessel, of providing substrate (i.e. food) to the microorganism (including oxygen for aerobic fermentation), for stirring it and for keeping it at the right temperature, pH, etc. Temperature control is especially critical for large volume fermentations, as metabolizing microorganisms produce a great deal of heat. Variations in layout include different sizing and spacing of the baffles (which ensure that the volume is mixed thoroughly by the stirring) and different types of impellors (stirrers). These can be a wide range of shapes and sizes: disc (Rushton) turbine, open turbine, marine impeller (i.e. like a ship's screw).

The other main variation between reactors is the gas injection mechanism. These are almost always via a sparger or diffuser (a pipe or plate with holes in it which shoots bubbles into the base of the reactor over a wide area at the base of the tank), but a wide range of shapes of pipes or plates have been used. The shapes—which include rings, crosses ("spiders") or dead-end pipes—must be selected for the particular shape and size of the reactor, and the amount of gas that has to be injected.

There is a great deal of expertise in designing a suitable bioreactor for culturing an organism or cell type. As a consequence, there are more companies specializing in bioreactor design, control and engineering than in recombinant DNA techniques and reagents, despite the much higher profile of "gene cloning."

Targeted drug delivery

This is using any method to deliver a drug to the site in the body where it is needed, rather than allowing it to diffuse into many sites. There are three approaches to such targeted drug delivery.

The first is to encapsulate the drug in something, usually a lipid coat (e.g. a **liposome**—see separate entry). The outside of the coat is itself coated with a material which binds to the target cells, either an antibody specific for those cells, or a glycoproteins, or a receptor molecule or ligand. The liposome travels round the blood until it finds its target, where it sticks to the cell and deliver the drug.

The second approach links the targeting mechanism directly to the drug. Here, the drug must either operate outside the cell, or be able to get itself into the cell on its own. Immunotoxins are examples of such chemical conjugates (see separate entry).

Problems with both these approaches is how to get the drug-carrier complex from the blood stream into the target tissue: unless the target is the endothelial cells of the blood vessels or a few cell types in liver, lung, or kidney, nothing as large as a liposome is going to be able to escape from the blood vessels to get to them.

The third approach is to make the drug as a "prodrug" which goes to all tissues of the body, but which is metabolized to the active drug only by one tissue because that tissue has a high level of an enzyme which cuts the prodrug into inert "carrier" and active drug. This is easier to do for tissues such as liver and kidney which have a battery of rather specific enzymes. Several drugs, such as the anti-viral drug Gancyclovir, are actually prodrugs that are activated by their target cells but remain less active in other cells.

Targets for drug discovery

Many small biotech companies are concerned with drug discovery, and often this relates to finding a new drug target. This is the molecule that the drug actually works on in the body. A drug is usually a small molecule, which acts by altering the function of another molecule in the body, usually a protein. We therefore want to find a target that is involved in the disease we are seeking to treat (so interfering with it reduces that disease), and is not involved in anything else (so side-effects are minimal), and then find a small molecule chemical that blocks its action. We also want one that is not like any other protein in the body. If it is very similar to another target, then the chances are that our drug will affect both our target and its similar "twin," causing side-effects.

Almost all known drugs affect proteins. A major thrust of the human **genome project** was therefore to find proteins that could be new drug targets, ones that no existing drug worked on. This has turned out to be much harder than it was initially believed, because it is very hard to work out whether a protein is actually involved in a disease or not. Thus, a major effort in current drug discovery is target validation, finding out whether a potential target is actually a good one—linked in with the mechanism of disease, specific to that disease.

Another requirement is that the target is "druggable." This means that its structure is such that it is possible to design a small molecule to block it. Some potentially promising targets have failed to lead to new medicines because they are "not druggable"—no one has found a way to make a small molecule that blocks them specifically.

Major targets for current drugs are enzymes, especially proteases, GPCRs, and other receptors.

Telomeres

Telomeres are unusual genetic sequences at the end of eukaryotic chromosomes. Having "raw" DNA ends at the end of a chromosome would be bad, as nucleases meant to defend the cell against virus DNA would break them down, gradually destroying the chromosome from the end, and there would be a chance that two ends would join up, forming an aberrant chromosome. So, special complexes of specific DNA complexed to proteins cap the end of the chromosomes—these are the telomeres.

The telomeres gradually shorten as cells age. This is probably a defence against cancer—when they have shortened below a certain critical point the chromosome is destroyed and the cell dies. The telomere theory of ageing suggests that aging is an effect of shortening the telomeres in all cells of the body, so they can no longer divide and repair themselves. One approach to "fundamentally" curing aging is therefore to stimulate an enzyme—telomerase—which rebuilds the telomeres. It is still too early to have any idea whether this is a sensible approach, let alone practical. However, it has been shown that putting telomerase gene (and hence enzyme) into cells is one component of immortalizing them.

Reversing this logic, telomerase is also a target for anti-cancer therapy. Around 85% of cancers have mutations which allow them to make telomerase, so the cancer cells can "get around" the growth limit that shortening telomeres would usually cause, and grow indefinitely. In theory, by blocking its action, you could in principle make cancer cells mortal again, so that their growth stops after a few cell divisions. Again, this is a research area still, with no evidence yet that it will work in the clinic.

Textiles

Biotechnological products are used in some areas of the textile industry to process cloth or cloth precursors. The majority of products are enzymes. The applications include:

Enzymatic pre-processing of fibers, such as the extraction of flax fibers from linseed using ligninases to break down the woody material binding the fibers together. Processing time can be reduced from days to hours by accelerating the release process in this way.

Amylases to desize cotton. "Size" is a coating of starch put on many fibers to stop them breaking during weaving. It has to be removed from the final cloth, which is usually done with acid or high temperatures. Amylase enzymes can be used instead. (Sizing was also used to make the cloth on the wings of early aeroplanes more rigid and airtight.) Other steps in cotton processing are also being tackled with enzymes, including softening fibers with pectase rather than with alkali.

"Pre-washed" jeans. "Stonewashed" jeans are actually washed with pumice stone to make them look worn. It is particularly suitable for processing denim, as the indigo dye used sticks mostly to the outside of the fiber, where stone washing wears some of it off, producing an authentic "worn" look. The same effect can be obtained by brief treatment with cellulase, to smooth exposed fibers and hence release dye. The majority of European and American jeans manufacturers now use "biostoning."

"Polishing" fibers with cellulases. Cotton fibers are covered with microscopic protruding fiber ends, which can catch and cause the cloth to go "blobby" with wear. Novo Nordisk has developed a cellulase product which polishes these fibers off, a process sometimes called biopolishing.

Bleaching, and bleach removal. Enzyme bleaching systems similar to those being tried for wood pulp bleaching have been applied to cloth. Generally, the systems are oxidase or peroxidase enzymes which generate peroxide radicals, that react with colored materials and oxidize them to colorless products.

Enzymes are also used extensively in detergent manufacture (*see* Laundering).

Biotechnology has also provided some limited new materials, such as the "chitopoly" polymer made of chitosan and another polymer. The chitosan is meant to have anti-fungal properties, useful for underwear. Agracetus is working to fill the hollow centres of cotton fibers with polyhydroxybutyrate (PHB) using enzymes, so as to produce an intimate, natural cotton/polyester mixed fiber. So far these have not hit the shops.

Therapeutic cloning

Therapeutic cloning is the name given to the still-unachieved goal of cloning tissues or organs from an individual for therapy. The object here is specifically not to "clone" an individual, but to generate a replacement organ that is genetically identical to them, so that it is not rejected when it is transplanted into them.

The concept is related to the idea of artificial tissues. However, whereas artificial tissue are usually built up from cells and synthetic matrices for them to grow in, a therapeutically cloned heart or liver would be grown as a complete organ.

At the moment, this is not possible without cloning an entire individual, and allowing the fetus to develop at least until the organs or tissues we want have differentiated (*see* Stem cells). So, until developmental biology has developed to be able to make a fertilized egg or a stem cell develop into a complete organ, but *not* develop all the rest of the person, therapeutic cloning is a theoretical idea.

Related to is it the use of embryo technology to generate stem cells— see separate entry.

In fact, a version of therapeutic cloning has already taken place. Several highly publicized instances have occurred where parents of a genetically ill child have conceived a second child *in vitro* and selected only those embryos that have tissue compatibility with their older sibling. When they are born, they are "custom made" tissue donors for the sick child. The parents do not love the new baby any the less, but even so some people feel that it is ethically questionable to "design" babies even for the best of motives.

Thermal sensors

Thermal sensors, sensors that detect tiny changes in heat or temperature, are well known in many applications. Such sensors are often used in gas chromatography systems to detect molecules emerging from the GC column.

There have been some attempts to harness thermal sensors as biosensors. Here, a probe detects the heat given out when an enzymatic reaction occurs. This could be much more flexible than enzyme electrodes, as, while relatively few enzyme reactions involve the transfer of electrons which could be picked up by an electrode, nearly all result in the release of heat. The problem is that, for small samples of dilute material, the amount of heat released is tiny, so very small and very sensitive heat sensors are needed.

Thermophile

A thermophile is an organism which grows at a higher temperature than most other organisms. Generally, as a wide range of bacteria, fungi and simple plants and animals can grow at temperatures up to 50 °C, "thermophiles" are considered to be organism which can grow above 50 °C. They can be classified fairly arbitrarily depending on their optimal growth temperature into "slight thermophiles" (50–65 °C), "thermophiles" (65–85 °C) and "extreme thermophiles" or "hyperthermophiles" (>85 °C).

Thermophiles and extreme thermophiles are usually found growing in very hot places: hot springs and geysers, and domestic hot water pipes for example. Deep sea exploration submarines have brought back microorganisms from the "black smoker" volcanic vents on the mid-ocean ridges which can apparently grow at over 100 °C, under pressure.

Many industrial processes could be catalyzed by enzymes if the enzymes could be speeded up, the diffusion effects that limit the chemistry could be overcome, and the cost of keeping a big reaction vessel cool could be reduced. If you can use a thermophilic enzyme, then you can heat the reactor up to speed up the reaction and diffusion, and do not need cooling. Heating up the reaction is also desirable because it reduces viscosity, and so reduces the amount of stirring and pumping energy needed, and the heat prevents other enzymes from working or (usually) contaminating organisms from growing in your reactor. It also helps removal of volatile products, like ethanol.

Enzymes from thermophiles also frequently exhibit increased stability to organic solvents. Thus, there is substantial interest in isolating these enzymes and using them for organic phase catalysis. Because the bacteria themselves are usually tricky to grow (and must be grown at high temperatures), once a suitable enzyme is identified, it is common to seek to "clone" its gene into a bacterium which grows at a more moderate temperature. This also means that they may be purified from all the other proteins in the bacterial cell simply by heating it up: all the other, non-thermostable proteins will precipitate, leaving a reasonably pure preparation of the target enzyme.

A range of thermostable enzymes are used in industrial processes. As with all research into isolating enzymes from bacteria, one key feature is to have a large number of diverse sources of candidate organisms to screen. This is why Iceland, with one of the world's densest concentration of different types of hot spring and geyser, has been the source of a majority of the publicly available thermophilic organisms in use.

Tissue culture

This is sometimes used interchangeably with **cell culture** (see separate entry). Strictly, it means the cultivation of tissues, that is multi-cell assemblies, outside the body. It uses very similar techniques and materials to cell culture.

Whole tissues are cultured when the scientist wants to examine the cell's interactions with their extracellular matrix (i.e. the stuff that supports the cells in the body), when they want to examine how cells interact, or when there is no known way of culturing those cells in isolation so that they maintain the properties they have in the body. Thus, liver slices are used quite widely to study hepatitis C virus biology, because there is still no known way to make HCV grow on isolated liver cells. The technology to culture liver cells in isolation and have them still behave like liver cells is still in development. Brain slices are also used to examine how nerves interact as a network.

The technology of tissue culture and cell culture overlaps with that necessary to build artificial tissues from isolated cells, discussed in the entry on **artificial tissues.**

Tissue repair and regeneration

This is making tissue grow and repair itself after injury or disease. It is a halfway house to making completely new, artificial tissues to replace damaged areas (*see* **Artificial tissues**), and is part of regenerative medicine (although regenerative medicine is sometimes used to include artificial tissues, stem cell technologies and bionics). It is related to "tissue engineering" although the latter is more concerned with the mechanical restructuring of tissue.

This confusing network of terms arises because this is not a well-defined or mature discipline, but a research topic.

Critical aspects of getting a tissue to re-grow and regenerate are:

- *Growth factors* (see separate entry). Each tissue needs its own specific cocktail of growth factors to start to regenerate. For many tissues, this cocktail has not yet been found.

- *Cells*. Often the body's own cells can be stimulated to some regeneration ("autologous" cells—cells that are the same as the patient's). Sometimes additional cells are needed. Genzyme has developed a treatment for joint damage based on injecting the patient's own cartilage cells into the space between the bone and the cartilage (the periosteum).

- *Supporting structure*. Only blood cells have no supporting structure around them: most other tissue types need to grow around and into a structure to regenerate. Thus, blood vessels have been grown around polymer tubes, and cartilage re-generated around a biodegradable polymer structure (this was the famous "mouse with a human ear" that attracted media ridicule—*see* **artificial tissues**). Polyurethane has been used as a structure for ligament repair: the cells from the surviving ligament gradually migrate into the polymer, break it down and replace it with authentic ligament.

Some of the early "artificial tissues" were in reality aids to tissue regeneration: the Apligraf from Organogenesis was more a source of growth factors than a replacement for skin. Most commercial skin replacements have a combination of tissue regeneration and tissue replacement effects.

Toxins

Living things make some of the most dangerous compounds known, like ricin (castor bean toxin) and pertussis (whooping cough) toxin. One molecule of *Botulinus* toxin, delivered to the inside of a nerve cell a billion times more massive than it is, will kill the cell, and less than a tenth of a microgram could kill an adult human: by comparison, the lethal dose of plutonium is guessed to be half a microgram. Such powerful poisons have potential uses, and biotechnology has the potential to make them relatively safely.

Toxins can be used on their own as therapeutics. *Botulinus* toxin is being developed as a way of blocking unwanted muscle spasm (marketed as "bottox," and abused as a cosmetic treatment for facial wrinkles). Clearly, it cannot be injected generally as other drugs would be—it would kill the patient. However, if minuscule doses are injected into the muscle they can paralyse that muscle. The amount of protein is so small that it escapes the notice of the immune system, and so the body does not make antibodies which could neutralize future doses. Allergan and Porton International produce commercial version of *Botulinus* toxin for this application.

Toxins can also be coupled to other things to give them a lethal "sting." **Immunoconjugates** are probably the best example—see separate entry.

Making such toxins is difficult, even with all the panoply of microbiological containment available (*see* **Physical containment**). People have tried cloning the genes for these protein toxins into bacteria to express them more efficiently (as they are normally present in extremely small amounts). Such scientists tend to find themselves in the middle of empty rooms whenever they talk about their ambitions at conferences.

Recently, the mass media have speculated that toxins could be used as biological warfare agents—the fact that an aspirin tablet-sized amount of botulinus toxin could kill the entire population of the UK sounds pretty devastating. In fact, making this would be very hard, as mentioned above, would almost certainly kill the terrorist before the toxin could be delivered, and would have to be injected individually into everyone in the UK to work (which is scarcely practical). But the threat of these agents alone is enough to have an effect, and encourage scientists and culture collections to keep a tight hold on the relevant microorganisms.

Transfection, transduction, transformation

These terms are all used to mean ways of getting DNA into cells, usually animal or bacterial cells. The meanings are different, depending on the type of cell being discussed.

Transfection. Strictly, this means carrying a piece of DNA into a cell as part of a virus particle. For mammalian and plant cells, it is used more generally to mean almost any way of getting DNA into a cell.

Transduction. A relatively specialist genetic technique, this means transferring a piece of DNA from one organism to another via natural DNA exchange processes. It occurs almost exclusively in bacteria, and is a method of genetically engineering large pieces of DNA such as the *Agrobacterium tumefaciens* Ti plasmid.

Transformation. For bacteria, this means getting the bacterium to take up DNA which the experimenter has added to its medium. Bacteria that have been treated to make them able to do this are called "competent." Demonstrating transformation was one of the key proofs that DNA was the genetic material: Avery and his colleagues showed that DNA could transform cells to a different genotype, but protein could not. For mammalian cells, transformation means turning the cell from one whose growth is limited by its neighboring cells into one whose growth is limited only by the media available to it. Transformation is a step in the development of cancer cells, and is also a crucial part of generating an "immortalized" cell line. Because these two meanings of "transformation" grew up beside each other, genetic engineers who are manipulating mammalian cells often say that they "transfect" the cells with DNA, rather than "transform" them.

Mammalian cells can be transfected by adding DNA to them as a calcium phosphate precipitate. The transfection can be Transient or stable. Stable transfection is when the DNA is spliced into the cell's own DNA, so that every "daughter" cell of the original transfected cell gets the same number of transfected genes in the same place in the chromosome. Stably transfected cell lines are permanently altered. Transient transfection, by contrast, is when the DNA is not integrated, but sits in the cell as separate molecules. Because they are not joined with the chromosomes, they are not duplicated when the cell divides, and so daughter cells will have less and less DNA in them. The DNA is also broken down by cellular enzymes, speeding this loss. However, it only takes a day or so to transiently transfect cells, but weeks to months to produce a stably transfected cell line.

Transgenic

A transgenic organism is one which has been altered to contain a gene from another organism, usually from another species. While this would suggest that any genetically engineered organism could be called "transgenic," the term is usually only applied to animals or plants. Bacteria and yeasts are always called "recombinant," and with plants you can use either term, or called them "genetically engineered" or "genetically manipulated."

Creating transgenic plants is a relatively mature science, and is discussed under **plant genetic engineering**.

Creating transgenic animals is more complex. Transgenic rats, mice, rabbits, cows, sheep, pigs, goats, and rhesus monkeys have been made (this last with only a marker gene for green fluorescent protein inserted into their chromosomes). The germ cells (i.e. the egg, sperm or newly fertilized zygote) must be altered—altering some of the cells in the adult (the somatic cells) only alter those cells, not the rest of the cells of the body and not any cells in the animal's descendants (although it may be valuable for other reasons—*see entry* on **gene therapy**). Thus, unlike plant genetic engineers who can regenerate a new plant from almost any cell in the old plant, animal genetic engineers must develop methods for getting DNA into the germ cells. There are several ways of doing this:

- *Microinjection*. This, the first successful method, simply injects the DNA into the egg's nucleus (diameter about 1/100 of a millimeter) with a very fine needle. Microinjection needs considerable skill, but is still by far the most common method. This is the only method which works for cows, sheep, goats, and pigs. (Pictures of this technique are stock footage for news film about any sort of animal gene technology: they usually show a large, blunt pipette, onto which the egg is held by suction, and a very fine pipette used to inject the DNA.)

- *Transfection*. This is chemical treatment of the egg with the DNA. While this works well for somatic cells, it is a bit dodgy for eggs. An Italian group claim that they had found a simple way of making sperm absorb DNA from solution, which if true will make creating transgenics much easier.

- *Electroporation*—see separate entry. Not very successful with animal cells, and not at all successful with eggs.

- *Using embryonic carcinoma cells* (EC cells). *See* entry on **chimeras**.

- *Retroviral vectors*. Some viruses, notably the retroviruses, can carry DNA into a cell and splice it into the cell's own DNA. There is a lot of interest in harnessing this ability to genetically engineer all sorts of animal cells.

- *Transomics*. This is an injection technique, but instead of injecting pure DNA, its practitioners dissect out sections of chromosomes under the microscope and inject them. As chromosomes are only a thousandth of a millimeter or so long, (and much thinner), this is not a profession for myopics or those with shaky hands.

Foreign genes introduced into transgenics are usually called exogenous (in animals) or ectopic (in plants) genes.

Transgenic animals: applications

Trangenic animal technology is used mostly to produce research results. The favored tools here are "knockout" and "knock-in" animals (see separate entry). There are three areas in which transgenic animal technology has been used to create biotechnological products, as opposed to research results.

The first is creating animal models for disease. This is probably the most successful application to date, discussed in a separate entry.

The second is as production systems for foreign proteins, usually therapeutics. This is known as **pharming** (see separate entry).

The third application area is in farm animal improvement. For example, 60% of the fixed cost of producing a pig is the cost of feed, so if a pig could be engineered to turn that feed into meat more efficiently, that would represent a substantial saving for the farmer. In principle, expression of a transgenic growth hormone gene in the pig should do this: however, early experiments showed that the side-effects of engineering GH genes into pigs or cattle outweigh the potential benefits. In addition, the debate over the use of injected BST suggests that, even if the genetic engineering is successful, it will be controversial on regulatory and social grounds. Engineering phytase, an enzyme not normally present in their guts that breaks down organic phosphates in the food (so they can use the phosphate in their diet more efficiently) has been more successful in research application in pigs. It should also mean that their feces contain less phosphate, which is a substantial pollution problem in farm sewage.

Other engineering of farm animals has looked at improving **wool quality** (see separate entry), and milk quality by introducing more milk protein into cows' milk, and engineering animals so that their organs are suitable for transplantation into humans (*see* entry on **Xenograft**).

One area that is becoming possible but is hotly argued is generating transgenic racehorses. Intra-nuclear sperm injection, the technique that injects a sperm head directly into donor egg's nucleus, was achieved successfully on horses in mid 2001, and there is only a small step from here to adding new genes to the horse. The opposition comes not primarily from animal activist groups (who prefer to target medical research), but from the racing industry who say that if they wanted to watch the products of human engineering race they would race cars instead.

Transgenic disease models

One application of transgenic animals is to model human diseases. Having an animal model of a disease can be very valuable, as it allows medics to study how the disease starts and how it progresses in well defined circumstances, rather than wait until the disease has caused serious damage and the patient feels ill enough to present themselves to a doctor. It also allows experiments that are not ethical or practical to do on humans.

However, quite a few human diseases are not mimicked accurately by any animal model: atherosclerosis is almost unknown in the animal kingdom, for example. Transgenic technology seeks to create animals, especially mice, which get diseases which are in some specific way characteristic of a human disease. These can then be used for screening for potential new therapies or drugs.

Animal models are always controversial, because they are never perfect mimics of human disease. So are you developing a treatment for a human disease, or a mouse disease? The "war on cancer" funded in the 1970s and early 1980s was very successful at finding cures for cancer in rats, but it also discovered that cancers in rats were not a very good model for how humans get cancer. Hence, the development of the "oncomouse" and similar transgenic models to try to mimic how human cancer developed. Many other transgenic models are now in use, but they are just models, not the full complexity of the human disease.

Transgenic plants: applications

Transgenic plants have been tested widely for agricultural use, and some have come into economic production. There were early concerns that the transgenic trait would reduce productivity, thus rendering the crop uneco-nomic, and the only way to test this is to grow the plant in real farming con-ditions in large amounts to see just what the yield is. In some cases it has been shown that the genetic engineering of the plant does make it unprofitable to grow compared to its unengineered counterpart, despite its theoretical superi-ority, but this is not always the case. For example, genetically engineered flax, containing the acetolactate syntase gene from *Arabidopsis* which allows it to be resistant to sulfonylurea herbicides, grows as well and as productively as the original strain, despite carrying the biochemical burden of the extra gene.

Typical applications of genetically engineered (transgenic) plants include:

Improved disease resistance in crops. Some virus resistance has been engi-neered into plants by giving them genes for virus coat proteins: the excess of the virus proteins in the plant prevents the live virus from replicating effec-tively. This is called pathogen-derived resistance (PDR), and was first shown with tobacco mosaic virus. Pest resistance has been very successfully engi-neered into plants by giving them the genes for chitinase (an enzyme that attacks insects' skeletons), or **Bacillus thuringiensis** toxin (B.t.k—*see* **Biopesticides**).

Protein production in plants. There has been a lot of work, and some progress, on using plants to produce recombinant proteins, particularly **anti-bodies**. See separate entry.

Herbicide resistance in crops. A common target is to engineer crop plants to be resistant to herbicides, so that fields can be sprayed with herbicides that will kill off weeds but not affect the crop plants. Concerns that this will lead to increased herbicide use have not been vindicated in practice, but neither have claims that this will reduce overall spraying by making it more targeted.

The problem with going beyond these simple biochemical changes to improving more complex properties of the plant is that plants are usually grown outdoors, where conditions are very variable. Biochemical fine tuning can be completely cancelled by variations in temperature, humidity, air quality, temperature or other factors. Prairie Plant Systems demonstrated an improved the yield on some transgenic tobacco, under conditions where all the environmental effects were completely controlled, by growing the plants in greenhouses down an abandoned mine: however this is not practical for larger-scale growth, such as agricultural applications.

Transmissible encephalopathies

Also called transmissable spongiform encephalopathies (TSEs), this is a generic term for a group of diseases that includes bovine spongiform encephalopathy (also "mad cow disease"), scrapie (affects sheep), Cruetzfeldt–Jacob disease (CJD), and Kuru (affect humans). They are slow, degenerative disease of the brain that end up with a brain with many spherical lesions in it that appear almost empty under the microscope, looking like a sponge, hence the name.

The diseases are caused by prions, proteins that aggregate, and can trigger aggregation of further prions, thus making the disease the only one known where a protein alone, with no nucleic acid, can transmit the disease from one individual to another. The prion protein is all alpha helix, and is soluble, before it aggregates, but switches to a different, partly beta sheet structure when it encounters another aggregated prion making it insoluble. The insoluble prion then forms long fibers in cells, triggering their death.

A "new variant" of CJD was identified in Britain in 1996 which may be linked to BSE: there have been 121 cases in Britain, 6 in France and, 1 in Republic of Ireland since the end of 2002.

The more common form of CJD causes 40–50 deaths a year in Britain. Despite Britain being the world center for the disease, research in Britain has been dogged by delay and incompetence, most notably in 2001 when the results from a five-year experiment to find if sheep's brains could carry BSE had to be scrapped because the researchers had used cows brains by mistake. However, BSE in the United Kingdom is not the only source of human TSE infection: Colorado elk suffer from a related TSE, and at least two people have died in Colorado from CJD since 1993 apparently contracted from eating elk meat.

The medical interest is that the disease this causes is lethal, irreversible, and (at the moment) incurable. The causative agent is very hard to destroy: boiling it, digesting it in acid or leaving it in the sun for a week seem to have little effect.

The encephalopathies started to cause concern in the biotech industry because of the possibility that the agent that causes the disease will get into biotech products produced from cultured cells. Many cell culture systems use fetal calf serum as part of the medium in which the cells grow. As it is not known how the prions are transmitted, or whether they could "grow" in cultured cells, it is hard to prove that the serum is not "infected," and any risk has to be avoided. The same concerns apply to collagen for medical use, as much collagen is made from cow hide.

Biotechnology is also working on tests for prions, all of which are antibody based, aiming to discriminate between the differently folded form of PrP. The aim is to make the tests able to detect the prion before symptoms appear, for controlling the disease in animals. There is little point is diagnosing it in people, because absolutely nothing can be done about it.

Transplant

Transplantation is putting a new organ, tissue, or (more recently) cell into a patient. Traditionally the new organ or tissue has come from another human being. Biotechnology approaches are allowing animal sources and synthetic tissues to be used.

The fundamental problem with putting an organ or tissue from one human into another (an allograft) is that of tissue rejection. Tissues are rejected when the recipient immune system recognizes them as carrying foreign proteins on their cell surfaces. In particular, most cells carry specific proteins called major histocompatability complex (MHC) proteins (human leukocyte antigens—HLA—is an older synonym) which vary hugely between people, so the chances that they are the same between any two individuals is tiny. The MHC proteins are concerned with the way the immune system detects foreign proteins, so if they themselves are foreign the receiving body has a strong reaction to them.

Some cells "present" such antigens much better than others, and so cause much stronger tissue rejection. Immune cells, especially dendritic cells, present most efficiently, and removing them from a transplanted tissue reduces the rate of rejection. Other approaches to reduce rejection include masking the donor's cell surface antigens with antibodies, irradiating the cells or tissues with UV light, or giving the recipient large doses of immunosuppressive drugs. This last course is the standard method, but has severe drawbacks, so the search is continuously on for better methods. Among them are genetic engineering approaches to removing the rejection antigens: transgenic animal technology has been used to create replacement organs which could in principle be transplanted into man. *See* **Xenograft**.

Biotechnology also uses transplant technology as a part of cell therapy, removing cells from a patient, engineering them, and then reintroducing them into the same person or into someone else.

There are ethical concerns around transplantation as around embryonic stem cells and xenotransplants. It is meant to be illegal anywhere to sell parts of your body (apart from blood and hair): however the demand is strong, as illustrated by an eBay "auction" of a kidney that topped $5.7millon in September 1999 before being revealed as a hoax. This has lead to many stories of the poor selling organs, and is one substantial drive behind developing non-human alternatives, such as xenotransplants or tissue regeneration technology.

Transporters

In a biochemical context, a transporter is a protein that carries molecules from one place to another, usually across the membrane of a cell. Transporters can be active—pumping a molecule across—or passive—allowing the molecule to diffuse across. The former often fall into a gene family of called ATP-binding cassette (ABC) transporters, that use the energy of ATP to concentrate molecules one side of a membrane.

Specific transporters of interest to biotechnology include:

- *P-glycoprotein (P-gp)*. Sometimes also called MDR (for multi-drug resistance), it is the main transporter responsible for pumping unwanted chemicals out from the intestinal lining and back into the gut, and out of cancer cells.

- *Bacterial ABCs*. Many bacteria gain resistance to drugs by acquiring or up-regulating ABC proteins that actively pump the drug out of the bacterial cell before it can do any harm.

- *Gut transporters*. There are a range of transporters in the gut wall that allow sugars and amino acids from the gut into the blood. If they could be used to pump drugs or proteins, these molecules would also be efficiently absorbed by the body.

- Specific cell surface receptors for molecules such as folic acid and ferritin (which carries iron into the cells).

The other type of transport system common in animal cells is receptor-mediated endocytosis. Here, receptors bind to their target molecule, then all cluster together on the cell surface and the cell takes in the whole patch of membrane, receptors, other molecules and a small portion of the extracellular fluid as well. The resulting little vesicle is called an endosome. Sometimes the receptors are "recycled" back to the cell membrane, sometimes they are broken down inside the endosome and have to be made anew. Such systems can also be harnessed—Ark Therapeutics has used the scavenger receptor on macrophages (which captures oxidized lipids from the arterial walls) to carry drugs or genes into cells.

Treatment protocol program

An FDA initiative to allow terminally ill patients to be given experimental drugs before they have cleared all the hurdles of full regulatory approval. This concept was driven to public notice by AIDS patients, who objected that the rate of approval of new drugs for AIDS was so slow that they would be dead before anyone got a drug on the market to cure their disease.

Triple DNA

Most introductory textbooks will tell you that RNA is "single stranded" and DNA is "double-stranded," that is that DNA consists of a helix of two strands wrapped round each other. However, it has been known for a while that RNA can be triple stranded, and recently triple stranded DNA has been demonstrated too (also sometimes called triplex DNA). It has several potential applications.

The "third strand" of triple DNA binds to the other two by specific base pairing, and so it can be used as a reagent which recognizes a specific DNA sequence. (The base pairing "rules" are rather different from those in normal double stranded DNA.) This can be used as a DNA probe, or to block gene activity. In this second role, a short oligonucleotide is made that binds to a gene, preventing the RNA polymerase from opening it and copying it onto RNA. This has the same effect as an antisense agent made against that gene product, but acts at the DNA level. So far, this has worked to a limited extent in the "test tube" only. This is an application of **aptamer** technology (see separate entry).

Binding a third strand to the double helical DNA can also force it into a left-handed helical form, called Z-DNA. This structure occurs naturally in gene promoters, and so making it appear (or disappear) may be another route to blocking gene function.

There are a number of other, related complex structures which have been made of from DNA, for various purposes. Chiron made branched polymers of DNA as an aid to increasing the sensitivity of hybridization assays. Nadrian Seeman has used oligonucleotides to make cage-like structures, opening the intriguing possibility of using DNA as a biomaterial, or for building nanotechnology structures.

Tumor marker

A tumor marker is any molecule which shows that a cancer is present. Usually each type of marker is produced by only a few types of cancer, so as well as showing that a cancer is present, it helps to identify the type of cancer, and hence the most suitable therapy. Tumor markers can be used for diagnosis or, potentially, as the targets for biopharmaceutical drugs such as immunotoxins.

Tumor markers fall into two categories. The first type are the products of oncogenes or tumor suppressor genes, and hence their presence represents part of the reason that the cell is a cancer cell to start with. The others are apparently incidental to the cause of the cancer, but are always found associated with a particular type of cancer. Such proteins are usually made in only a few cell types in the healthy body, but cancer cells make them in much larger amounts [such as prostate-specific antigen (PSA)] or in inappropriate places (such as the hCG, usually only made by the placenta in early pregnancy, but also produced by some cancers). Quite a few markers, like CEA, are proteins normally found only in the fetus (sometimes called oncofetal proteins). Some are identified solely because a monoclonal antibody that binds to tumors has those proteins as an antigen. A number of them are glycoproteins or carbohydrates: cancer cells often add sugar units to proteins in a slightly different order from normal cells, so creating different glycoforms of those proteins. Polymorphic Endothelial Mucin (PEM) is an example—the protein is covered in sugar residues in normal tissue, but exposes some of its polypepide chain in cancers.

Using tumor markers for diagnosis and prognosis can be tough, as they are only *indicative* of the cancer, not an infallible sign. Substantial efforts in genomics, especially expression profiling, is aimed at finding better, more specific markers.

Tumor markers are also targets for tumor vaccines (also called cancer vaccines), aiming to get the body's own immune system to recognize the cancer and eliminate it. *See* entry on **cancer**.

Two hybrid

This is a widely used method of finding if two proteins interact *in vivo*. In summary, two hybrid genes are made that contain a "bait" protein tied onto one protein, and a library of "fish" proteins tied onto another. The proteins are such that if they are brought together, they activate a gene—there are several versions of essentially the same idea using different proteins. So if the "bait" hooks a "fish," the target marker gene is turned on in that cell, and the experimenter knows that the two fusion proteins are binding together.

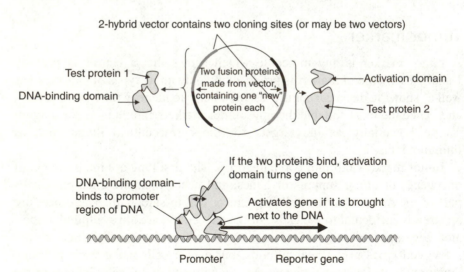

Because the two proteins are expressed at higher levels than is normal inside their original cell, and other modulatory proteins are not around, 2-hybrid methods can generate quite a lot of "false positive" results. Nevertheless, it is a powerful tool to discover proteins that are interacting with each other. Several academic groups, as well as the company Hybrigenics, are building up systematic interaction "maps" that show how many proteins interact with each other. This is an extension of proteomics technology, taking the basic identification of proteins into the realm of systems biology.

Two-hybrid is also used to identify compounds that disrupt (or trigger) protein–protein binding. A pair of interacting proteins is put into a 2-hybrid system, and then the assay carried out under conditions where the two should interact, and the experimenter looks for cells where they do *not* give a signal, that is where something is stopping the interaction.

More complex, three-way versions of the system ("tribrid" or "three hybrid" systems) have been developed to look at complex protein assemblies, but are very tricky research tools to get to work. To an extent, it is expected that 2-hybrid methods will be replaced by protein chip technology when that becomes more reliable: then the proteins binding to a specific "bait" will be detected directly, as DNA binding to DNA is detected on DNA chips now.

Vaccinia virus

Vaccinia viruses are DNA viruses from the same family as cowpox and smallpox. However, it is a safe virus to work with, and so has been used for several biotechnology applications.

Vaccinia has been used as the basis of an expression vector system. It has a lot of DNA, quite a bit of which can be removed using suitable genetics. Thus, quite large foreign genes can be spliced into it, and then the recombinant virus can be used to infect a wide range of cells, allowing the biotechnologists to choose the cell most suitable for the process. They can be used to make more than one protein at once in a cell, which can be useful for making proteins with more than one polypeptide chain (multi-subunit proteins).

Vaccinia has also been used as the basis for "live virus" vaccines. It does not itself cause severe disease, and because it can infect a wide range of species it can be used to produce a range of animal vaccines, which are the first targets of this sort of technology. Provisional approval for a field trial of a live vaccinia virus vaccine was granted in the United States in 1990.

Foreign genes are usually introduced into vaccinia by recombination, rather than by isolating the vaccinia DNA and manipulating it *in vitro*. This is because the vaccinia virus is too large to be manipulated conveniently.

Cowpox and raccoon pox viruses, which share some of the useful characteristics of vaccinia, are being looked at as alternative vector systems.

Vaccines

Vaccines are preparations which, when given to a patient, elicit an immune response which subsequently protects the patient from a disease. Usually, the vaccine consists of the organisms that causes the disease (suitable attenuated or killed) or some part of it. Attenuation of a virus (or bacterium) is manipulating it so that it does not lose its ability to grow in culture, but looses some or all of its ability to cause disease in animals. Usually bacteria, and to a degree viruses, slowly lose their ability to colonise living things (and hence to cause disease) as they are cultured outside the body.

There are a range of biotechnological approaches to produce vaccines.

- *Viral vaccines*. These are vaccines that consist of genetically altered viruses. See separate entry.

- *Enhanced bacterial vaccines*. These are bacteria that have been genetically engineered to enhance their value as vaccines. Intervet markets a genetically engineered *E. coli* as a vaccine for pigs, which is meant to express proteins that stimulate the immune response.

- Recombinant DNA techniques can also be used to generate attenuated strains, by deleting pathogenesis-causing genes, or by engineering the protective epitope from a pathogen into a safe bacterium (*see* Live vaccines).

- *Biopharmaceutical vaccines*. Proteins, or sections of proteins, which are the same as the proteins in a virus' or bacterium's wall, can be made by recombinant DNA methods as vaccines. This is a standard biotechnological route, and has the advantage that there is no possibility that the resulting vaccine will contain any live virus particles. Peptide vaccines are often incorporated by genetic engineering into a larger carrier protein to improve their immunogenicity (i.e. how well they cause the body to become immune), or their stability. Examples of such vaccines include HBV, HVC, flu. Rhein Biotech does is a major producer of these.

- *Multiple antigen peptides (MAPs)*. Developed by J.T. Tam, these are peptide vaccines which are chemically stitched together (usually onto a poly-lysine "backbone"). This means that several vaccines can be delivered in one shot.

- *Poly-protein vaccines*. This is a similar idea to MAPs, but here a single protein is made by genetic engineering in which the different peptides form part of a continuous polypeptide chain.

- *DNA vaccines*. Here, the DNA is delivered into cells, which then make the protein. The aim is to make the protein slowly over a substantial period, to allow the immune system to "learn" its structure, rather than give it all as one shot. This is particularly valuable (in principle) when the protein normally elicits a poor immune response. Other genes can be stitched into the DNA, to make proteins that stimulate the immune response (egg cytokines). Cell GeneSys' DNA cancer vaccine for lung cancer is in phase III trials using this approach. This could also be considered a version of gene therapy.

Vaccines have also been produced by a range of non-standard methods. The conventional method is to produce the vaccine from an attenuated version of the disease-making organism, grown in cultured cells or hens eggs. Alternatives include:

- Producing single protein vaccines in recombinant microorganisms.

- Producing vaccines in transgenic plants. Plants will not have any immune response to a human virus, and will not get ill from them. They can produce lots of the protein and could, in principle, be used as food to provide an oral vaccine. This is now being done in quite large scale by companies such as EPIcyte (using corn), and Meristem (using tobacco) (*see* **Protein production in plants**).

There has been a rapid growth in biotech vaccine production since September 2001, driven by fears of a biological terrorism attack. Companies such as Powderject and Acambis have been given substantial contracts to make large supplies of vaccines agents such as smallpox and anthrax, to be stored in case of an attack. How they would be delivered to the population is less clear.

Another challenge for biotechnological vaccine production is a vaccine for AIDS. Developing a therapeutic vaccine for something that destroys the immune system is difficult. Current development targets include the gp120 and p24 and p55 proteins of the virus' core, but usually include several proteins or peptides in one vaccine in order to increase the chance that the immune system will "see" the vaccine. A prophylactic vaccine— one that protects people who do not have an HIV infection from catching the virus—should be easier to develop, but is much harder to test—who are you going to give it to, and what are you going to do to show that they are protected? Immune Response Corp. was embroiled in publicity in 2001 when it cancelled its AIDS vaccine trial after it was not showing sufficiently good effect. Some of the researchers involved said that Immune Response was refusing to let them publish the results in the scientific literature, although Immune Respone had announced why the trial was being pulled in 1999, and published more data on it when all the researchers had finished analyzing it.

Vector

A vector is something that carries something else: a disease vector is an organism that carries the disease from person to person, like mosquitoes carry malaria. A vector in biotechnology is usually a DNA segment that carries another piece of DNA in a clone, generated using recombinant DNA techniques.

DNA does not replicate itself: it needs a battery of enzymes that replicate it when the cell needs a new copy, and make sure that only one copy of the whole molecule is made at a time. To allow this to happen the DNA must contain a "start here" signal, called the origin of replication, that the enzymes recognise. A unit of DNA that has an origin of replication (and a signal to stop replication at the other end, if that is necessary) is called a replicon. As most bits of DNA do not contain an origin, they must be given one: this is done by splicing them together with an origin-containing bit of DNA, called a vector. Vectors can be thought of as little replicons into which we can add other DNA. The added bit of DNA is called the "insert."

This is the basic function of vectors. To make them convenient to use, they have a range of other properties.

Most cloning vectors are episomes, that is, they are genetic elements which can be replicated separately from the host cell's chromosome (i.e. the rest of its DNA). Episomes can be plasmids (small loops of DNA with no function that is deleterious to the cell) or persistent viruses (bits of DNA with the potential of coding for virus particles, such as the m13 phage) (*see* the entry on **Plasmids**). A few vector types integrate their DNA into the chromosomes of their targets. Examples are YIPs (Yeast integrating plasmids), a plasmid that inserts itself into the chromosomes of yeast, and retrovirus vectors for mammalian cells. Other yeast cloning vectors, such as the YEPs (yeast episomal plasmids) based on the "2-micron plasmid" remain separate from the chromosomes like a bacterial plasmid does.

The DNA replication parts of a vector are specific to the organism it is intended for—a bacterial plasmid will not (usually) replicate in mammalian cells. You can construct a vector such that it has the DNA replication origins for two organisms—for example, for *E. coli* and mammalian cells, so that the vector can be manipulated in *E. coli* (which is easy to grow) and then transferred directly to mammalian cells (where we want to do the experiment). Such two-host vectors are called binary vectors or shuttle vectors.

Vectors contain a range of genetic elements to make cloning with them easier. These can include:

- *Selectable genes*. These code for something that allows the cell to survive an otherwise hostile environment. A common one is a gene for resistance to an antibiotic: growing the engineered organism in the presence of antibiotic will select those organisms which contain the vector (and hence whatever genes we have spliced into the vector).

- *Polylinker*. This is a piece of DNA made to contain many restriction enzyme sites, so the vector can be cut at that one place for splicing in other genes.

- Specialist origins than allow many copies of the vector to exist in the cell, such as high copy-number or runaway replication plasmids.

- *Promoters, Enhancers, leader peptides* . These are elements to help you express a gene that is cloned in the vector—*see* **Expression systems**.

Because there are so many possible vectors which can be assembled out of these components, some vector systems are made not as complete vectors but as cassette systems, where different selectable genes, origins, etc. can be slotted together to make the vector of your choice.

A critical feature of a vector is the amount of foreign DNA that it can hold. The practical limits (in thousands of base pairs—kilobases) are:

Type of vector	Typical maximum "insert" size
"Conventional plasmid vectors"	15
Lambda phage vectors (including "phagemids")	20
Baculovirus	20
Cosmids	45
P1 bacteriophage vectors	85
BACs (bacterial artificial chromosomes)	300
YACs (yeast artificial chromosomes)	2000
MACs (mammalian artificial chromosomes)	at least 1000

Vectors for gene therapy

A major barrier to getting gene therapy into the clinic is a lack of effective ways of delivering genes to cells in living patients. This requires a "vector" to carry the DNA into the cell, as DNA on its own gets into mammalian cells very inefficiently.

There are three approaches: viral vectors, chemical vectors, and physical systems.

Several viruses have been engineered to act as gene therapy vectors, with varying success. Here you splice the gene to be used into the vector DNA, and then viral infection of the cell carries the "new" gene in with it. Usually the viruses are replication deficient, that is, the engineer has removed a number of genes that are essential for the virus to grow, both to make the safer and to make room for the "new" DNA. However, they still need to be grown in the laboratory, so growing them requires either a helper virus or a helper cell line. Either provide the function of the missing genes, the former in another virus (which need not be the same type of virus), the latter engineered into the cells. It is important that the genes from either helper virus or cells do not share any common sequence with the gene therapy virus, as otherwise recombination between them can cause the virus to regain its missing genes from the helper, and become a replication competent virus again.

The most commonly discussed are:

- *Retroviruses.* Retroviruses efficiently transport their RNA into cells, copy their RNA into DNA, and then insert that DNA into the cell's chromosome. However, engineering suitably safe vectors and manufacturing them is difficult, as the genetics of the cells you grow the virus on can readily generate replication-competent viruses. A further practical difficulty is that retroviruses do not naturally infect cells that are not dividing such as nerve or muscle cells.

- *Adenovirus*. Adenoviruses can incorporate their DNA into the cell's nucleus without inserting it into the chromosomes. However, the genes that are engineered into them are not expressed permanently in the receiving cells. Adenoviruses can also provoke a strong immune reaction, especially in people who already have antibodies to the "wild" virus (which causes cold). In September 1999, immune reaction to an adenovirus gene therapy killed Jesse Gelsinger, the first recorded death from a gene therapeutic. Ovine adenovirus may be an alternative, as people are unlikely to have antibodies to this sheep virus.

- *Adeno-associated virus (AAV)*. These are vectors based around the virus particle from adenoviruses, but with no viral genes. The foreign DNA has the ends of the Adenovirus DNA spliced onto its ends can be "packaged" into otherwise empty virus shells. AAV virus-based therapeutics for cystic fibrosis, which take the "correct" CF gene directly to the lungs of affected patients in a spray, are in the clinic.

- *Herpes viruses*. The herpes viruses carry DNA into specific types of cell, there to lie low for months or years. Engineered versions have been made to carry foreign DNA into cells either in replication-deficient viruses or in otherwise "empty" virus coats. Again, this is largely a research-stage idea.

Chemical vectors. These wrap the DNA into liposomes or complex it with cationic (positively charged) lipids such as lipofectin to speed the DNA into cells. This gets round the safety problems of viruses, but is very much less efficient—under 0.1% of cells affected as compared to over 10% for some adenovirus systems. Examples are:

- *Using liposomes*. DNA that has been encapsulated into liposomes and injected is taken up by the liver and, to a lesser, extent, by the spleen, and any genes it carries are expressed briefly. Liposomes can also be targeted to other tissues (*see* entry on **Liposomes**). DNA can also be complexed with negatively charged lipids such as Lipofectin: the two molecules form nanometer-scale clumps, which are taken up by cells.

- *Using cationic lipids*. Positively charged lipids will form many-molecule aggregates with nucleic acids that are fairly lipid soluble and electrically neutral. These are taken up by cells much more readily than "naked" DNA. The same approach has been used to get antisense oligonucleotides into cells, using a lipid formulation called lipofectin.

- *Use of peptides*. Rather than use whole viruses, you could complex DNA to proteins that bound to specific cells. For example, Genzyme are trying to link peptides with the RGD sequence in them to DNA. RGD binds to Integrins (*see* **Cell adhesion molecules**).

Mechanical methods. The obvious way to get DNA into something is by injection. Simply injecting DNA complexed with calcium phosphate into liver or muscle causes some cells to take the DNA up and express the genes in it.

Other mechanical methods include electroporation and biolistics. **Electroporation** is not used much for gene therapy because the process kills most of the cells (see separate entry). **Biolistics** is used mainly to transform individual cells, but it can be used to put DNA into cells that are still part of an animal. *See* entry on **Biolistics**. A related technology is the various forms of needleless injection, such as that pioneered by Powderject to deliver DNA vaccines.

Vertical integration

A must term for management consultants, this means a company that can perform all the parts of development, production, and sale of something. In the pharmaceutical industry, a vertically integrated company would be one which researched, developed, manufactured, marketed, and sold a drug.

There is a major difference in philosophy between companies that seek vertical integration and those that do not. The latter see themselves as providing a service for larger, "mainline" pharmaceutical companies: they discover or invent drugs, develop new ways of delivering them or provide research or development capabilities for making the drugs. By contrast, some biotechnology companies feel that it is their destiny to become large drug companies, doing everything from drug discovery to knocking on GP's doors. They want to become FIPCOs—fully integrated pharmaceutical companies. A less ambitious goal is to become a FIDDO (fully integrated drug discovery and development organization), leaving the manufacture, marketing, etc. of the product to established pharmaceutical companies.

Outside healthcare, in areas such as environmental cleanup or specialist chemicals synthesis, the same criteria do not apply, as the biotechnology companies act as service suppliers for other companies and individuals in many industries: no biotechnology company making food products wants to run high-street supermarkets, for example.

A related idea is the virtual company. This is a company that assembles its capabilities by alliance or contract with other organizations, rather than building it "in-house." Thus, for example, a pharmaceutical company could license the world-class research of a biotechnology start-up, get pre-clinical development and clinical trials performed by clinical research organizations (CROs), have manufacturing done by a contract chemical manufacturing company, even hire an off-the-shelf sales force. The company's value lies in its ability to put this consortium together, make it all work, and fill in the gaps, and pay for it all. This is no mean feat, and so very few truly virtual companies exist in biotechnology.

Veterinary biotechnology

Veterinary biotechnology, the application of biotechnology to animals, is usually a spin-off of human healthcare applications of biotechnology. An exception is in genetics, where applications are substantially advanced over human uses because no ethical issues over rights to information.

Applications of DNA technology in the veterinary field include:

- *DNA fingerprinting of animals for identification*. This is done for paternity identification, and for identifying expensive animals. *See* **DNA fingerprinting**.

- *DNA mapping for breeding*. Unlike human breeding, animal breeding is directed by humans, who can therefore select genetic traits based on detailed gene maps. This can be in farm animals, for desirable productivity traits, or in pets for size, shape, behavior, or coat colour. VetGen Inc. is offering genetic testing services to pet breeders for this type of selection.

- *Pet genetic disease testing*. Many of the inbred animal strains used as pets have a high load of genetic disease, as do racehorses. DNA screening can identify carrier individuals, or potential sufferers.

- *Rare breed identification*. This is a specialist version of DNA fingerprinting, and is used both for valuation and for smuggling prevention. A related application is using DNA fingerprinting to identify illegal smuggling of endangered species, to check that ivory comes from elephants that have been killed as part of a legal "cull" and similar conservation applications.

The other headline-grabbing application is cloning animals, both for agricultural purposes and as pets. This is still a very new area of science, and hence both uncertain and expensive, but this does not deter some pet owners (*see* **Animal cloning**).

Viral vaccines

Also called live virus vaccines, these are vaccines that consist of whole viruses rather than dead ones or separated parts of viruses (another name for a complete, infectious virus particle is "virion"). Clearly, the virus itself cannot be used, as that would simply give you the disease, so the virus has to be "attenuated" to reduce its ability to cause disease (its pathogenicity). This can be done by growing it for years in cell culture, but can also be achieved by one of two genetic engineering methods.

The first is to genetically engineer the disease virus so that it is harmless, but can still replicate (albeit inefficiently, sometimes) in cultured animal cells. The genetic engineering route seeks to delete whole genes from the virus, and so make sure that the attenuated virus has no chance of mutating back to a "wild-type," pathogenic virus.

VIRAL VACCINES

The other approach is to clone the gene for a protein from the pathogenic virus into another, harmless virus, so that the result "looks" like the pathogenic virus but causes no disease. Vaccinia and adenoviruses have been used in this way, notably to make rabies viruses for distribution with meat bait: a trial of such a vaccine was carried out in summer 1990 in the United States. A similar vaccinia virus engineered to produce the haemagglutinin and fusion proteins of rindepest virus (a disease of cattle) has been developed successfully by University of California at Davis.

Walking

There are several techniques known as "gene walking" or "chromosome walking." They are all methods for cloning large regions of a chromosome. The diagram illustrates the basic idea. Starting from a known site, a gene library is screened for clones that hybridize to DNA probes taken from the ends of the first clone. These clones are then isolated, and their ends used to screen the library again. These clones are then isolated and their ends used, . . ., and so on. This can go on as long as is needed to get from where you are (usually at a "linked marker") to where you want to be.

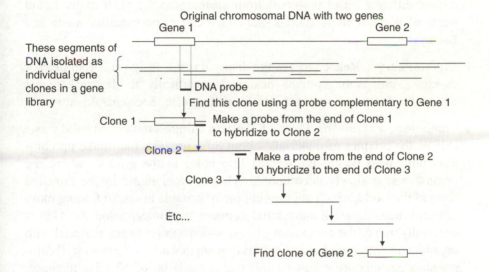

There are variants called "gene jumping" or "chromosome jumping" that allow some of the intermediate steps to be cut out: these rely on rearranging the original chromosomal DNA during cloning.

To "walk" a chromosome quickly, it is useful for the clones to each cover an extremely large amount of DNA, as otherwise each "step" will only cover a small amount of the genome. Thus, cosmid vectors (which hold 40 000 bases of foreign DNA per clone) and YAC vectors (which can hold up to a million bases) are preferred.

Increasingly, "walking" techniques are only used for large eukaryotic genomes of exotic species. For common genomes, you look up the location of the gene in a genome database, and for bacterial genomes it is almost quicker to sequence the whole genome at random than to try to do directed cloning of this type.

Waste disposal

Waste disposal is a major concern of environmental biotechnology (the other major concern being bioremediation—the removal of environmental poisons). Biotechnology addresses waste disposal in many ways.

- *Sewage disposal*. All sewage disposal is through biotechnological routes to degrade the organic matter. In crowded areas of the world like Holland and Japan, sewage disposal is a major problem. See separate entry.

- *Industrial waste disposal*. Biotechnology has had a substantial impact on the cleaning of contaminated waste water from industrial processes. This has a quite different set of problems from cleaning up the other major liquid waste of Western civilization–sewage (*see* entries on **Industrial waste** and **Sewage**).

- *Digestor design*. Related to sewage disposal is the design and operation of **digestors**, bioreactor systems designed specifically to destroy organic material rather than convert it into something else. See separate entry.

- *Composting and landfill*. Much of Western culture's municipal solid waste (MSW, the mixture of materials, about 80% organic, that make up common household waste) is dropped into holes in the ground, where the organic matter slowly decomposes. This is problematic in the crowded parts of the world, which are running out of suitable holes, so finding more efficient methods are a substantial concern for biotechnologists. This is especially true of the composting stage, which applies to any material with organic content. Such landfill material generates a lot of gas as it decomposes, a gas mixture called landfill gas is made up of 50–65% methane, 30–40% CO_2, and small amounts of nitrogen, hydrogen, and volatile organic chemicals. Large sites can generate $1000\,m^3$ of landfill gas per day, which represents a substantial explosion risk if it is not released. It can be used as a source of biogas fuel, but must be scrubbed first to remove hydrogen sulfide and other smelly and poisonous trace gases.

  A better use of the organics in MSW would be composting them, both to produce biogas more efficiently and so that the reduced volume of compost could be used for fertilizer. The technology to do this is well developed (although some such wastes have too high a metal content, from batteries, electronics, inks and other sources, to be usable). The barriers to making it happen are primarily social and organizational.

- *Tests and assays*. Biotechnology can provide a range of biologically relevant tests for toxicity, BOD, and trace chemicals that have potent biological effects, like hormones or drugs. Using biology-based toxicity tests has been standard in the water industry for years, but they often rely on whole trout as a sensor, which is not very compact or cheap. Ones based on bacterial biosensors are in development in many places. Biotech can also faster and on-the-spot tests using ELISA and biosensor technology, rather than sending samples for testing in distant reference labs.

Wood

Most tree biotechnology addresses food production via fruit trees, or **wood pulp** production (see separate entries). Some efforts have gone into engineering forest trees that are felled for wood. Forest trees have long life cycles, so traditional breeding takes decades, or centuries. With increasing economic pressure on natural forests and slower-growing species, biotechnology seeks to provide "high value" forest trees. Some examples are:

More easily farmed variants on standard trees. An example is introducing herbicide (glyophosate) resistance, so young plantations can be sprayed with broad spectrum herbicides to kill off undesirable tree seedlings. Insect resistance via Bt proteins has also been engineered into trees.

Trees with altered morphology (shape). An example is introducing a gene called rolC from *A. tumefaciens* into Aspen trees, which results in much denser branching and smaller, densely packed leaves. Unfortunately, all the genes tried so far have undesirable effects (such as these) as regards the lumber properties of the wood. A combination of "rol" genes, however, has a more beneficial effect of making the trees grow faster when young.

Such trees are created through:

Micropropagation (see separate entry)—growing a very large number of identical trees from a single "optimal" individual. Poplar and eucalyptus can be propagated almost indefinitely by cuttings: however, many other species can also be grown using more advanced tissue culture techniques. This allows unique variants to be propagated into a whole forest very quickly.

Transgenic trees. Transgenic spruce, larch, poplar, eucalyptus, and aspen trees have all been grown, using biolistics or *A. tumefaciens* to get genes into the trees. Several "morphology" genes have been engineered into forest trees, as well as herbicide resistance. Pine trees have to be transformed with biolistics, as *Agrobacterium* cannot grow on this type of tree (gymnosperms—trees that do not produce flowers).

There have been almost no field trials of transgenic trees.

Wood pulp and paper

Worldwide about 17% of felled wood is used to make paper, about the same as is used as sawn construction wood. Wood processing has attracted increasing attention from biotechnology in part because traditional methods produce a lot of effluent that is considered to be environmentally unfriendly and in part because wood is a biological material that is well suited to processing by biological means.

Nearly all the bioprocessing of wood is aimed at paper production, which takes wood chips and turns it, via wood pulp, into clean, white cellulose for making paper. The same technology could be applied to recycling cellulose-based waste, which comprise about 40% of all solid municipal waste (*see* **Waste disposal**). Wood pulping is a very large-scale process, a typical plant using 750 000 tonnes of wood to produce 300 000 tonnes of dry pulp a year, so process economics are key.

The six areas on which biotechnology has focused are:

Tree genetics. Breeding trees is very hard—in the past, people have found a tree that happens to be suitable, and cloned it by micropropagation. Genetic mapping technology allows you to test a seedling for the desired combination of genes, long before they make themselves felt in the adult tree. One of the traits being mapped is the amount of lignin in the wood, which can be reduced by selective breeding or by "knocking out" one of more of the genes involved in lignin production. So far, this has had limited success, as knocking out lignin synthesis also affects many other aspects of the plant's biochemistry. However, a poplar tree strain announced in 2002 from a consortium of European scientists reduced cinnamyl alcohol dehydrogenase (involved in making lignin), and was easier to pulp without showing slower growth or more susceptibility to pests.

Other fiber sources can be used, especially straw, which is a waste-product from cereal crops and hence easy to breed (short life cycle) and cheap. However, it produces inferior quality fibres, and more waste products.

Preprocessing. This is the removal of pitch and resins from wood. Pectinase could be used to help loosen the bark from trees, but is rarely economic. The wood from most trees contains a substantial amount of complex, oily chemicals which preserve the wood from being attacked by insects and bacteria. This has to be removed, and can be by "fermenting" wood pulp with microorganisms which grow on pitch or digesting it with lipases that break the pitch down into water soluble materials.

Pulping. Usually wood chips are converted to pulp mechanically or using chemicals. Enzyme methods are being investigated. The object here is to break down the non-cellulose materials that hold the cellulose fibres together. Several fungi that make ligninases are known, and these soften the wood, reducing the power needed in the mechanical pulpers. At the moment such methods are used in conjunction with mechanical mashing.

Fiber modification. The nature of paper depends strongly on the sort of fibers that make it up. The cellulose fibres can be modified by trimming off surface irregularities. Generally, biotechnological methods are too expensive for this use in the paper industry.

Biobleaching. The color of paper is extremely important. Wood is colored because of a large number of compounds that permeate the fibers, mostly lignin. Ligninases can be used to bleach pulp without resort to the chlorine and chlorine oxides usually used in the paper industry. Xylanases are also used here: these break down polysaccharides other than cellulose and so release colored materials trapped in the pulp. (It is important that the xylanases be free of any contaminating cellulase, as this would break down the cellulose as well.) Novozymes market such a xylanase preparation, called Pulpzyme. Thermophilic xylanases able to work at 80 °C are being developed for this application, both to speed up the process and to avoid having to cool the wood pulp down before treating. Mannanases have also been tried in this role, and ligninases have tried to digest lignin directly.

Waste disposal. Producing new paper and recycling old paper generates a lot of waste water that contains a lot of carbon compounds. These can be a substantial pollution problem, raising the BOD of the waste water to unacceptable levels (*see* **Sewage treatment**). Biological treatment of wood pulp waste is a way of removing reducing these effluent problems. The same technologies have been tried for processing other agriculturally derived waste, such as straw. They are mainly made of the same cellulose and hemicellulose as wood, with small amounts of lignin, and can be broken down by a combination of grinding, steam or hot water processing, enzyme treatment, or direct bacterial breakdown. Fungal breakdown is the normal mode of breakdown of woody materials in European countries, and many bioprocessing approaches to both woody waste treatment and to bleaching use direct fungal fermentation to break down lignin.

Wool

One of the targets of the application of genetic engineering to farm animals is to improve the quality or quantity of wool that sheep produce. This is a complex problem, but one that a number of research groups are working on, especially in Australia, which produces a substantial fraction of the 2 billion kilograms of wool produced annually worldwide.

Research on improved wool production focuses on:

- Inserting the gene for growth hormone in sheep. This has been tried, and seems to produce an increase in wool production, although no one is sure why.

- Inserting new genes for keratins into sheep. There are several types of keratins in wool, and altering the ratio of them may improve the quality of the wool. This is an empirical approach, because it is not clear what effect any particular gene insertion will have on the wool even if it makes protein in the right cells and at the right time.

- Inserting the genes for improved synthesis of cysteine into transgenic sheep. Keratin, the protein in wool, has a lot of cysteine, which is a limiting factor in the rate of growth of wool. Sheep cannot usually make cysteine themselves since they lack the relevant enzymes, so the engineering aims to give the sheep enzymes from bacteria which can make cysteine from sulfides generated in the rumen. This has not been very successful, and has been supplanted by

- Engineering feed plants. An alternative method to getting more cysteine into sheep is to engineer the plants they eat to have more cysteine. The problem here is that the rumen bacteria break down a lot of the cysteine in the food, and so improving the forage plants may not improve wool output. Some storage proteins from peas are proof against rumen breakdown, and may be suitable.

- Engineering rumen bacteria. An alternative route is to manipulate the rumen bacteria to convert cellulose in the feed into chemicals that the sheep can use more efficiently, or to make more essential amino acids, and especially cysteine, available to the sheep. This research is still in a very early stage, in part because, to model accurately what the bacteria do, you need something very like a sheep's stomach as an incubator.

Xenobiotics

A xenobiotic is a chemical that would not normally be found in a given environment, and usually means a toxic chemical which is entirely artificial such as a chlorinated aromatic compound or an organomercury compound. However, formally speaking, all chemically synthesized drugs are xenobiotics.

Biotechnology brushes with xenobiotics in three areas. First in determining their toxicity and effects on living systems. Second, biotechnology has developed methods for removing them through bioremediation or enzyme-based degradation. Lastly, a range of biotechnological products aim to replace the compounds which, if they get out of their target site, are classified as xenobiotics. Among these are the chemical herbicides and pesticides which biocontrol agents and biopesticides hope to replace.

Nearly all drugs are xenobiotics, and so much of healthcare biotechnology is concerned with them as drugs rather than as chemicals. Proteins are not xenobiotics, of course. A substantial element of how xenobiotics interact with mammals is their metabolism, particularly that mediated by a group of enzymes called P450s, which catalyze the first stage of attack on a wide range of different xenobiotics.

Xenograft

A xenograft is a graft of a tissue which comes from another species. A related idea is a xenotransplant, where a whole organ from a non-human species is transplanted into a human: the two terms are often used interchangeably. The first such transplant (from chimps into humans) was done in 1984, transplanting a baboon heart into a human. They are medically attractive because there is always a shortage of human tissue and organs for transplant, although baboons are not appropriate donors for more than research purposes.

However, there are substantial technical problems with this approach. Our immune system would violently reject such a graft (especially because of the alien major histocompatability antigens—MHC—on them, discussed in the entry on **transplants**), an exaggerated response called hyperacute rejection. Strategies have to be developed to prevent this. These include:

- Creating transgenic animals that show the human transplantation antigens, not the animal ones.

- Creating animals that show no transplantation antigens at all, by gene knockout. A favored route for this is knocking out the galactosyl transferase that is responsible for adding sugar residues to the cell surface proteins in pigs that are the primary triggers of hyper-acute rejection in primates. Imerge BioTherapeutics and PPL Therapeutics cloned pigs with such knockouts in 2002. Both this and the previous approach can be used to generate cells or tissue from grafting, of whole organs for transplantation.

- Encapsulating the organ or tissue so that the human immune system does not "see" it.

- Inducing tolerance (also called "anergy") in the human, so that they come to accept the foreign tissue as their own. One of the routes to engineering this is to make pigs express molecules such as CD59, a human protein that stops parts of the immune system attacking those particular cells.

So far, none of these have been successful in the clinic, although all have had some research success.

The most developed practical human xenografts come from pigs, although most of the research work is done in other species for convenience. For organs such as the heart, the donating animal has to be similar is size to man, and with a similar physiology: pigs and baboons are favorites, and as pigs are much easier to breed they are the focus of much of the recombinant DNA work.

There are also general practical objections centring around the concern that the xenografted tissue will carry new infections, particularly new viruses, into the human population. The historical example behind this case is probably AIDS, which some scientists believe spread from being a relatively contained disease in monkeys to a pandemic in man. With the acceptance that pigs are the best source of donor material, such debate has focused on porcine endogenous retroviruses (PERVs), possible retroviruses that are natural to pigs, cause no disease in the animal, but when introduced into man could start to grow and cause new diseases. Such viruses have been found, and can be induced to grow from transplanted tissues in mice, but no PERV growth has been observed in a primate (monkey or ape), so whether this could ever occur in man had not been resolved by the end of 2001.

There is also a substantial ethical debate about xenografts, which offends many religious groups, especially when it uses pigs.

YACs

YACs are yeast artificial chromosomes, and are cloning vectors which are gaining a lot of use in the human genome project. They consist of those pieces of DNA which define the ends (telomeres) and the middle (centromere) of a chromosome in yeast. Both these elements are needed to allow a chromosome to be replicated in yeast cells. If there is no telomere then the ends of the chromosome are liable to be broken off, or to join onto other chromosomes, and if there is no centromere, then newly made chromosomes will not be pulled into the new cells during cell division. In addition, there is an origin of replication so that the DNA is replicated.

These elements are placed in a single DNA fragment, which can be used as a vector to clone foreign DNA into yeast. The advantage of a YAC is that there is no effective limit to how big the piece of DNA can be. Thus, while conventional bacterial cloning using bacteriophage or plasmids is usually limited to cloning foreign DNA fragments of a few tens of thousands of bases long, YACs can and have cloned fragments millions of bases long. This makes mapping whole genomes of DNA much easier, as the whole genome map has to be assembled from far fewer YAC maps, and also makes cloning very large genes such as the gene for muscular dystrophy (which is at least 2 million bases long) more straightforward.

YACS are technically difficult to use, and a common problem is that DNA from two different regions of the original genome can end up in one YAC, giving the erroneous appearance that they were originally linked together, a result called a chimeric clone.

BACs (bacterial artificial chromosomes) and MACs (mammalian artificial chromosomes) use the same concept with the genetic elements from their respective species (*see* entry on **MACs**). Other vectors capable of cloning large pieces of DNA and the bacterial vectors based on cosmids and P1 (*see* entry on **Vectors**).

Yuk factor

A flippant term for the very real observation that the public, and indeed many scientists, judge the ethical acceptability of experimental procedures and biological manipulations in accordance to a scale of personal distaste. Thus, the creation of the first cloned carrot in the 1960s was greeted with amusement in the press, while the creation of the first cloned frog in the early 1970s was treated with interest and some caution and the creation of Dolly, the cloned sheep, in 1997 resulted in widespread alarm. Similarly, tests that rely on newts have less negative impact on public relations than those on rats, and rats are considered more acceptable than rabbits or dogs.

In general, this reflects a concern for animals that look or behave more like human beings. The ultimate public condemnation is therefore reserved for the potential scientific interference with human fetuses or children. In public debate, the yuk factor is sometimes a deciding one: much of the opposition to Monsanto's promotion of BST as a biopharmaceutical to boost the milk production of dairy cattle was based not on arguments about farm economics but on the feeling that it must be horrible for the cow to be turned into a milk-producing machine.

This is a very real scale of values, and one that many scientists do not take seriously enough (hence their calling it a "yuk factor" rather than a "value scale"). It cannot be acceptable for society to support a scale of values that treats chimpanzees the same way as bacteria, if only because the jump from how we treat chimps and how we treat each other is so much smaller than that between chimps and bugs. However, going to the extreme of beating up the CEO of animal-based research companies with baseball bats (as happened to the CEO of UK-based Huntingdon Life Sciences) is taking the ethical argument to extremes that few would support.

Zoonosis

Zoonosis is infection of one animal by an organism that usually infects another species, for example, infection of humans by hantanaviruses (which usually only infect rodents). It is an increasing concern for biotechnology for three reasons.

New diseases. Many of the "new" diseases that are claimed to be a major health threat to Western cultures, such as Ebola, Marburg and Lassa viruses, are probably zoonoses. They are "natural" diseases of tropical primates, but can also infect humans. With increasing travel between tropical and temperate zones, it is likely that more such cross-species infections will turn up. West Nile virus is probably basically a bird disease that sometimes accidentally infects humans.

AIDS. The reason that this may be important is that AIDS, and possibly Lyme disease, may originally have been diseases of other species which jumped the species barrier to infect humans as zoonoses, but which subsequently became "naturalized" human illnesses. AIDS in particular is similar to a range of illnesses that are endemic to old world monkeys. Understanding how this jump happened could both give insight into AIDS biology and understanding of whether diseases such as Ebola, which are now very rare, could become common among humans.

Xenografts. One way that AIDS was transmitted was by the most common transplant procedure—blood transfusion. There is concern that grafting animal organs (xenografting) could transmit animal infections to humans through a route that would otherwise be impossible for the bacterium or virus to traverse. Viruses are the major concern here, as once introduced into a host they might mutate very rapidly to give rise to a human version of a pig or goat disease. The possibility that a version of Cruezfeldt–Jacob disease has been passed onto humans from cows enhances people's concern about this type of zoonosis.

Further reading

There are a libraries of books on specific technical aspects of biotechnology. Most books in a bookshop will be arguing for or against biotechnology rather than presenting ungarnished facts. For a general introduction to most areas of biotechnology as an industrial process, see

- Basic Biotechnology: ed Colin Ratledge and Bjorn Kristiansen (Cambridge University Press, 2001)

For an authoritative overview of the drug discovery process, see

- In quest of tomorrow's medicines, by Juergen Drews (Springer Verland, 2003)

and for a characteristically opinionated support for the genomics agenda, see DNA: the secret of life bt James Watson and Andrew Berry (Knopf, 2003) Journals providing technical reviews of biotechnology at several depths are:

- Nature Biotechnology (Nature publishing Ltd – www.nature.com)
- Trends in Biotechnology (Elsevier – www.elsevier.com)

The former also has industry overviews on what companies are doing and review and opinion pieces on topical issues of the day.
The web is avast resource of information, mis-information and downright lies. Reliable sites are:

- The NCBI (www.ncbi.nlm.nih.gov) and the EBI (www.ebi.ac.uk) for updates on genome projects and other matter genomic and bioinformatic
- Biospace (www.biospace.com) and Bio (www.bio.com) for industry news (usually replicated from company press releases)

and of course, this book's own web site at

www.biotech-atoz.com

Index

INDEX

INDEX

shelf life – of sensors 68
shigella toxin 223
shuttle vector 369
signal amplification (for DNA) 125
signal peptide 331
  removal 298
signal transduction (in cells) 244, 318
signalling cascade – repoter genes 322
silica
  as immobilization medium 146
  in HPLC 210
  solid phase 335
silk 54
  genetic instability in bacteria 189
simulation (of biological systems) 347
SINE 179
single cell protein 329
single chain antibody (SCA) 117
single domain antibody 18
single locus probe – in DNA fingerprinting 125
single nucleotide polymorphism 333
single strand conformation polymorphism
    (SCCP) 260
SIP (sterilization in place) 104
site-directed mutagenesis 304
size exclusion chromatography 101
skin substitutes 25
slot blot 74
smallpox 51
  vaccine 368
smart drugs 23
SMB chromatography 103
SNP 333
  as polymorphism 296
  for DNA fingerprinting 125
soil – microorganisms in 30, 344
soil amelioration 334
solar cells 57
solar energy 335
solid phase 335
solid phase (chromatography) 101
solid phase chemistry 107
solid phase synthesis of peptides 281
solvents – used for bioconversion 37
somaclonal variation 336
  and genetic instability 190
somatic gene therapy 47, 183
SOP 194
Southern blot 73
soy bean 112
  as fermentation feedstock 149
soy sauce 168
space shuttle 302
sparger 8, 348
species hybridization 213
speigelmer 22

spider diffuser 348
spiders silk – as a biomaterial 54
spidroin 54
spinosyn 58
Spirulina 9, 53, 330
splicing 186
splitting (cell culture) 83
sports and biotechnology 336
SPR 267
SSCP 260
stabilization of enzymes – by immobilization 147
stable cell line 88
standard equipment 337
standard operating procedure 194
starch
  and glycosidases 198
  as fermentation feedstock 149
  as fermentation feedstock 161
  processing 297
    amount of enzymes used 144
Starlink corn 182
stationary phase 87, 101
stearic acid 237
stem cell 338
  growth factors 204
  and chimeric animals 96
  and embryo technology 140
  in bone marrow 269
stereospecific 71
sterile insects as biocontrol 49
sterile methods (fermentation) 160
sterile plants 292
sterility – in cell culture 83
sterilization (of equipment) 159, 340
sterilization in place (SIP) 104
steroids – production by bioconversion 37
stirred tank bioreactor 341
stock (shares) 78
stoichiometry – in metabolism 246
stonewashing jeans 351
stop transfer peptide 332
storage compartment of plant 305
strain (cultivar) 342
strain development (improvement) 343
strain improvement and darwinian cloning 119
strain isolation 344
  for environmental biotechnology 66
strategic alliance 9
straw 378
streptavidin 33, 70
streptokinase 72
streptomyces 249
structural genomics 302
structure-activity relationship 317
structure-based drug design 317
  and X-ray crystallography 302